Materials Design and Manufacturing: Theory and Practice

Materials Design and Manufacturing: Theory and Practice

Edited by **Gerald Brooks**

WILLFORD PRESS

New York

Published by Willford Press,
118-35 Queens Blvd., Suite 400,
Forest Hills, NY 11375, USA
www.willfordpress.com

Materials Design and Manufacturing: Theory and Practice
Edited by Gerald Brooks

© 2016 Willford Press

International Standard Book Number: 978-1-68285-023-7 (Hardback)

Printed in the United States of America.

Contents

Preface

There have been rapid advancements in the field of materials designing and their applications in manufacturing processes. The topics that have redefined this field such as various computational and data management techniques, integrating different toolsets, managing technical challenges in data curation and model verification, emerging applications of advanced materials, etc. have been presented in this book. This text will serve as a reference to a broad spectrum of readers.

Significant researches are present in this book. Intensive efforts have been employed by authors to make this book an outstanding discourse. This book contains the enlightening chapters which have been written on the basis of significant researches done by the experts.

Finally, I would also like to thank all the members involved in this book for being a team and meeting all the deadlines for the submission of their respective works. I would also like to thank my friends and family for being supportive in my efforts.

Editor

Novel microstructure quantification framework for databasing, visualization, and analysis of microstructure data

Stephen R Niezgoda[1,2]*, Anand K Kanjarla[3] and Surya R Kalidindi[4]

*Correspondence:
niezgoda.s@gmail.com
[1] Materials Science and Technology
Division, Los Alamos National
Laboratory, Los Alamos NM 87545,
USA
[2] Present Address: Department of
Materials Science and Engineering,
Department of Mechanical and
Aerospace Engineering, The Ohio
State University, Columbus, OH
43218, USA
Full list of author information is
available at the end of the article

Abstract

The study of microstructure and its relation to properties and performance is the defining concept in the field of materials science and engineering. Despite the paramount importance of microstructure to the field, a rigorous systematic framework for the quantitative comparison of microstructures from different material classes has yet to be adopted. In this paper, the authors develop and present a novel microstructure quantification framework that facilitates the visualization of complex microstructure relationships, both within a material class and across multiple material classes. This framework, based on the stochastic process representation of microstructure, serves as a natural environment for developing relational statistical analyses, for establishing quantitative microstructure descriptors. In addition, it will be shown that this new framework can be used to link microstructure visualizations with properties to develop reduced-order microstructure-property linkages and performance models.

Keywords: Microstructure; Two-point correlations; Microstructure database; Structure-property relationships; Reduced-order models

Background

It is well understood that advances in the development of materials with enhanced performance characteristics have been critical in the successful development of advanced technology, and are important drivers for continued economic prosperity. It is also widely recognized that modern materials cannot be understood through their chemistry and bulk processing alone, but that the key to continued material development lies in understanding and optimizing the myriad details of the hierarchical three-dimensional (3-D) internal structure, or microstructure, which spans several disparate length scales (from the electronic to the macroscale). While developing microstructure/processing/property linkages is the central theme in materials science, as a field we are only beginning to develop the tools necessary to truly explore and harness multi-scale microstructure sensitive design and manufacturing. Given the complexity of microstructure in virtually all engineering and natural materials, the traditional approaches to materials development, relying on combinatoric experimentation guided by engineering principles and physical intuition, have only categorized and exploited a small handful of the readily accessible material microstructures.

In recent years, impressive advances have been made in materials characterization as well as the development of sophisticated physics-based multi-scale modeling and simulation tools. Through these advents, materials science is undergoing a transition from a data-limited field to a data-driven but analysis-limited field. It is now possible to automate acquisition of large experimental and simulation datasets, and our ability to generate data is rapidly outstripping our ability to process it. Radically new approaches are required to integrate the looming deluge of data from advanced simulation and characterization techniques into *useful materials knowledge*. While this data crisis is a significant technical challenge, it is also an opportunity to explore completely novel inverse approaches to material design and deployment while at the same time reducing our dependency on slow combinatoric experimental approaches. This realization has been highlighted via the Materials Genome Initiative for Global Competiveness [1], the DOE Needs Reports on Computational Materials Science and Chemistry [2] and the continued growth of Integrated Computational Materials Engineering (ICME) [3]. It is widely realized that the community must develop a Materials Innovation Infrastructure (MII) or Materials Innovation Ecosystem [1,2] to exploit fully advanced simulation and coupled experiments, improve predictive capabilities, and provide the design, certification and monitoring tools for rapid and holistic materials development and deployment.

While the need for and end goals of a practical MII are well articulated, its requirements and components are largely undefined. It is envisioned that a successful MII framework will accelerate the development and deployment of materials by reducing/replacing expensive and evolutionary empirical experimentation with revolutionary optimization of structure and processing through simulations that are verified and validated with key experiments. The absence of a general framework for categorizing and visualizing the space spanned by collected microstructure data, understanding the statistical properties and variability inherent in materials, and efficiently integrating digital materials data with the disparate components of multi-scale modeling frameworks is a significant barrier to the development of a practical MII and to the overall goals of the Materials Genome Initiative.

The creation and curation of large scale materials databases has been widely cited as a critical required component for the acceleration of materials development and deployment [1,2] and has been a recognized need by the materials community since the 1970's [4]. Prior efforts at the creation and maintenance of large scale materials databases, such as the efforts of ASTM Committee E-49 for the Computerization of Materials Property Data [5] and the National Materials Property Data Network [6] (which was operated commercially from the mid 1980's until 1995), were largely focused on chemical composition, average properties and effective material response. A shortcoming of these efforts, largely due to the limited computing power of the times, was that details of the materials internal structure or microstructure was not considered and captured alongside the composition and properties. Understanding the relationship between structure, properties and processing is the central theme in materials science and engineering, and it is critical that future materials databases effectively capture all three legs of the triangle if such databases are to be successfully applied to materials design applications. While a significant amount of development, across the entire materials community, is required to even outline the structure and specific requirements of such a database, this paper focuses on one critical area; namely, the definition and description of a computational framework and tools for the visualization, analysis, and quantitative comparison of microstructure datasets (collected across several different "materials") consisting of ensembles of 2D or 3D micrographs collected for each material.

The work addresses several aspects necessary for the development and deployment of successful materials databases, or more exactly microstructure databases. In particular, we seek to address the following questions: 1) What is the appropriate way to represent the microstructure for inclusion into the database? 2) How do we describe the inherent structural variability in a material, and how do we connect this variability to variability in properties? 3) How do we quantitatively compare ensembles of material volumes from different samples or material systems? 4) How do we visualize the relationship between materials and the span of collected microstructure data? 5) How do we determine if additional characterization would add to our knowledge base or if it would be redundant? In this work we demonstrate a reduced-order microstructure visualization space, built upon a stochastic process representation of microstructure. It is envisioned that this visualization space will serve as the interface between the large scale database of microstructure data and the user, and will provide tools that will allow the user to interact with microstructure data in new ways that go far beyond those accomplished traditionally by visual inspection of micrographs or three-dimensional data sets.

Previous work from the authors and others laid the groundwork for reduced-order representation and categorization of microstructure data, on which many of the ideas presented in the manuscript build. Sundararaghavan and Zabaras proposed the construction of a material library from a reduced-order representation of digital micrographs, and demonstrated the application of supervised learning techniques such as support vector machines for material classification [7,8]. Ganapathysubramanian and Zabaras explored more advanced non-linear dimensionality reduction schemes and used these representations to construct inputs to stochastic multi-scale models [9,10]. Niezgoda [11] and Niezgoda, Yabansu and Kalidindi [12] formalized the description of microstructure via statistical metrics into a stochastic process interpretation, and developed the basic theory for the microstructure visualization space presented in this work. Additionally, they developed relationships between the observed variance in microstructure for a material ensemble and the corresponding ensemble variance in properties/performance and proposed applications of the microstructure visualization space in materials design and manufacturing quality control and process monitoring.

While the previous work largely focused on developing the microstructure space and visualizations for ensembles of a single material or microstructure class, this work is focused on the extension of the above concepts to the simultaneous categorization, analysis and visualization of multiple materials, the development of descriptive and relational statistics that facilitate the exploration of microstructure relationships between different materials, and the exploration of microstructure/property relationships in the microstructure visualization space. The main purpose of this manuscript is to present the philosophy of the visualization space and the mathematical formulation that underlies its construction, followed by some simple examples of the type of microstructure analyses and property explorations that can be performed in the space. The examples are chosen to show the utility of the proposed space, rather than to suggest a specific analysis procedure or regime. The main idea is that the proposed visualization space is a platform upon which a large suite of analysis tools can be built, and where the analysis can be tailored for a specific material system and design problem.

The following key ideas and features behind the proposed microstructure visualization space will be explored in detail throughout the remainder of this paper: 1) The understanding that microstructure posses an inherent randomness suggests that that the mathematics of stochastic processes is a natural description of the multi-scale variability of materials. 2) While powerful, such an approach leads to significant abstraction of the data and the resulting data-set lies on the surface of a complex high-dimensional manifold which is not intuitive to many materials, design, or manufacturing practitioners. In order to simplify the presentation of the data a reduced-order representation of the microstructure data is required. Here we apply principal component analysis to construct the reduced order representation. 3) The reduced order representation forms the basis for the visualization space. Each sample is represented by a single point in the space. Points that are near each other correspond to samples with similar structures, disparate points correspond to samples with drastically different structures. 4) The space spanned by all the characterized samples from a material or material class gives an indication of the microstructure variability within that material. This scatter in the visualization space can be directly tied to the scatter in properties or response of the material. 5) Comparison tools based on relational statistics can be easily developed for further analysis in the space. Additional tasks such as automated classification or clustering analysis are naturally formulated in the visualization space. 6) Explicit mapping of properties or performance characteristics into the visualization space can be performed to formulate invertible structure/property linkages.

Methods

Microstructure data generation and modeling methodology

For this study ensembles of 8 different material classes were generated by thresholding random fields. The 8 material classes were all porous composites with varying degrees of anisotropy in pore shape, connectivity, and spatial distribution. In order to prevent variations in porosity between classes from exerting an overwhelming influence on property calculations and to highlight the higher-order effects of the spatial distribution of pores, the volume fraction of pores was controlled to be nominally 25% and normally distributed with a standard deviation of $\sigma = 8.7\%$ across all material classes. $64 \times 64 \times 64$ arrays were populated by sampling from a uniform distribution $\mathcal{U}(0, 1)$, the field was then locally averaged by circular convolution with an anisotropic 3-D Gaussian filter with a diagonal covariance matrix, Σ. The resulting periodic random field was thresholded, with the threshold value sampled from a normal distribution, $\mathcal{N}(\mu = 0.25, \sigma^2 = 0.0075)$. Representative realizations of the 8 material classes are shown in Figure 1. For each class two ensembles of 50 volumes were generated. The first ensemble was used to create the microstructure database and calibrate the classification and property models, while the remainder were reserved for validation.

The mechanical response of each volume element was evaluated using an image based fast Fourier transform (FFT) elastic-viscoplastic model, that directly uses the three dimensional voxel material volumes as input. This type of model was first developed by Suquet and collaborators [13], for nonlinear composites, and further developed by Lebensohn and collaborators for polycrystals [14]. For this work we adopted

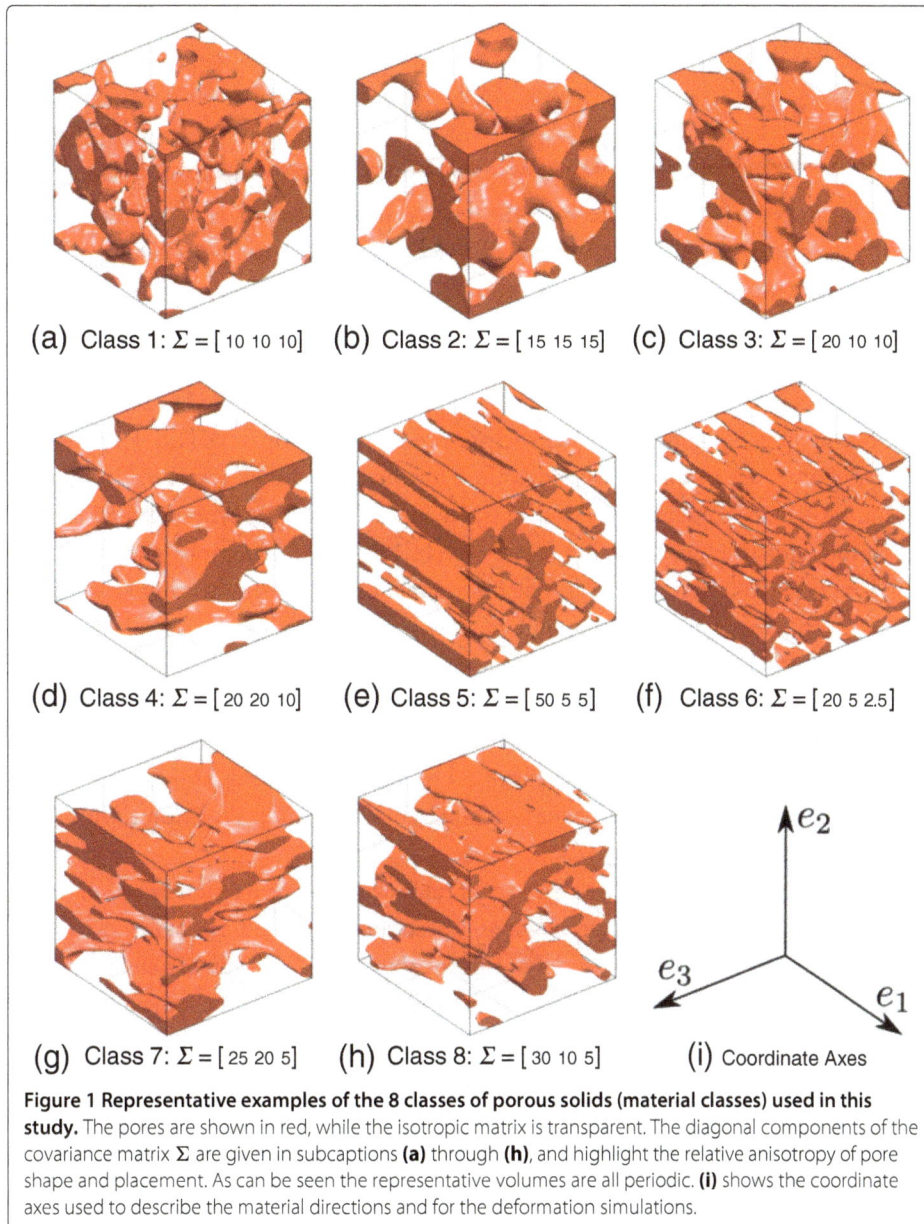

Figure 1 Representative examples of the 8 classes of porous solids (material classes) used in this study. The pores are shown in red, while the isotropic matrix is transparent. The diagonal components of the covariance matrix Σ are given in subcaptions **(a)** through **(h)**, and highlight the relative anisotropy of pore shape and placement. As can be seen the representative volumes are all periodic. **(i)** shows the coordinate axes used to describe the material directions and for the deformation simulations.

the elasto-viscoplastic formulation of Lebensohn and colleagues [15,16], to generalized power-law composites. The FFT-based algorithm computes a compatible strain-rate field that minimizes the average work rate under the constraints of the constitutive relation and stress equilibrium. The FFT method is predicated on the observation that the mechanical response of a heterogeneous non-linear medium can be calculated as a convolution between a linear reference material and a polarization field, which describes the non-linearity and heterogeneity of the actual composite material response. The FFT name comes from the fact that the resulting stress equilibrium equations take a computationally convenient form when cast in terms of Fourier transforms.

The matrix material is considered as elastically and plastically isotropic with a Young's modulus of 40GPa and a Poisson's ratio of 0.3. The matrix had an initial yield strength,

σ_0, of 45 MPa. A visco-plastic constitutive model was adapted to relate the stress to the plastic strain rate as

$$\dot{\varepsilon}_{ij}^p = \frac{3}{2}\dot{\varepsilon}_0 \left(\frac{\sigma_{ef}}{\sigma_0}\right)^n \frac{\sigma'_{ij}}{\sigma_{ef}} = C^* \left(\sigma_{ef}\right)^{n-1} \sigma'_{ij} \tag{1}$$

where $\dot{\varepsilon}_{ij}^p$ is the viscoplastic strain rate, σ'_{ij} is the deviatoric stress tensor, $\dot{\varepsilon}_0 = 1$ is a reference strain rate, and σ_{ef} is the scalar Von Mises effective stress. C^* can be considered as an effective plastic modulus given by $C^* = \frac{3}{2}\dot{\varepsilon}_0(\sigma_0)^{-n}$.

Stochastic process representation of microstructure

The mathematical formalism of stochastic processes has been demonstrated to be a powerful tool for the quantification of microstructure variability and variance [12]. Here we present a brief overview of the theory as required for the applications presented in this manuscript; for a more complete overview of the theory of stochastic processes and the application to microstructure modeling please see [11,12]. Consider a probability space defined by the ordered triplet $(\Omega, \mathcal{F}, \mathcal{P})$. The first element Ω is termed the sample space and is a non-empty set of experimental outcomes or observations, individually denoted as ω. $\mathcal{F}$ is the set of all theoretically possible events and is formally defined as a Borel σ-algebra [17], and $\mathcal{P}$ denotes the standard probability measure on $\mathcal{F}$. A random variable $\mathbf{x}$ is then defined as a function, with domain $\mathcal{F}$, which maps to each experimental outcome ω a number $\mathbf{x}(\omega)$ such that the axioms of probability are satisfied. The probability distribution function (pdf), $f(\mathbf{x})$, associates each possible experimental outcome with a probabilty such that $\mathcal{P}\{x_1 \leq \mathbf{x} < x_2\} = \int_{x_1}^{x_2} f_{\mathbf{x}}(x)dx$.

A stochastic process, $\mathbf{x}(t)$, is by extension a set of rules which assign a function $x(t, \omega)$ to every experimental outcome ω of experiment Ω. These rules take the form of associated probability distributions. To completely determine the statistical properties of a stochastic process the nth order distribution function $f(x_1, \ldots, x_n; t_1, \ldots, t_n)$ must be known for all t_i, x_i, and n. For virtually all processes (and microstructures) this is impossible. Instead, attention is usually restricted to lower order statistical measures such as the mean and autocorrelation [12,18].

The microstructure of virtually all materials exhibit rich details that span several length scales. A microstructure constituent that can be assigned a distinct local structure can be considered to posses a distinct local state. The local state is denoted h and can be considered an element of a local state space, H, that identifies the complete set of local states that could theoretically be encountered. For example, the local state at point x in a polycrystalline metal sample may be defined by the thermodynamic phase ρ, the local lattice orientation g, the state of dislocation α etc. In this way the local state may be considered a random vector $\mathbf{h}(x) = [\rho, \mathbf{g}, \boldsymbol{\alpha}, \ldots]$, and assuming a stationary process the local state distribution $f_{\mathbf{h}}(h)$ gives the volume density of material points in a volume with local state h. A common example of local state distribution is the well known orientation distribution function used to quantify preferred crystallographic orientations in polycrystalline materials [19].

In terms of the probability space defined above, the set of all possible events $\mathcal{F}$ is the set of all possible spatial arrangements of local states $h \in H$. The sample space Ω consists of an an ensemble of microstructure realizations, where each microstructure realization is considered an experimental outcome ω. The microstructure can then be interpreted as a stochastic process $\mathbf{m}(x, h)$, sometimes termed the microstructure function, that assigns a local state distribution field $m(x, h, \omega)$ to every realization. It is important to note that the stochastic process $\mathbf{m}(x, h)$ cannot be observed directly. The microstructure function is best understood as a series of higher order probability distributions which describe how the local states are placed in a material relative to each other. Instead, the local state fields $m(x, h, \omega)$ can be observed for an ensemble, and can be used to estimate the microstructure statistics.

Statistics of the microstructure

Consider a material system where the local state is defined by a combination of k features of interest, thus the local state is described by the random vector $\mathbf{h} = [\boldsymbol{\beta}_1, \ldots, \boldsymbol{\beta}_k]$. The probability of finding local state $\mathbf{h}$ at material point x is in a region of local state space $\mathcal{H}$ is given by

$$
\begin{aligned}
\mathcal{P}\{\mathbf{h}(x) \in \mathcal{H}\} &= \int_{\mathcal{H}} f(h, x) dh \\
&= \int_{\mathcal{H}} f(\beta_1, \beta_2, \ldots, \beta_k, x) d\beta_1 d\beta_2 \ldots d\beta_k
\end{aligned}
\tag{2}
$$

where the PDF $f(h, x)$ is refered to as the first order density of the microstructure function, and can be interpreted as the spatially resolved volume fraction of local state $\mathbf{h} = h$.

Estimation of 2nd order PDFs from characterized microstructure realizations, is often impractical and estimation of higher order PDFs ($n \geq 2$) is often impossible. Fortunately, a framework of hierarchical spatial statistics of the microstructure is available in the literature in the form of n-point correlation functions [20-25]. Local state distributions, $f(h)$, are often termed one-point statistics, as they reflect the probability density of finding a specific local state at a randomly selected point in the microstructure.

Expanding on this idea, the two-point correlation is the joint density of occurrence of local state h_1 and h_2 at material points separated by the vector r

$$
f_2(h_1, h_2 | r) = E\{\mathbf{m}(x, h_1)\mathbf{m}(x + r, h_2)\}
\tag{3}
$$

The microstructure statistics must be estimated from an ensemble of characterized realizations. Consider a ensemble of P volumetric realizations $\Omega = (\omega_1, \ldots, \omega_P)$, where the pth realization has an associated local state field $m(x, h, \omega_p)$. The two point correlation for the microstructure can be estimated as

$$
f_2(h_1, h_2 | r) \approx \widehat{f_2}(h_1, h_2 | r) = \langle f_2(h_1, h_2 | r, \omega_p) \rangle
\tag{4}
$$

$$
= \frac{1}{P} \left[\sum_{p=1}^{P} \frac{1}{vol(\omega_p | r)} \int_{x \in \omega_p | r} m(x, h_1, \omega_p) m(x + r, h_2, \omega_p) dx \right]
$$

where $\omega_p | r = \{x | x \in \omega_p \cap x + r \in \omega_p\}$. $f_2(h_1, h_2 | r, \omega_p)$ is the estimate of the two-point correlation obtained from a single realization or volume. Equation (4) is readily identified as a convolution integral which is readily computed via Fourier transform techniques [22, 25]. Three-point and higher order correlations are defined analogously, however, for the remainder of this paper the discussion will be limited to 2-point correlations.

Results and discussion

The microstructure space

Principal component representation

The set of physically meaningful n-point correlation functions lie in a very high dimensional space that is not amenable for efficient computation, visualization, analysis, or other design purposes [25]. For even the simple material systems considered here, where the local state is either pore or solid, and restricting the statistical description to two-point correlations, some form of reduced order representation of the microstructure statistics is required. While there are numerous approaches to dimensionality reduction including spectral methods, manifold learning approaches, metric multidimensional scaling, among others [26], the requirements of the proposed microstructure space places strong constraints on the type and complexity of the dimensionality reduction scheme. In order to determine an appropriate low dimensional representation the following factors were considered: 1) The approach must be computationally robust and insensitive to the type or structure of the underlying data. 2) The representation must be able to be updated in real time as new information or datasets are added to the system 3) The representation must be invertible in real time, so that the microstructure space can be used to explore new microstructures in both an interpolative or extrapolative manner. 4) In order to facilitate the computation of descriptive and relational statistics the representation should be an orthogonal decomposition of the data so that each dimension in the reduced frame can be considered as independent variables. Based on these requirements and considerations of computational complexity, and the excellent performance in previous work [12], dimensionality reduction via principal component analysis (PCA) was chosen for this study.

PCA is a linear approach to dimensionality reduction which can be understood as a coordinate transformation that maps a set of (possibly) correlated variables onto a new set of orthogonal (independent) variables. This is most easily understood as projection of a high dimensional dataset onto a new orthogonal coordinate frame where the axes are defined by the directions of highest variance [27]. PCA as been likened to a shadow of the data cast from its most informative projection. One key strength of PCA is that it forms a natural basis for the exploration of microstructure variance as the principal directions align with the directions of highest variance in the microstructure data [12]. A significant drawback to using PCA is that, as the microstructure data (2-point correlations) is inherently non-linear (meaning the data naturally lies on an embedded curved manifold in the high dimensional space rather than on a hyper-plane), the resulting representation will not necessarily be the most compact or efficient. However it will be shown that when the different materials are of the same "family", such as the various classes of porous composites explored here, a suitably compact representation can be developed and useful visualizations and maps can be developed in just a few dimensions. If the range of structures were to be expanded to include multiple material "families" then more advanced dimensionality reduction approaches would be required. However, in practice, for most materials design applications the materials system is fixed as a design constraint and the goal is optimizing processing or chemistry to achieve a target microstructure or properties. For these cases the PCA representation is expected to perform quite well.

Consider an ensemble of P realizations or volumes from a single material class (in multiple volumes cut from a single large sample or multiple samples with the same nominal processing history) denoted $\omega_1, \omega_2, \ldots, \omega_P$ each with associated two-point correlation estimate $f_2(h_1, h_2|\omega_p)$. The PCA representation of the correlations measured from the pth member of the ensemble can be written as

$$f_2(h_1, h_2|r, \omega_p) = \sum_{j=1}^{P-1} \alpha_j^p \phi_j + \widehat{f_2}(h_1, h_2|r) \tag{5}$$

where $\widehat{f_2}(h_1, h_2|r)$ is the ensemble average of the overall microstructure statistics (see Equation (4)). ϕ_j are the orthogonal principal component vectors and α_j^p are the corresponding weights or the PCA representation of the pth member of the ensemble. Mathematically the decomposition consists of a few basic steps

1. Mean center the data

$$\Phi^p = f_2(h_1, h_2|r, \omega_p) - \widehat{f_2}(h_1, h_2|r) \tag{6}$$

2. Compute the covariance matrix of the mean centered data

$$C = \frac{1}{P} \sum_{p=1}^{P} \Phi^p \left(\Phi^p \right)^T \tag{7}$$

3. Perform an eigenvalue decomposition

$$C\phi_j = b_j \phi_j \tag{8}$$

4. Project Φ^p onto the eigenvectors

$$\alpha_j^p = (\phi_j)^T \Phi^p \tag{9}$$

In practice, it is not necessary to explicitly construct the covariance matrix C. Instead, for large datasets, an algorithm called the method of snapshots is used to reduce the computational burden [28]. Additionally, it is possible to incrementally build the basis (realization by realization) by taking advantage of the understanding that any portion of new data not in the span of the eigenvectors must be orthogonal to the current basis [29]. Building the basis in this manner has the advantage that addition of new realizations does not require the re-computation from scratch. Assuming no linear dependencies, the rank of C is $P - 1$, implying that the maximum number of parameters necessary to represent the data is approximately the number of members in the ensemble. The eigenvalues of the decomposition are an indication of the significance of that principal component (degree of variance in the data). By taking only the components with the highest eigenvalues it is often possible to approximate the data in a handful of parameters.

For this study, two PCA decompositions were performed to generate a two tiered representation. First an intra-class decomposition was performed on the members of each material class. The intra-class PCA representation will be used to explore microstructure variance within a material system, and to identify interesting or representative members of a specific class. Then an inter-class PCA decomposition was performed over all the realization of all classes. The inter-class representation will be used to explore relationships between the individual classes and to develop property relationships across multiple material systems. The inter-class representation was truncated to 50 principal components, primarily for computational reasons. However, it will be shown that for this example fewer than 20 components are sufficient for the classification and property modeling examples presented here. 50 realizations from each materials class were used to build the inter-class representation, the additional 50 members were reserved to validate the representation and microstructure classification scheme.

Visualizations in the microstructure space

The inter-class PCA weights, α_j^p, are an explicit representation of the individual microstructure realizations in an "optimal" orthogonal reference frame defined by the eigenvectors of the data covariance matrix. This reference frame spans the space of the collected microstructure datasets (up to the error introduced by truncation), and is a natural setting in which to visualize and explore the range of data collected, the relationships between datasets and the variability within material classes. By projecting the microstructure data onto the first three eigenvectors, microstructure maps can be constructed. Other operations such as descriptive and relational statistical analysis and the development of homogenized property relationships can be performed in this microstructure map space. As such, these maps and visualizations are a central result of this work. The projection of the first eight microstructure classes onto the inter-class PCA basis is shown in Figure 2. As a reminder, each data point in the figure corresponds to a single material realization. Convex hulls bounding the realizations of each microstructure class are plotted to help visualize the differences between the volume of space occupied by the members of each class.

An advantage of the PCA representation is that the axes are naturally ordered by significance in that the first principal component is the direction of highest variance and the 2nd is the direction of highest variance orthogonal to the first and so on. By choosing the first three components or directions for visualizing the projected microstructure data we are selecting our vantage point so that we are capturing the view which captures the most variability in the collected microstructure data. This does not, however, guarantee that the view is optimal for separating the different realizations by microstructure class. As can be seen in Figure 2 the different microstructure classes are indeed heavily overlapped in three dimensions, indicating that more terms are needed to delineate the different microstructure classes. It will be shown for this particular example that 15 principal components are sufficient to build a robust classification model for these eight material classes (see Section "Support vector machines for classification of microstructures").

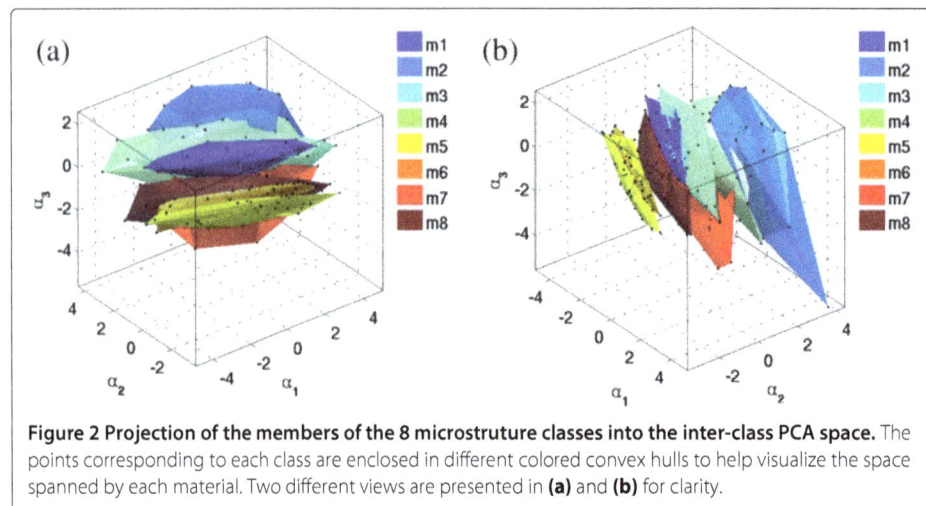

Figure 2 Projection of the members of the 8 microstruture classes into the inter-class PCA space. The points corresponding to each class are enclosed in different colored convex hulls to help visualize the space spanned by each material. Two different views are presented in **(a)** and **(b)** for clarity.

Figure 2 is a key result of this work. At a glance it shows the relationships between the different microstructure groups. Distances in this space (defined below) serve as an indication of how similar or different structures are. Material classes whose bounding hulls are close (or overlapping) share more common features than those classes whose bounding hulls are well separated. This can be seen in Figure 2(b) which shows the hulls bounding classes 5 and 6, which share a highly anisotropic elongated pore shape, are overlapping and are well offset from the classes 1 and 2 which have an isotropic pore shape. The center of mass of the cloud of points for each class, serves as an estimate of the average or representative material volume for that class. The volume of the hull bounding a microstructure class also serves as a qualitative measure of the inherent structural variability of the members of a material class relative to the other classes. Again, the figure shows that the classes 5 and 6 exhibit less structural variability than the other material classes. Microstructural clustering or the uneven dispersion of points within the hull, can be also readily observed. While not the focus of this paper, the scatter in microstructure realizations collected for a single class allows the ready identification of structural outliers or realizations expected to exhibit performance or property values far from the mean. Previous work from the authors focused on intra-class variance and the development of tools for analysis of multiple ensembles of a single microstructure class [11,12].

Sundararaghavan and Zabaras also used PCA for reduced order representation for the construction of microstructure libraries [7,8]. In that work they chose to perform the decomposition on the characterized microstructures directly, rather than working with statistical descriptors as we do here. While the motivations behind this work and theirs are substantially different, it is worthwhile briefly discussing some of the advantages and disadvantages of each approach. PCA is a linear transformation and thus effective dimensionality reduction can only be accomplished if the data can be approximately fitted to an embedded linear manifold (hyperplane) in the high dimensional space. Unfortunately, as the ongoing work of Zabaras group has demonstrated, the underlying data in microstructure datasets often lies on an embedded highly non-linear surface and large numbers of principal components must be kept for reasonable representations or non-linear data mappings must be employed [10]. Some basic properties of the n-point correlations, including having a natural origin at $r = 0$ and translation invariance, greatly reduce this non-linearity and adequate representation can be produced with only a few principal components. For comparison, a PCA decomposition was done directly on the 3-D microstructure datasets and the resulting projection into the PCA space is shown in Figure 3. As can be seen in the figure, the hulls bounding the microstructure classes are heavily overlapped and the relationships between the classes is not readily apparent from a view of the first three dimensions of the space. The eigenvalues of the decomposition give an indication of how significant each principal component is to the representation of the data-set, and the rate of decay of the eigenvalues gives an indication of how many terms must be kept for accurate representation. Figure 4 shows the eigenvalues for the first 50 principal components for the inter-class PCA, performed on the spatial correlations (described earlier), and the PCA performed on the microstructure images. For the inter-class representation constructed on 2-point correlations there is a sharp drop in the eigenvalues within the first 10 principal components while there is no such drop in the representation constructed from the images themselves. The slow decay is highlighted by the ratio between eigenvalues, for the decomposition on the structures

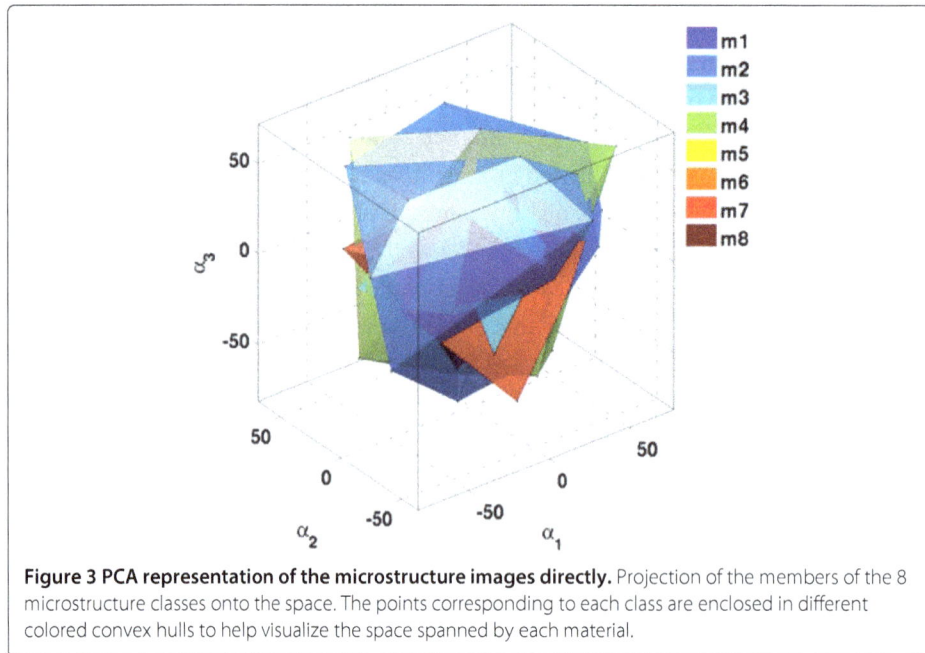

Figure 3 PCA representation of the microstructure images directly. Projection of the members of the 8 microstructure classes onto the space. The points corresponding to each class are enclosed in different colored convex hulls to help visualize the space spanned by each material.

$\alpha_1/\alpha_{50} \approx 3$, while for the decomposition on the statistics $\alpha_1/\alpha_{50} \approx 65$. These results indicate a much stronger concentration of the representational power in the first few principal components when the microstructure statistics are used to construct the reduced order representation.

Working in the correlation space does, however, add a significant abstraction. When working directly with the microstructure realizations, once the basis is defined, new realizations can be created for any point in the spanned space by simple linear combinations

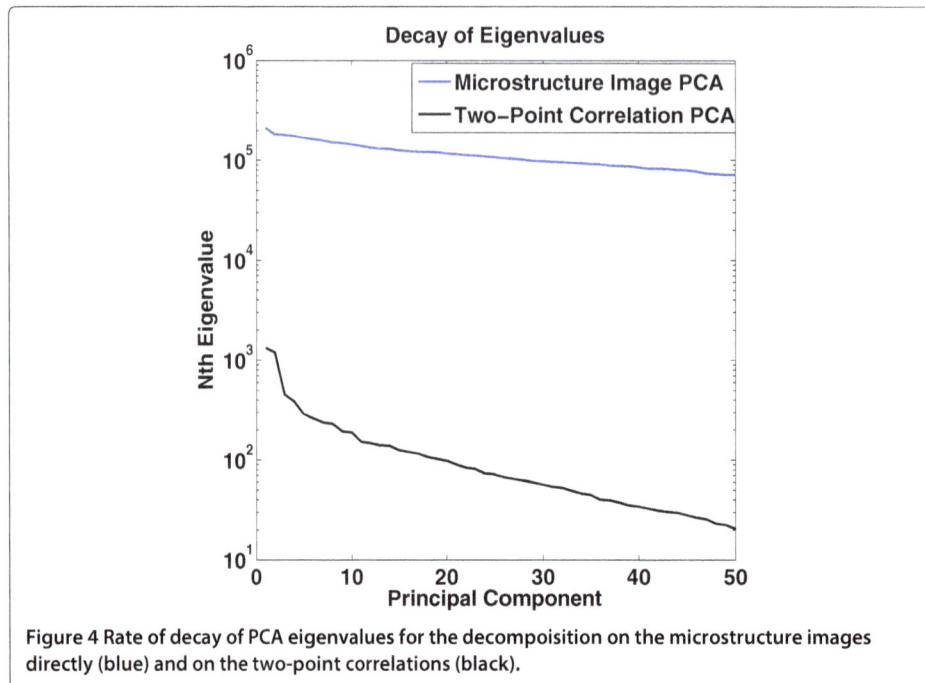

Figure 4 Rate of decay of PCA eigenvalues for the decompoisition on the microstructure images directly (blue) and on the two-point correlations (black).

of the eigenvectors. When working with the correlations each point in the spanned space represents a set of potential microstructure correlations, and new microstructure realizations must be reconstructed from the statistics. Reconstruction from statistics is an active area of research, and significant advances have been made in recent years [30,31]. In the opinion of the authors the benefits of a more compact linear (PCA) representation and a cohesive framework based on the formalism of stochastic processes outweigh the added abstraction.

Relational statistics in the microstructure space

In order to effectively utilize the PCA space for microstructure quantification or comparison of structures, a measure of distance in the space must be introduced. For intra-group comparison between individual points (i.e. how similar two points are) the Euclidean distance is adequate. However in this work we are interested in comparisons between groups of points and for the application of microstructure classification we are interested in the question of how close are the clouds of points representing two material classes or how close is a new realization from an unknown class to the cloud of points from a known class. For this we apply the Mahalanobis distance, which gauges the similarity between an unknown sample (or sample set) to a known or classified sample set [32]. Consider a multivariate vector $x = (x_1, \ldots, x_n)^T$, the Mahalanobis distance between x and a distribution of vectors $f(y)$ with mean μ_y and covariance matrix C_y is defined as

$$D_M\left(x, f(y)\right) = \sqrt{\left(x - \mu_y\right)^T C_y^{-1} \left(x - \mu_y\right)} \tag{10}$$

The feature of the Mahalanobis distance is that it takes the shape of the known probability distribution into account, in that all points that lie on the same iso-probability surface of the known distrubtion will have the same Mahalanobis distance to that distribution regardless of the shape of the distribution[a]. The Mahalanobis distance can be generalized to a similarity measure between two multivariate vectors x and y, where y comes from a distribution $f(y)$ with covariance C_y as

$$d(x, y) = \sqrt{(x - y)^T C_y^{-1} (x - y)} \tag{11}$$

When computing distances in the PCA space the individual components of x and y, x_i and y_i are independent (orthogonal) variables and the covariance matrix is by definition diagonal. In this case $d(x, y)$ becomes the normalized Euclidean distance, $d(x, y) = \sqrt{\sum_i (x_i - y_i)^2 / \sigma_i^2}$, where σ_i is the standard deviation of $f(y_i)$. In the limit that the covariance matrix is the identity matrix (e.g. for the standard multivariate normal distribution) the Mahalanobis distance reduces to the standard Euclidean distance.

When working with ensembles of characterized material volumes in the design space, the first natural question to ask is whether two samples (material volume ensembles) truly posses different microstructures or have the same microstructure but different structural variance. In other words, how likely is it that the two ensembles share the same mean or that same realization can serve as a representative volume for both ensembles. For quality control and process monitoring applications it is important to have a quantitative measure of drift in the manufacturing process or be able to compare material from different processing lines. It is a natural question to ask if process paths 1 and 2 are indeed producing material with the same microstructure, and if not how different are they? If they are confirmed to posses the same microstructure, then analysis and comparison of the variability of the ensembles may be carried out to determine the likely effects on properties or performance (see [12]).

The standard tool for hypothesis testing on the means of multivariate data is the 1-way multivariate analysis of variance (1-way MANOVA). The 1-way MANOVA tests the null-hypothesis that the means of each group lie in the same d-dimensional subspace. Choosing $d = 0$ tests that the means are the the same or that the ensembles posses the same microstructure. By extension, choosing $d = 1$ tests that the means lie on the same line through the PCA space, etc. For a full discussion of the 1-way MANOVA see standard multivariate statistical texts such as [32]. The major assumptions made in performing an MANOVA analysis is that the different classes have the same covariance matrix, which is clearly not the case in our data. However, MANOVA analysis has been shown to be robust to a violation of this assumption provided the groups are all well behaved unimodal distributions [32]. The basic notion of the MANOVA analysis is to transform the original n-dimensional dataset into a new variable on a d-dimension subspace, termed the canonical variable, that maximizes the separation between the classes, then to perform a statistical test on the ratio of within group scatter of the canonical variable to the total scatter across all groups.

If we consider a dataset of K material volume ensembles where $\Omega_k = \{\omega_1^k, \ldots, \omega_P^k\}$ denotes the P_k members of the kth ensemble. By projecting onto the PCA space each ω_p^k is represented by the J weights ${}^k\alpha_j^p$. For compactness, let A_{pk} denote the J dimensional column vector of PCA weights for the pth member of the kth ensemble. Further let $\bar{A}_k$ indicate the mean vector of PCA weights over all P_k members of ensemble k. The within group scatter is characterized by the intra-class sum of squares and cross product (SSCP) matrix, W, defined as

$$W = \sum_{k=1}^{K} \sum_{p=1}^{P} (A_{pk} - \bar{A}_k)(A_{pk} - \bar{A}_k)^T \tag{12}$$

The between group scatter is characterized by the inter-class SSCP matrix defined as

$$B = \sum_{k=1}^{K} P^k (\bar{A}_k - \bar{A})(\bar{A}_k - \bar{A})^T = \sum_{k=1}^{K} P_k \bar{A}_k \bar{A}_k^T \tag{13}$$

where $\bar{A}$ is the mean vector of PCA weights across all K ensembles which is by definition the zero vector. The total scatter is simply $T = W + B$. The test statistic for the 1-way MANOVA can be chosen as Wilk's lambda $\Lambda = \frac{|W|}{|T|}$ or Pillai's trace $Tr\left(WT^{-1}\right)$. Wilk's lambda is a multivariate extension of the F test, while Pillai's trace (also F distributed) is more robust to violations of the assumptions (such as our case).

The representation of the data in terms of the canonical variables can be found by performing an eigenvalue decomposition of $W^{-1}B$ then projecting the PCA representation onto the eigenvectors. Let $L = $ eigenvectors $\left(W^{-1}B\right)$, then the canonical representation can then be found as $c_{pk} = LA_{pk}^T$.

The canonical representation is useful for visualizing clustering between the groups. Figure 5 shows that microstructure classes can be easily separated by projection onto the first two canonical variables. In other words there exists a plane in the 50 dimensional PCA space, that when the microstructure data is projected onto it there is virtually no

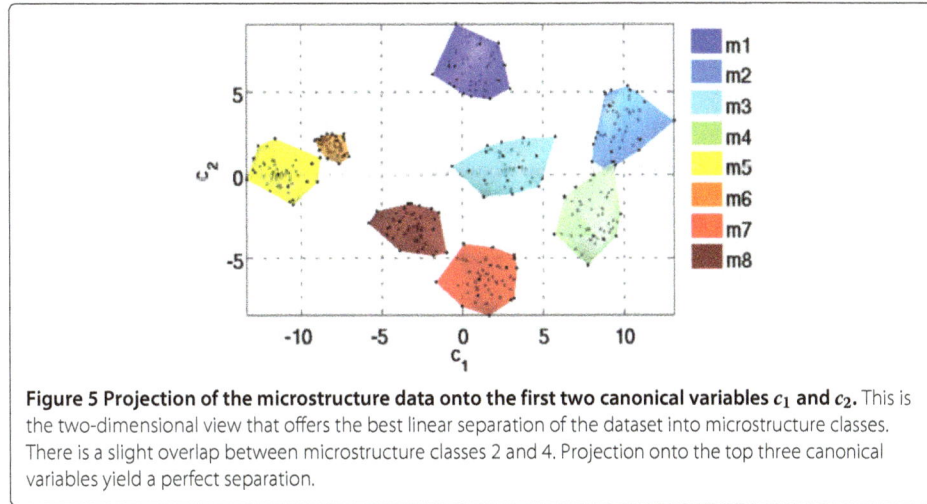

Figure 5 Projection of the microstructure data onto the first two canonical variables c_1 and c_2. This is the two-dimensional view that offers the best linear separation of the dataset into microstructure classes. There is a slight overlap between microstructure classes 2 and 4. Projection onto the top three canonical variables yield a perfect separation.

overlap between the classes. Projection onto the first three canonical variables yields a perfect separation, for this dataset.

That a perfect separation can be achieved by only three canonical variables does not imply that the data lies on a hyperplane in the 50 dimensional PCA space. The results from the 1-way MANOVA indicate that at 5% significance we can reject the null hypothesis that the means of the 8 microstructure classes lie in a 6 or lower dimensional subspace ($p = 0.048$ for $d = 7$), but that they may lie in a 7 dimensional space. Performing the test pairwise between the groups indicates that we can state with statistical certainty that the means of the 8 classes are all different or that the microstructures of the 8 ensembles are all distinct ($p \approx 0$ for $d = 0$ for all pairwise tests).

This notion of Mahalanobis distance can be used to further explore the relationships between the various groups. A key question to ask is what is the expected distance between a randomly selected member of class a to class b. By computing this expectation for all combinations of classes the entire dataset (8 material classes) can be broken up into groups and subgroups of classes that share similar features by a hierarchical cluster analysis. Hierarchical cluster analysis is a technique to build a multi-level hierarchy of groups and subgroups that are easily represented by a tree-like graph. The analysis is performed by first computing the distance between the different groups, in the case the distance measure is the expected Mahalanobis distance between a randomly selected member of class a and the cloud of points corresponding to class b, $\overline{D}_M(a, b) = E\{D_M\left(x \in a, f(y \in b)\right)\}$. Then the two closest groups are combined into a larger group. The procedure is then repeated until all the smaller groups have been combined into a single large class.

The results of the hierarchical cluster analysis are shown in Figure 6 as a cluster tree. The cluster tree is interpreted by moving up on the y-axis, which shows the expected Mahalanobis distance, the height of the inverted U connecting two groups is the expected distance between those groups. Note that $\overline{D}_M(a, b) \neq \overline{D}_M(b, a)$ due to the differences in covariance between the groups, for plotting the cluster tree the minimum of the two expectations is taken. Groups that are linked lower down are more similar, than groups that are linked higher up. For example classes 5 and 6 are more similar than 7 and 8, and the expected distance between a group formed from classes 1–4 and the group consisting

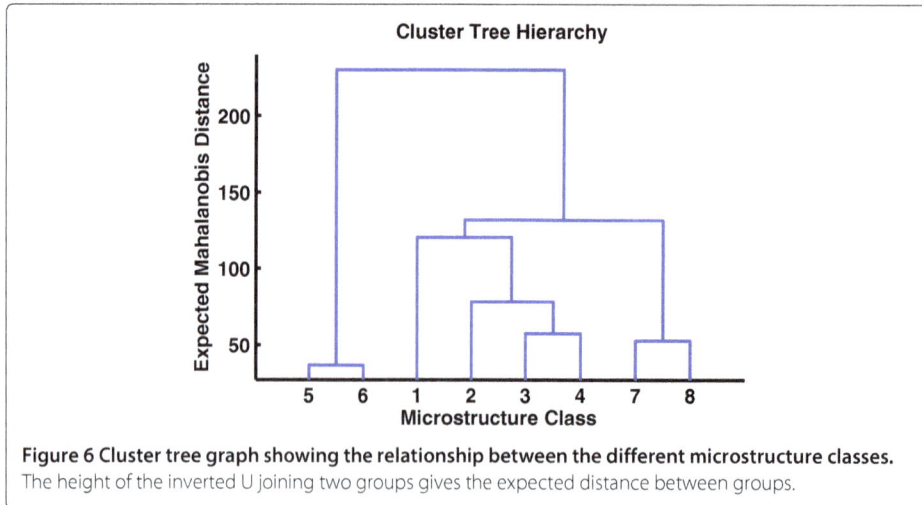

Figure 6 Cluster tree graph showing the relationship between the different microstructure classes. The height of the inverted U joining two groups gives the expected distance between groups.

of classes 7 and 8 is approximately 135. While the cluster tree graph is a useful visualization of the relationships between the microstructure classes, it is up to the end user to decide on the number of clusters or groups in the dataset or where to cut the tree to define clusters. For this example a likely pruning scheme is to break the data into three groups: the first consisting of classes 5 and 6, the second containing classes 1–4, and the third classes 7 and 8. It is interesting to note that this grouping naturally follows the degree of anisotropy in the pore shape as seen in Figure 1. Classes 1–4 all have a low degree of anisotropy with a covariance ratio of 2 or less between the highest and lowest direction, while the other two groups are more highly anisotropic with 5 and 6 having highly elongated pores along 1 primary axis, while classes 7 and 8 have pores elongated along two directions.

The PCA or visualization space serves as a natural setting for a wide range of different statistical analyses on the microstructure data. As mentioned above the center of mass for each class can be thought of as the average or representative realization for that class. A natural question to ask is "How well is the mean known given the amount of characterization for this class?" or equivalently "If the experiment were repeated and a different ensemble collected how different would our estimate of the average structure be?" Given the orthogonality of the principal component vectors, common tools of statistical inference can be applied treating each principal component as independent. This allows us to use readily available univariate statistical tools and descriptors without concerning ourselves with correlations between the different directions. For example the questions asked above can be readily answered by computing confidence regions about the mean of each microstructure class. 95% Confidence regions are shown in Figure 7(a). This gives us a formal probabilistic approach to selecting realizations to serve as representative volumes elements (RVE) for the ensemble. The confidence region is interpreted not as a probability on the true microstructure mean (which is unknowable) but rather on the process by which we are estimating the average structure. If we were to validate the confidence regions by collecting a very large number of ensembles and estimating the ensemble mean, 95% of the time the calculated mean would lie within this box. Under this interpretation any material volume lying in our confidence region is a potential RVE for that

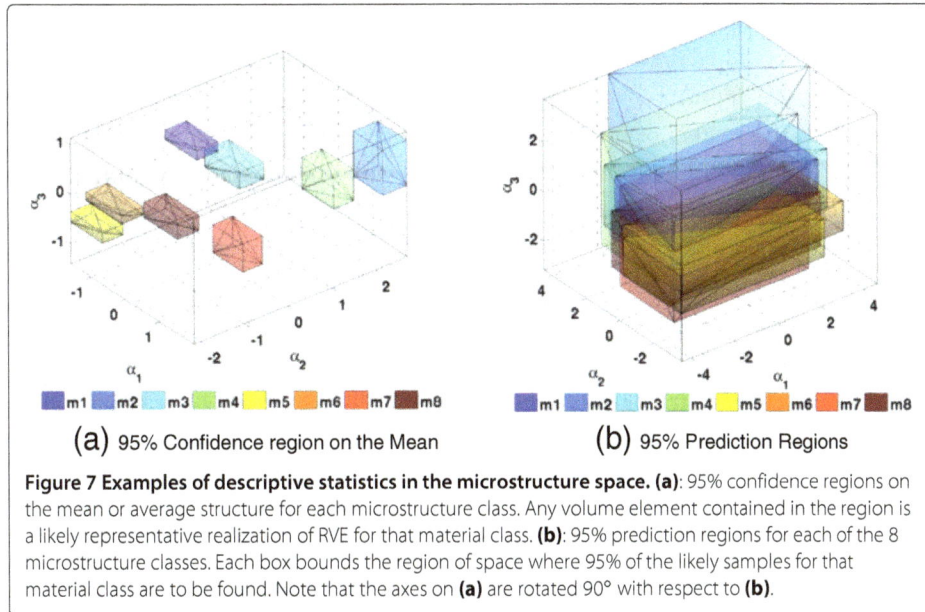

Figure 7 Examples of descriptive statistics in the microstructure space. (a): 95% confidence regions on the mean or average structure for each microstructure class. Any volume element contained in the region is a likely representative realization of RVE for that material class. **(b)**: 95% prediction regions for each of the 8 microstructure classes. Each box bounds the region of space where 95% of the likely samples for that material class are to be found. Note that the axes on **(a)** are rotated 90° with respect to **(b)**.

microstructure class. The RVEs shown in Figure 1 were chosen as the material realization closest to the center of the confidence region.

In a similar manner, we may wish to have an estimate on the range of each microstructure class or given the variability observed in a class how likely are additional samples to fall within some region. Prediction bounds and related measures can also be readily calculated in the PCA space. Examples of 95% prediction regions are shown in Figure 7(b). Simple analysis such as this give a way of sampling representative volumes from a fuller range of likely structures for modeling and simulation [33] or more advanced intra-class statistical analysis.

Support vector machines for classification of microstructures

Projection onto the canonical variables showed that the microstructure classes can be well separated in a low dimensional subspace, therefore it should also be possible to develop a robust classification scheme to sort new microstructure realizations into their appropriate classes. For the example classes presented here where there are clear microstructural differences between the classes, classification could be accomplished by simply assigning a new realization to the nearest class, the class which minimizes the Mahalanobis distance. While this simple approach will work for the "toy" examples given in the paper, we wish to apply a more formal probabilistic approach to classification that is able to handle the general case when the data may not be linearly separable. Following Sundararaghavan and Zabaras, we adopt a kernel support vector machine (SVM) classification scheme [7,8]. Unfortunately, a mathematical treatment of SVM classification, particularly for the multi-class examples of interest here are beyond the scope of this manuscript. For the interested reader a full treatment of SVM classification and regression can be found in standard references such as [34], here we will provide a descriptive overview of the technique and a classification example.

If we consider the projection of the microstructure classes onto the canonical variables (see Figure 5), and we wish to develop a linear categorization rule that assigns a new point to either microstructure class 1 or 2 it is obvious that there are numerous possible lines that could be drawn which would partition the space between these two classes. The intuitive approach to finding an "optimal" partition would be to imagine a line connecting the centers of mass of each class and draw the partition normal to this line at the midpoint. In performing this operation we would be finding a partition that approximately maximizes the distance from each class to the partition. SVMs are supervised machine learning models for binary linear classification, that implement, in a more theoretically rigorous manner, the basic approach just described. They attempt to find the separating hyperplane through the data such that the distance from it to the nearest data points on each side (the support vectors) is maximized. Such a plane, provided it exists, is called the maximum-margin hyperplane. There are however many cases where such a plane does not exists, such as when the data is non-linearly separable or when where is noise in the data and the outliers from each class overlap. In the case of outliers where the data is nearly linearly separable, a penalty is introduced on points on the wrong side of the margin boundary that increases with the distance from it. This is known as a soft margin SVM and was introduced in 1995 by Cortes and Vapnik, and revolutionized the machine learning field [35]. Previous to soft margin approaches the options for nearly linear or fuzzy classification were limited to neural networks or other unsupervised techniques that have a strong tendency to overfit the data. The soft margin concept allowed for an optimal classification without overfitting, in the sense that the margin is maximized, even when a clean linear separating hyperplane did not exist. In instances where the data is not linearly separable, such as a quadratic separating surface, an approach termed kernel SVM introduced by 1992 by Boser, Guyon and Vapnik, is used to map the data points to a higher dimensional space where they are linearly separable [36]. The approach relies on the so called kernel trick, where it was realized that if the optimization was formulated properly the mapping to the higher dimensional space only appears as a dot product, implying that the mapping never needs to be explicitly performed thus keeping the dimensionality of the optimization equation low. For multi-class problems such as our example, the problem is reduced to multiple binary classification problems by developing pairwise classification for all combinations (1-against-1) or by considering one class against the remainder of the data set (1-against-rest) [37].

The open-source software library LIBSVM was used to create a classification on the data in the PCA space [38]. Performing the classification in terms of the canonical variables would have been more efficient as the data is 100% linearly separable in three dimensions. However for this example we are interested in exploring how many principal components are really necessary to capture the range of microstructures in the dataset. The 50 realizations from each of the 8 classes that were used to construct the PCA representation were once again used as the training dataset. As we know *a priori* that the data is linearly separable the applied kernel is simply the identity matrix. The 1-against-1 approach was found to perform better as the individual classifications remained linear rather than needing to apply a non-linear kernel for 1-against-rest. Validation was performed by projecting the 50 members of each set held in reserve into the PCA space (the PCA basis was not updated, the weights of the new data given the current basis were computed) and exercising the classification model assuming no prior knowledge of correct class membership. Classification models were constructed for increasing numbers of principal components. The accuracy of the classification model for both the training and validation sets is shown in Figure 8.

As can be seen in the figure, the accuracy of the classification rapidly increases with additional principal components, by 15 components the accuracy of the training set is 100% at the accuracy of the validation set saturates around 92%. That the classification accuracy of the validation dataset does not approach 100% implies that 1) there are outliers in the validation dataset that are being misclassified and 2) the additional material realizations contain some useful microstructure information that is absent in the original dataset and is not captured in the original PCA decomposition. By examining the PCA representation of the training data and the projection into this space of the validation set, it can be seen that the hulls bounding the points do not entirely overlap and indeed there are outliers which skew the shape of the bounding convex hull. An example of this projection for microstructure class 1 is shown in Figure 9.

A key advantage of using SVMs for classification is that when new points are classified they are assigned a probability of membership for each existing class. This probability can be used to construct visualizations and maps in the design space on class membership. Such a map giving the probability of class membership over a region in the PCA space is shown in Figure 10. The probability of class assignment gives additional information concerning the state of the classification model and the sufficiency of the microstructure data collected. Virtually all of the misclassified points in the validation dataset were assigned a significant probability of belonging to two or more microstructure classes. When adding a new ensemble to the system the classification probabilities are a powerful tool in determining if the ensemble belongs to a currently defined class or a new class. While not discussed in detail here, the probabilities can also be used as weights when assigning average properties or other metrics to the various classes. In effect this is one method of handling the outliers of a microstructure class or of estimating the tails of the performance distribution, given the current state of the microstructure knowledge.

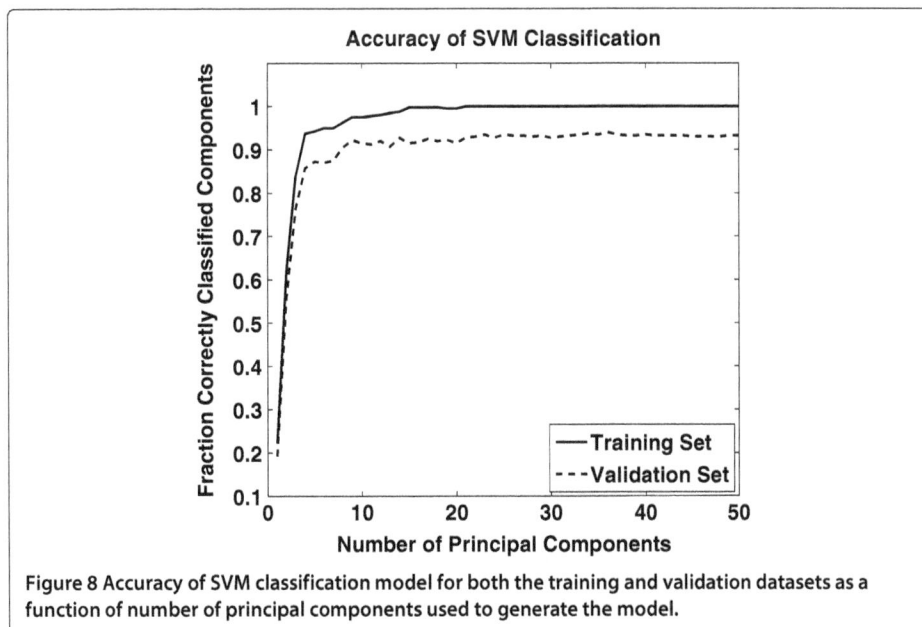

Figure 8 Accuracy of SVM classification model for both the training and validation datasets as a function of number of principal components used to generate the model.

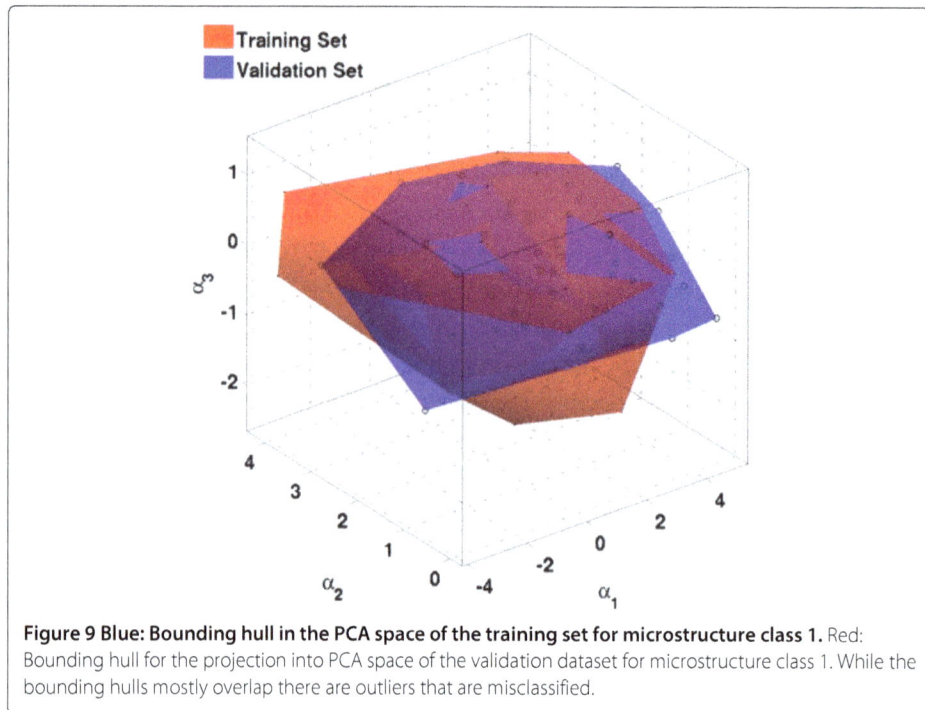

Figure 9 Blue: Bounding hull in the PCA space of the training set for microstructure class 1. Red: Bounding hull for the projection into PCA space of the validation dataset for microstructure class 1. While the bounding hulls mostly overlap there are outliers that are misclassified.

Connection with properties and performance

The above discussion was focused solely on the exploration and visualization of the microstructural relationships between multiple materials classes in an interactive materials design space. In this section we link the materials design space with properties. Being able to connect the interactive materials data with properties and response data is a critical task if the materials design space concept is to be applied in a practical manner. The idea of generating invertible mappings between a microstructure space and a property space is not new. The Microstructure Sensitive Design framework [22,39],

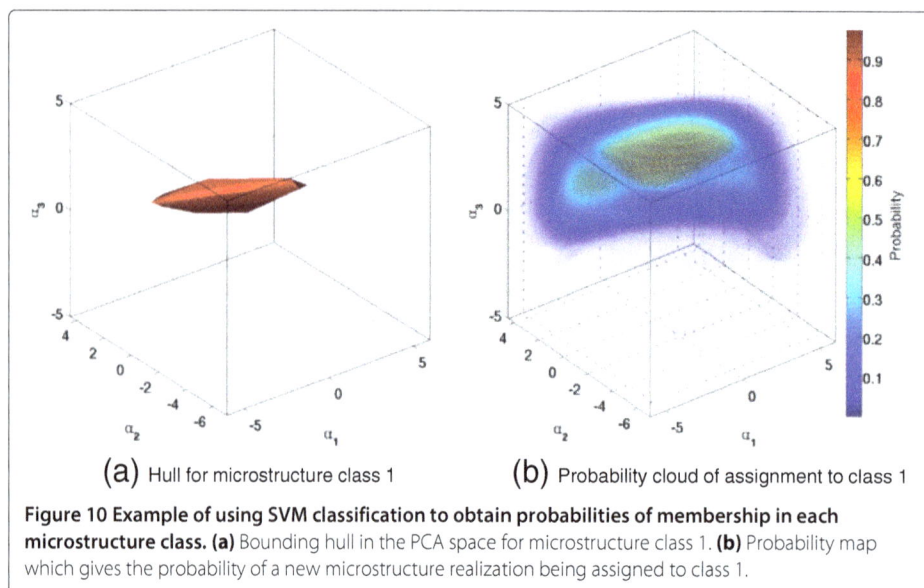

(a) Hull for microstructure class 1   **(b)** Probability cloud of assignment to class 1

Figure 10 Example of using SVM classification to obtain probabilities of membership in each microstructure class. (a) Bounding hull in the PCA space for microstructure class 1. **(b)** Probability map which gives the probability of a new microstructure realization being assigned to class 1.

for example, was predicated on delineating a microstructure design space termed the microstructure hull that bounded the complete set of theoretically feasible microstructures of a given class and then projecting microstructures from the design space to a property space, typically in the form of homogenization relationships for the delineation of property closures. The main difficulty of such an approach is the construction of the microstructure hull, typically only first order microstructure descriptors such as volume fraction or crystallographic texture could be used. As was described above the space of two point correlations for even the simple porous composite material system here, renders the problem intractable for higher order microstructure descriptors [25]. Here we are restricting the problem from the complete set of theoretically possible structures to the space of microstructures available through either characterization or digital simulation. Instead of delineating property closures, or strict bounds on theoretical property values, we are exploring and mapping the property space that is currently achievable.

The successful production of property models directly in terms of distributions of the microstructure process or microstructure function is important if such structural models are to be incorporated by industry and the larger ICME or materials design community. In the long term, it is hoped that the microstructure design space described here will find applications in quality control and design certification, where current decisions are made on measured property or experimental response values rather on the microstructure details. Moving toward microstructure based acceptance criteria will require the widespread construction and validation of the types of linkages presented here.

As described in Section "Microstructure data generation and modeling methodology", the mechanical response of each realization for the 8 microstructure classes was simulated using elastic visco-plastic FFT modeling. The resulting distributions in yield stress and effective elastic modulus are shown in Figure 11. In previous work Niezgoda et al. observed a nearly perfect linear relationship between a measure of microstructure variance and the observed variance in properties for multiple ensembles of the same material class [11,12,40]. In that work the material system was a low to moderate contrast two phase composite, and while the ensembles all had large differences in structural variance,

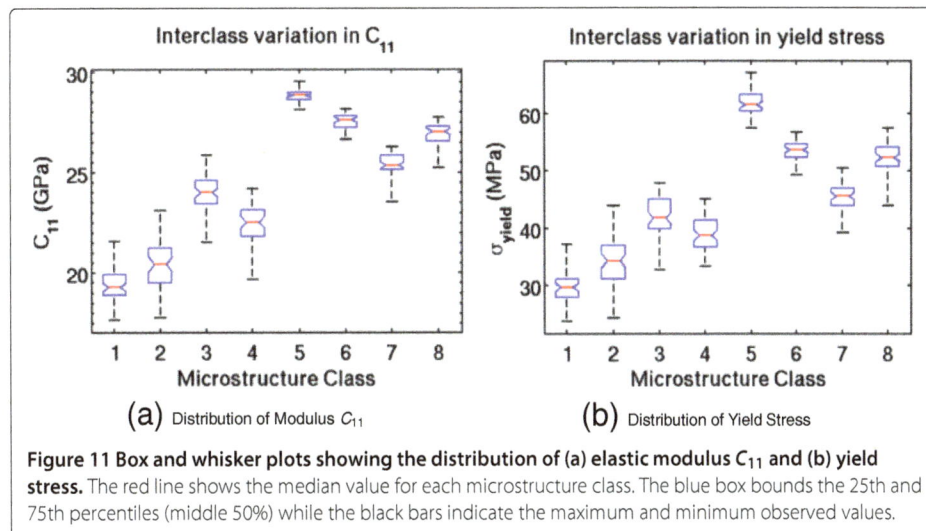

Figure 11 Box and whisker plots showing the distribution of (a) elastic modulus C_{11} and (b) yield stress. The red line shows the median value for each microstructure class. The blue box bounds the 25th and 75th percentiles (middle 50%) while the black bars indicate the maximum and minimum observed values.

the multiple ensembles all had the same mean or average structure. Here we extend that observation to effectively infinite contrast porous solids and show that the relationship between structural variance and property variance shows the expected linear trend across multiple material classes.

The descriptor of microstructural variance found to be most useful was the total variation defined as

$$VAR_{tot}^p = b_1^p + b_2^p + \cdots + b_J^p \tag{14}$$

where b_j^p are the eigenvalues of the intra-class PCA decomposition for the pth microstructure class. The total variation can also be approximated for a given class from the inter-class PCA representation by $VAR_{tot} \approx \sum_{j=1}^{J} var(\alpha_j^p)$ where the variance is calculated over the members of the pth specific class. For each class the total variation was computed from the intra-class PCA representation, and the relationship with property variance is shown in Figure 12. As can be seen in the figure there is a strong correlation between microstructure variance and property variance. The observed effect is not as strong as seen in previous cases, however this example was deliberately set up as a worst case in that the contrast between matrix and pore was effectively infinite and that the microstructures spanned a wide range of anisotropy in pore shape and distribution.

Ashby charts or maps are a well known design tool for visualizing relationships between properties, performance and design constraints such as cost. Such maps have proved invaluable in guiding material selection based on macroscale properties or performance criteria. An ongoing goal of the authors is to link microstructure and microstructure variance into similar type maps to extend the utility to the realm of microstructure sensitive design and design of materials/microstructures. In order to develop such tools, properties and performance must be described as a function of the microstructure, or in our case the PCA representation of microstructure. Here we we perform a proof of concept example, by developing linear homogenization relationships in the microstructure design space. A linear model was fit between effective modulus and yield strenth and the PCA weights of each realization using a weighted least squares approach. The eigenvalues from the inter-class PCA representation were chosen as the least squares weights. Weighting the least squares regression in this manner enforced the assumption that the significance of

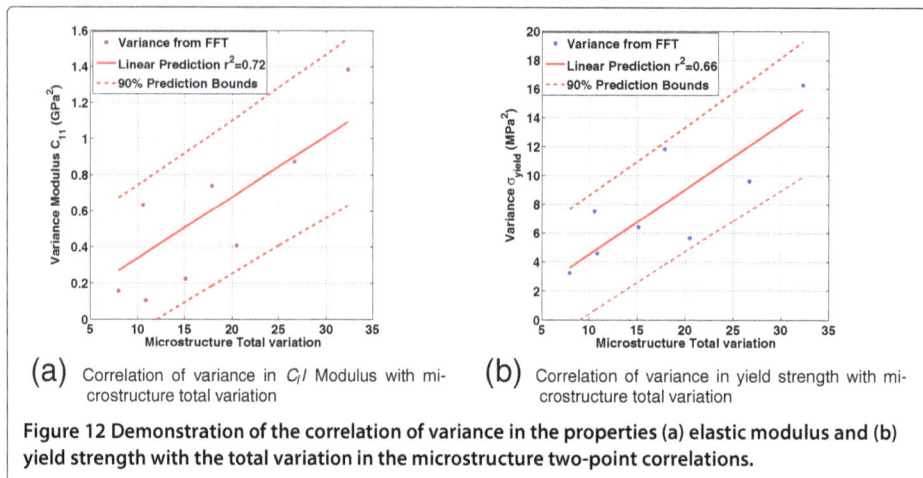

(a) Correlation of variance in $C_i l$ Modulus with microstructure total variation

(b) Correlation of variance in yield strength with microstructure total variation

Figure 12 Demonstration of the correlation of variance in the properties (a) elastic modulus and (b) yield strength with the total variation in the microstructure two-point correlations.

each successive principal component on properties will decrease, and add a physical basis for the simple linear model. In effect we are assuming the directions of greatest variability in the microstructure will have the largest influence on the change in properties. The results of the fit are shown in Figure 13, which shows that the simple linear model does an excellent job of capturing properties across the complete range of material classes.

Once an appropriate homogenization scheme is developed, property maps can be constructed by projecting the property values into the PCA space or onto the canonical representation. The projection of properties into the canonical microstructure representation is shown in Figure 14. The canonical representation was chosen as the cloud of realizations for each microstructure class are significantly overlapped in the PCA space but are well separated in the canonical representation. In contrast to Ashby charts and similar maps, constructing property maps in the microstructure space allows for the visualization of the range of average properties for a microstructure class but also to see the expected range of properties for individual microstructure realizations within a given class. Such maps offer a quick but powerful visualization of the effect of structure on properties and gives the user a means of rapid identification of microstructure realizations of interest.

Conclusions

This work was largely driven by the need, within the materials community, for effective and efficient microstructure databases and microstructure based design and development tools. In this paper, we presented a novel framework for the visualization and analysis of microstructure data which fills several of the critical technological gaps limiting the development of large scale microstructure design libraries. This work directly targets the following key technological questions:

1. What measures of the material microstructure best capture its salient features?
2. How do we describe the inherent variability in a material and how do we link this variability with scatter in properties or performance?

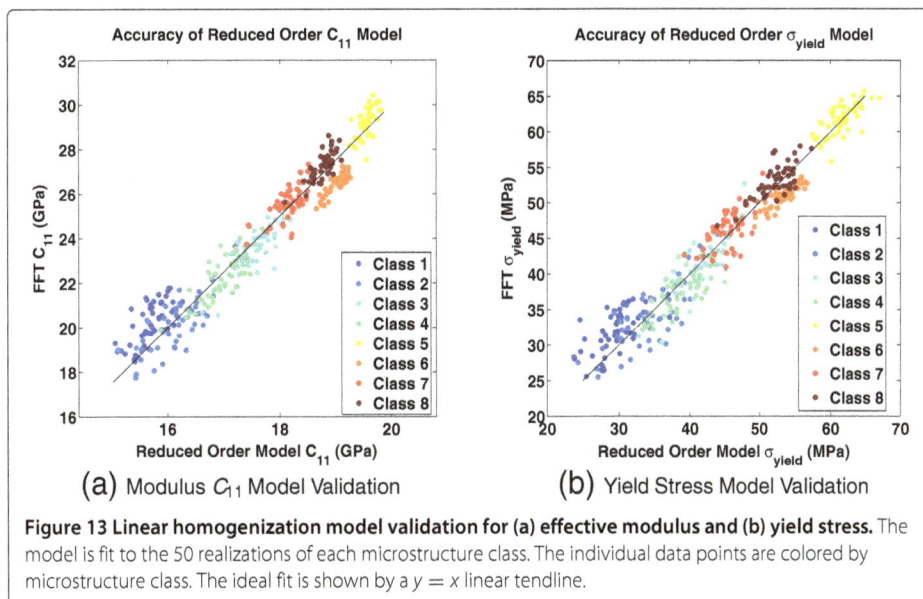

(a) Modulus C_{11} Model Validation   (b) Yield Stress Model Validation

Figure 13 Linear homogenization model validation for (a) effective modulus and (b) yield stress. The model is fit to the 50 realizations of each microstructure class. The individual data points are colored by microstructure class. The ideal fit is shown by a $y = x$ linear tendline.

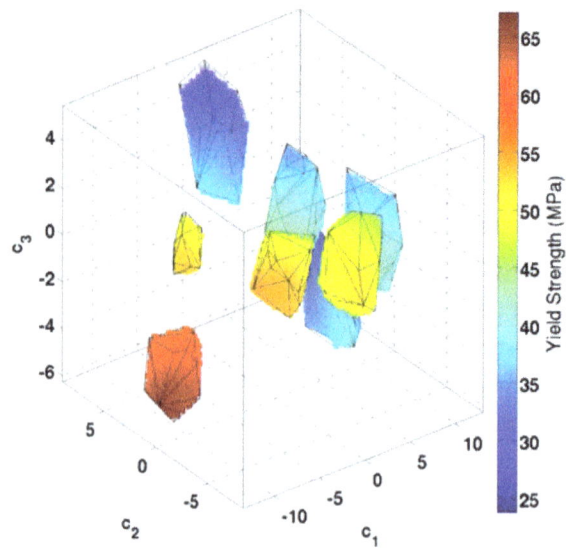

Figure 14 Projection of predicted yield stress values onto the top three canonical variables showing 1) the how yield stress varies in the microstructure space 2) average property values for each microstructure class and 3) the expected range of properties within a class. The class identification for each region can be cross referenced with Figure 5.

3. How do we quantitatively compare materials from different classes or material systems?

4. How do we place metrics on the quality or sufficiency of the microstructure data collected, or how do we place measures on how well we have characterized a particular material?

5. What objective data-driven reduced-order measures of the material microstructure are best suited for establishing the invertible structure-property relationships needed for materials design?

While a concerted effort from the larger materials community is required if such databases are to ever come to fruition, the authors believe this work represents a critical first step in their development.

In this work we developed several proof-of-concept tools and visualizations for the analysis of microstructure data. While the case studies presented were largely "toy-problems" or highly simplified examples they demonstrate the main features of the microstructure space and the range of tools that can be developed and the breadth of analyses that can be performed. While it is still in its infancy, we believe that this work represents a completely new approach to handle the analysis and visualization of microstructure data. In particular this work offers for the first time, an objective framework for the quantitative comparison of microstructure between differing material classes. These simple results potentially open completely new avenues for the exploration of structure-processing-property relationships. A key difficulty in incorporating microstructure evolution into design frameworks such as Microstructure Sensitive Design has been the lack of a suitably defined microstructure space for higher order statistical microstructure descriptors. In future work, the authors hope to explicitly incorporate microstructure evolution maps as processing path-lines or flow vectors through the microstructure space. Such maps could find great utility in process monitoring and material quality control applications, in addition to providing novel structure-processing-property visualizations to the design community.

The work presented here also has immediate relevance to the ongoing efforts in multi-scale modeling and characterization. If useful multi-scale structure/property/processing knowledge are to be developed, an automated analysis built on a rigorous mathematical formalism is necessary. Such formalism should handle a wide variety of data types across multiple length-scales. Such an approach will enable scientists and engineers to model the intrinsic randomness of microstructures, relate it to property variance, and treat it as a materials design parameter for verification and validation. Given the vast separation between the macro-, meso-, and atomic length scales of interest in most materials design problems, it should be clear that microstructure datasets collected at differing length scales are best integrated in a stochastic manner. Using this formalism a natural multi-scale representation of microstructure emerges where the structure at lower length scales is described in terms of conditional probability densities given the structure at higher length scales. Such a method aligns naturally with emerging multi-scale modeling techniques based around stochastic partial differential equations [41-43] and provides an elegant framework for the inclusion of materials uncertainty into material model uncertainty quantification and into component level design validation and verification.

The need to move beyond design based on average or effective properties is driving the integration of simulation and experiment within frameworks such as ICME and is a central theme of the Materials Genome Initiative. As time goes on this need is likely to become more acute. In the opinion of the authors, the integration of the disparate techniques and data paths required for ICME and the MGI necessitates a corresponding evolution of the way we think about microstructure data. It is the hope of the authors that this work will help jump-start discussions within the field of materials science and engineering, and pave the way for future developments.

Availability of supporting data

Computer code to create the simulated digital micrographs and to reproduce all of the case studies presented in this manuscript are available directly from SRN (Niezgoda. s@gmail.com).

Endnote

[a]This is only strictly true for Gaussian or near-Gaussian "well behaved" unimodal distributions. However, this will be the case for all microstructure classes examined here, and is expected to hold for virtually all real material systems.

Competing interests
The authors declare no competing interests.

Authors' contributions
SRN developed the case studies, performed the analysis and composed the manuscript text. AKK performed the FFT micromechanical simulations and discussed the results and manuscript drafts. SRK discussed the results and manuscript drafts. All authors read and approved the final manuscript.

Acknowledgements
S.R.N. acknowledges partial funding for this work from the Los Alamos National Laboratory LDRD program, Project 20110602ER. LANL is operated by LANS, LLC, for the National Nuclear Security Administration of the U.S. DOE under contract DE-AC52–06NA25396. S.R.K acknowledges funding from ONR award N00014–11-1–0759 (Dr. William M. Mullins, program manager).

Author details
[1]Materials Science and Technology Division, Los Alamos National Laboratory, Los Alamos NM 87545, USA. [2]Present Address: Department of Materials Science and Engineering, Department of Mechanical and Aerospace Engineering, The Ohio State University, Columbus, OH 43218, USA. [3]Department of Metallurgical and Materials Engineering, Indian Institute of Technology Madras, Chennai, 600036, India. [4]George W. Woodruff School of Mechanical Engineering, Georgia Institute of Technology, Atlanta, GA 30332, USA.

References
1. Materials Genome Initiative for Global Competitiveness. Tech. rep., National Science and Technology Council (2011)
2. Crabtree G, Glotzer S, McCrdy B, Roberto J (2010) Computational materials science and chemistry: accelerating discovery and innovation through simulation-based engineering and science. Tech. rep. U.S. Department of Energy, Office of Science
3. Allison J (2011) Integrated computational materials engineering: A perspective on progress and future steps. JOM J Minerals, Metals Mater Soc 63(4): 15–18
4. National Academy of Sciences (US) Committee on the survey of materials science and engineering (1975) Materials and man's needs: materials science and engineering – Volume II, The Needs, Priorities, and Opportunities for Materials Research. The National Academies Press, Washington
5. Rumble J (1991) Standards for Materials Databases: ASTM Committee E49. In: Kaufman J, Glazman J (eds) Computerization and networking of materials databases, Volume 2. ASTM International, pp 73–83
6. Freiman S, Madsen LD, Rumble J (2011) Perspective on materials databases. Am Ceramics Soc Bull 90(2): 28–32
7. Sundararaghavan V, Zabaras N (2004) A dynamic material library for the representation of single-phase polyhedral microstructures. Acta Materialia 52(14): 4111–4119
8. Sundararaghavan V, Zabaras N (2005) Classification and reconstruction of three-dimensional microstructures using support vector machines. Comput Mater Sci 32(2): 223–239
9. Ganapathysubramanian B, Zabaras N (2007) Sparse grid collocation schemes for stochastic natural convection problems. J Comput Phys 225: 652–685
10. Ganapathysubramanian B, Zabaras N (2008) A non-linear dimension reduction methodology for generating data-driven stochastic input models. J Comput Phys 227(13): 6612–6637
11. Niezgoda SR (2010) Stochastic representation of microstructure via higher-order statistics: theory and application. PhD thesis, Drexel University
12. Niezgoda SR, Yabansu YC, Kalidindi SR (2011) Understanding and visualizing microstructure and microstructure variance as a stochastic process. Acta Materialia 59(16): 6387–6400
13. Moulinec H, Suquet P (1998) A numerical method for computing the overall response of nonlinear composites with complex microstructure. Comput Methods Appl Mech Eng 157: 69–94
14. Lebensohn RA (2001) N-site modeling of a 3D viscoplastic polycrystal using fast Fourier transform. Acta Materialia 49(14): 2723–2737
15. Lebensohn RA, Kanjarla AK, Eisenlohr P (2012) An elasto-viscoplastic formulation based on fast Fourier transforms for the prediction of micromechanical fields in polycrystalline materials. Int J Plast 0: 59–69
16. Kanjarla AK, Lebensohn RA, Balogh L, Tomé C N (2012) Study of internal lattice strain distributions in stainless steel using a full-field elasto-viscoplastic formulation based on fast Fourier transforms. Acta Materialia 60(6–7): 3094–3106
17. Billingsley P (2012) Probability and measure, 3 edition. Wiley, New York
18. Papoulis A, Pillai SU (2002) Probability, random variables and stochastic processe. McGraw-Hill Education, Boston
19. Bunge HJ (1982) Texture analysis in materials science: mathematical methods. Butterworths, London
20. Tewari A, Gokhale AM, Spowart JE, Miracle DB (2004) Quantitative characterization of spatial clustering in three-dimensional microstructures using two-point correlation functions. Acta Materialia 52(2): 307–319
21. Torquato S (2001) Random heterogeneous materials: microstructure and macroscopic properties. Springer, New York
22. Fullwood DT, Niezgoda SR, Adams BL, Kalidindi SR (2010) Microstructure sensitive design for performance optimization. Prog Mater Sci 55(6): 477–562
23. Huang M (2005) The n-point orientation correlation function and its application. Int J Solids Struct 42(5–6): 1425–1441
24. Adams BL, Morris PR, Wang TT, Willden KS, Wright SI (1987) Description of orientation coherence in polycrystalline materials. Acta Metallurgica 35(12): 2935–2946
25. Niezgoda SR, Fullwood DT, Kalidindi SR (2008) Delineation of the space of 2-point correlations in a composite material system. Acta Materialia 56(18): 5285–5292
26. Lee JA, Verleysen M (2007) Nonlinear dimensionality reduction. Springer, New York
27. Fukunaga K (1990) Introduction to statistical pattern recognition. Academic Press, Boston
28. Turk M, Pentland A (1991) Eigenfaces for recognition. J Cogn Neurosci 3: 71–86
29. Li Y (2004) On incremental and robust subspace learning. Pattern Recognit 37(7): 1509–1518
30. Yeong CLY, Torquato S (1998) Reconstructing random media. Phys Rev E 57: 495
31. Fullwood DT, Niezgoda SR, Kalidindi SR (2008) Microstructure reconstructions from 2-point statistics using phase-recovery algorithms. Acta Materialia 56(5): 942–948
32. Stevens JP (2001) Applied multivariate statistics for the social sciences. Lawrence Erlbaum, Mahwah
33. Qidwai SM, Turner DM, Niezgoda SR, Lewis AC, Geltmacher AB, Rowenhorst DJ, Kalidindi SR (2012) Estimating the response of polycrystalline materials using sets of weighted statistical volume elements. Acta Materialia 60(13): 5284–5299
34. Steinwart I, Christmann A (2008) Support vector machines. Springer, New York
35. Cortes C, Vapnik V (1995) Support-vector networks. Mach Learn 20(3): 273–297

36. Boser BE, Guyon IM, Vapnik VN (1992) A training algorithm for optimal margin classifiers In: Proceedings of the fifth annual workshop on Computational learning theory, 5. ACM, Pittsburgh, pp 144–152
37. Hsu CW, Lin CJ (2002) A comparison of methods for multiclass support vector machines. IEEE Trans Neural Netw 13(2): 415–425
38. Chang CC, Lin CJ (2011) LIBSVM: a library for support vector machines. ACM Trans Intell Syst Technol 2(3): 27
39. Adams BL, Henrie A, Henrie B, Lyon M, Kalidindi SR, Garmestani H (2001) Microstructure-sensitive design of a compliant beam. J Mech Phys Solids 49(8). 1639–1663
40. Niezgoda SR, Turner DM, Fullwood DT, Kalidindi SR (2010) Optimized structure based representative volume element sets reflecting the ensemble-averaged 2-point statistics. Acta Materialia 58(13): 4432–4445
41. Karakasidis T, Charitidis C (2007) Multiscale modeling in nanomaterials science. Mater Sci Eng: C 27(5): 1082–1089
42. Xu X (2007) A multiscale stochastic finite element method on elliptic problems involving uncertainties. Comput Methods Appl Mech Eng 196(25): 2723–2736
43. Li Z, Wen B, Zabaras N (2010) Computing mechanical response variability of polycrystalline microstructures through dimensionality reduction techniques. Comput Mater Sci 49(3): 568–581

2

Phase field simulation of dendrite growth with boundary heat flux

Lifei Du[*] and Rong Zhang

* Correspondence:
dulifei@mail.nwpu.edu.cn
Key Laboratory of Space Applied
Physics and Chemistry-Ministry of
Education, School of Science,
Northwestern Polytechnical
University, Xi'an 710072, China

Abstract

Boundary heat flux has a significant effect on solidification behavior and microstructure formation, for it can directly affect the interfacial heat flux and cooling rate during phase transition. In this study, a phase field model for non-isothermal solidification in binary alloys is employed to simulate the free dendrite growth in undercooled melts with induced boundary heat flux, and an anti-trapping current is introduced to suppress the solute trapping due to the larger interface width used in simulations than a real solidifying material. The effect of heat flux input/extraction from different boundaries was studied first. With heat input from boundaries, the temperature can be raised and the dendritic morphology changed with gradient temperature distribution caused by the heat flux input coupling with latent heat release during the liquid-solid phase transition. Also, the concentration distribution can be also influenced by this irregular temperature distribution. Heat flux extraction from the boundaries can decrease the temperature, which results in rapid solidification with small solute segregation and concentration changes in the dendrite structures. Also, dendrite growth manner changes caused by undercooling variation, the result of competition between heat flux and latent heat release from phase transition, are also studied. Results indicate that heat flux in the simulation zone significantly reduces the undercooling, thus slowing down the dendrite formation and enhancing the solute segregation, while large heat extraction can enlarge the undercooling and lead to rapid solidification with large dendrite tip speed and small secondary dendrite arm spacing, while solute segregation tends to be steady. Therefore, the boundary heat flux coupling with the latent heat release from the solidification has an effective influence on the temperature gradient distribution within the simulation zone, which leads to the morphology and concentration changes in the dendritic structure formation.

Keywords: Computer simulations; Metals and alloys; Rapid solidification; Microstructure; Diffusion

Background

The mechanical properties of many materials have a significant relationship with the microstructure formation process [1], but in practice, it is difficult to observe the microstructure formation process with experimental methods, and computer simulations can visually show the phase transition process and provide much more information to calculate many other parameters that are related to the mechanical properties with the data achieved from simulations. For decades, the phase field method has become a popular technique to model various types of complex microstructure changes qualitatively, such as solidification, spinodal decomposition, Ostwald ripening, crystal growth and recrystallization, domain micro-structure evolutions in ferroelectric materials, martensitic transformation, dislocation dynamics, and crack propagation [2,3]. The phase field model has been used for computing solidification morphologies to avoid the explicit boundary tracking needed to solve the classical sharp interface model. With this advantage, phase field methods have attracted considerable interest in the last decades as a means of simulating the solidification process. In phase field models for solidification process, a variable $\phi(r, t)$, called the phase field order parameter, is introduced to indicate the physical state of the system in time and space. This order parameter $\phi(r, t)$ is assigned the value 0 (or −1) in the bulk solid phase and 1 in the bulk liquid phase, and it changes smoothly between these values over a thin transition layer that plays the role of the classical sharp interface. The governing equations for the growth of a solid phase from a liquid need to be derived from the irreversible thermodynamic law and conservation laws for both mass and energy, and the resulting equations need to be applicable to the entire space being modeled without any discontinuities between the various phases present. It also must be possible to determine the parameters in the governing equations from classical thermodynamic and kinetic quantities.

In the past decades, many researches have studied the solidification process using the phase field method [4,5]. Kobayashi [6] developed a simple phase field model for one-component melt growth including anisotropy and used this model to study the formation of various dendritic patterns. He found that the qualitative relations between the shapes of crystals and some physical parameters and noises gave a crucial influence on the side branch structure of dendrites. The first phase field model for alloys was developed by Wheeler et al. [7,8], called the WBM model. Kim et al. [9,10] presented another model for alloys by adopting the thin-interface limit, which is known as the KKS model. Karma [11] presented a phase field formulation to quantitatively simulate microstructural pattern formation in alloys, and the thin-interface limit of this formulation yielded a much less stringent restriction on the choice of interface thickness than previous formulations and permitted one to eliminate non-equilibrium effects at the interface. Dendrite growth simulations with vanishing solid diffusivity showed that both the interface evolution and the solute profile in the solid were accurately modeled by this approach. Recently, solidifications with forced flow or convection were studied in binary alloys in 2D and 3D [12-16]; solidifications of multi-component and multiphase were also studied using phase field methods [17-19].

As known, the undercooling in the liquid melt has a significant effect on the microstructure forming process and thus affects solute diffusion to change the concentration distributions within the structure. In industry, a cooling/heating system can be applied on the chilling wall during the casting process to control the solidification. So, in this study, we introduce complex boundary conditions in order to present the different cooling/heating conditions to find the relation between the structure forming and these solidification conditions. Experimental and numerical investigations have been presented in the past decades to study the heat transfer problems during solidification [20-24] and find that heat flow can significantly affect the temperature distribution and thus determine the quality of the solidification process. But few studies have been carried out to study the heat flux at the boundaries and its influence on phase transition using phase field methods [25]. Especially, the effect of heat flux on microstructures and distributions of temperature and concentration still need to be investigated in detail, for temperature field has a directional influence on the solidification process with undercooling. So, in this study, the boundary heat coupling with latent heat release from solidification is the main factor affecting dendritic structure forming in undercooling melts. Different dendrite patterns are obtained by changing the boundary heat flux at different boundaries and directions. Distributions of the concentration change as the result of temperature change caused by different boundary heat flux couplings with latent heat release are also given and analyzed in detail.

Methods

Model description and simulation method

The phase field model for non-isothermal simulation of binary systems [26,27] is implemented to study the flow field effect on the concentration and temperature distributions during dendritic growth. The main governing equations are listed below:

- Phase field equation

$$\frac{\partial \phi}{\partial t} = M_\phi \bar{\varepsilon}^2 \left[\nabla \cdot \left(\eta^2 \nabla \phi \right) - \frac{\partial}{\partial x} \left(\eta \eta_\beta^{'} \frac{\partial \phi}{\partial y} \right) + \frac{\partial}{\partial y} \left(\eta \eta_\beta^{'} \frac{\partial \phi}{\partial x} \right) \right] - M_\phi ((1-x_B)H_A + x_B H_B) \tag{1}$$

- Diffusion equation

$$\frac{\partial x_B}{\partial t} = \nabla \cdot D \left[\nabla x_B + \frac{V_m}{R} x_B (1-x_B)(H_B(\phi, T) - H_A(\phi, T)) \nabla \phi + \vec{j}_{at} \right] \tag{2}$$

- Temperature equation

$$c_p \frac{\partial T}{\partial t} + 30 g(\phi) \Delta \widetilde{H} \dot{\phi} = \nabla \cdot K \nabla T \tag{3}$$

In this model, the phase field variable ϕ varies smoothly between 0 in the solid phase and 1 in the liquid phase as we assumed, x_B is the mole fraction of solute B in solvent A, and T is the temperature. ε is the coefficient of gradient energy, which is determined by the interfacial energy. The anisotropy is included in the system because the phase change kinetics depends upon the orientation of the interface. Here, we introduce the anisotropy $\varepsilon = \bar{\varepsilon} \eta = \bar{\varepsilon}(1 + \gamma cos\kappa\beta)$, where $\bar{\varepsilon}$ is related to the surface energy σ and interface thickness λ, γ is the magnitude of anisotropy in the surface energy, κ specifies the mode number, and the expression $\beta = \arctan((\partial\phi/\partial y)/(\partial\phi/\partial x))$ gives an approximation of the angle between the interface normal and the orientation of the crystal lattice. $\vec{j}_{at}$ is the anti-trapping current introduced by Karma [11] to suppress the solute trapping due to the larger interface width used in simulations in order to get a more quantitative prediction.

The formulations included in these governing equations are the following:

$$H_A(\phi, T) = W_A g^{'}(\phi) + 30 g(\phi) \Delta H_A \left(\frac{1}{T} - \frac{1}{T_m^A} \right) \tag{4}$$

$$H_B(\phi, T) = W_B g'(\phi) + 30g(\phi)\Delta H_B \left(\frac{1}{T} - \frac{1}{T_m^B} \right) \tag{5}$$

$$c_p = (1-x_B)c_A + x_B c_B \tag{6}$$

$$\Delta \tilde{H}' = (1-x_B)\Delta H_A + x_B \Delta H_B \tag{7}$$

$$K = (1-x_B)K_A + x_B K_B \tag{8}$$

$$\vec{j}_{at} = a\lambda(1-k)\left[\frac{2x_B}{1+k-(1-k)\phi} \right] \frac{\partial \phi}{\partial t} \frac{\nabla \phi}{|\nabla \phi|} \tag{9}$$

where $g(\phi) = \phi^2(1-\phi)^2$ and W_A, W_B are constants, T_m^A and T_m^B are the melting points of pure A and pure B, respectively. ΔH_A and ΔH_B are the heats of fusion per volume, c_A and c_B are their heat capacities, and R is the gas constant. $p(\phi) = \phi^3(10 - 15\phi + 6\phi^2)$ is a smoothing function, chosen such that $p'(\phi) = 30\,g(\phi)$. The diffusion coefficient is postulated as a function of the phase field variable, $D = D_S + p(\phi)(D_L - D_S)$, where D_S and D_L are the classical diffusion coefficients in the solid and liquid, respectively. a in Equation 9 is the anti-trapping coefficient and needs to be adjusted to fit the solid concentration of the sharp-interface solution, and in the present calculations, $a = 1/\sqrt{2}$ [28]. $k = cs/cl$ is the partition coefficient, where c_L (c_S)is the concentration on the liquid (solid) side of the interface.

Governing equations are solved using the standard finite difference methods with the tridiagonal matrix algorithm (TDMA), and time stepping is by explicit Euler scheme. All simulations are carried out in a $1{,}200\Delta x \times 1{,}200\Delta y$ simulation box with $\Delta x = \Delta y = 0.96\lambda = 4.6 \times 10^{-8}$ m, and the time step is chosen as $\Delta t = 1.0 \times 10^{-8}$ s. All simulations are carried out for Ni-Cu binary alloy, and the physical parameters are listed in Table 1. The initial concentrations of the melt, x_0, and the initial temperature, T_0, are chosen to make the whole region initially contain supersaturated (0.86) and undercooled ($\Delta T = 20.5$ K) melt. Zero-Neumann boundary conditions for ϕ and x_B are imposed at all boundaries. The boundaries with different heat fluxes are chosen in the temperature-field calculations, and the boundary heat flux can be introduced as $Q = \lambda_i(\partial T/\partial x)$; the density of the heat flux at boundaries is the control parameter, which determines the magnitude and direction of the heat flux.

Results and discussion
Free dendrite growth in undercooling melt
Figure 1 gives the morphology, concentration, and temperature distribution maps at $t = 2.0$ ms during the free dendritic growth in an undercooled melt without heat flux. Typical dendrite microstructure forms with secondary arms due to the random noise introduced in the simulation. The whole dendritic structure and distributions of temperature and concentration have the approximate rotational symmetry. The primary arms have the largest growth speed and the secondary arms' growth was restrained because of high temperature due to the latent heat release during the liquid-solid phase transition. The solute level between the secondary arms is higher than that inside the solid arm because of solute segregation, and the cells of high concentration formed inside of the dendrite structure because rapid solidification in undercooled melts prevents the

Table 1 Physical parameters used in the simulations [7,26,27]

Parameter	Value
Melting temperature of Ni	1,728.0 K
Melting temperature of Cu	1,358.0 K
Latent heat of Ni	2.35×10^9 J/m^3
Latent heat of Cu	1.728×10^9 J/m^3
Heat conductivity of Ni	84.0 W/mK
Heat conductivity of Cu	200.0 W/mK
Specific heat of Ni	5.42×10^6 J/m^{3}K
Specific heat of Cu	3.96×10^6 J/m^{3}K
Diffusion coefficient of the liquid phase	1.0×10^{-9} m^2/s
Diffusion coefficient of the solid phase	1.0×10^{-13} m^2/s
Mole volume of alloy	7.42×10^{-6} m^3/mol
Surface energy of Ni	0.37 J/m^2
Surface energy of Cu	0.29 J/m^2
Interfacial kinetic coefficient of Ni	3.3×10^{-3} m/sK
Interfacial kinetic coefficient of Cu	3.9×10^{-3} m/sK
Interface thickness	4.9×10^{-8} m
Amplitude of the noise fluctuations	0.4
Model of anisotropy	4
Magnitude of anisotropy	0.04

diffusion of the solute. The low solute level presented in the middle of each dendrite arm is also caused by the rapid phase transition with undercooling.

The concentration and temperature profiles along the vertical lines labeled in Figure 1b are plotted in Figure 2. The solute diffusion layer is thinner in front of the primary arm tip than that beside the dendrite arms. The solute segregation is reduced in front of the primary dendrite arm tip compared with that beside the arms because sidebranching increases the temperature beside the primary arms as shown in Figure 2. The temperature distribution profiles indicate that the temperature along the primary arms is unsteady because of the side-branching effect, and latent heat release during the liquid-solid phase transition is the main factor to affect the temperature distribution. The solid phase has higher temperatures because the rapid solidification due to large undercooling prevents the heat diffusion. The temperature gradient formed due to the latent heat release changes the solute segregation and irregular microstructure formation. The solute diffusing inside the solid-liquid interface is affected by the temperature level and thus determines the solute level inside the solid dendrite.

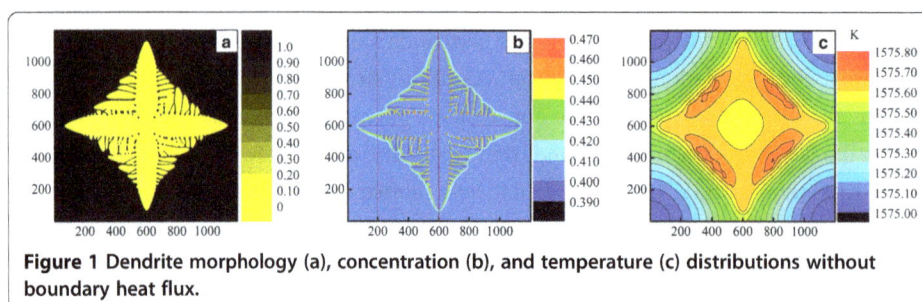

Figure 1 Dendrite morphology (a), concentration (b), and temperature (c) distributions without boundary heat flux.

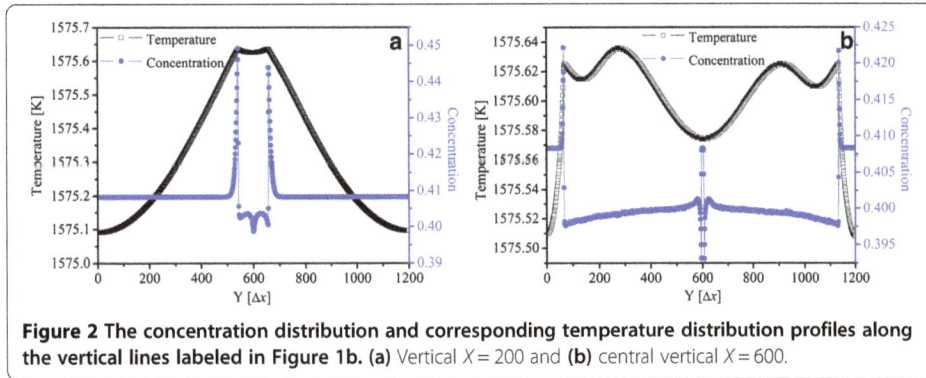

Figure 2 The concentration distribution and corresponding temperature distribution profiles along the vertical lines labeled in Figure 1b. **(a)** Vertical $X = 200$ and **(b)** central vertical $X = 600$.

Dendrite growth with boundary heat flux

Different heating/cooling conditions during the solidification can significantly change the temperature distributions, thus affecting the solute diffusion and dendrite structure formation. In the following simulations, boundary heat flux is introduced to present different heating/cooling modes during rapid dendrite growth in undercooling melts. Latent heat release during phase transition is also considered in these non-isothermal simulations.

The morphology, concentration, and temperature distribution maps at $t = 2.0$ ms during dendritic growth with heat flux input/extraction of $|\lambda_i| = 10 \times 10^{-3}$ W/m^2 from the north (top) boundary are shown in Figure 3. With heat flux input, temperatures near the north boundary are raised, and this prevents dendrite arm growth and side-branching close to the north boundary. The highest temperature zone finds its position near the primary dendrite arm tip growing towards the north boundary due to the heat flux input coupling with the latent heat release. At the same time, with heat extraction from the north boundary, the primary arm growing towards the north has a high velocity, and sidebranching beside this primary arm is enhanced due to the large undercooling caused by the heat extraction. The length of the primary arm growing towards the north boundary with heat flux input is 12% less than that with heat flux extraction. The temperature map in Figure 3f shows that the lowest temperature zone

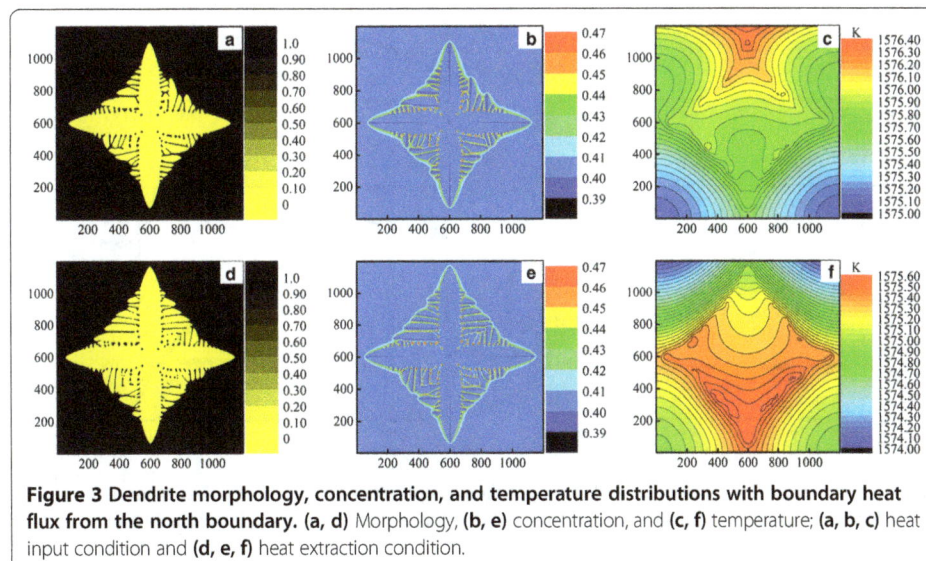

Figure 3 Dendrite morphology, concentration, and temperature distributions with boundary heat flux from the north boundary. **(a, d)** Morphology, **(b, e)** concentration, and **(c, f)** temperature; **(a, b, c)** heat input condition and **(d, e, f)** heat extraction condition.

is located at the corners near the north boundary, and the temperate gradient has an opposite direction compared with that of heat input. The concentration distributions have a similar map with heat input and extraction in these two cases, but the solute diffusion is enhanced by high temperature, which shows a thicker solute diffusion layer with slow dendritic structure formation, while large temperature undercooling prevents solute diffusion and leads to an enhanced structure formation with thinner solute diffusion layers.

The concentration distribution and corresponding temperature distribution profiles along the vertical lines are plotted in Figure 4. The solute level along primary dendrite keeps decreasing with the distance from the dendrite center, and this effect is enhanced with the heat flux input from the north boundary. While with the heat flux extraction from the north boundary, an opposite temperature gradient was formed; thus, the solute distribution along the primary arm still keeps decreasing with the temperature rising. That is because the heat flux input from the boundary significantly enlarges the solute segregation near the boundary and leads to an efficient solute diffusion, thus resulting in low solute distribution with high temperature. With heat extraction from the boundary, we find that the solute distribution along the primary dendrite arm near the north boundary keeps steady unlike that without heat flux; this is because the heat extraction partly balances the latent heat release from the liquid-solid phase transition, which keeps the temperature almost unchanged at the solid-liquid interface and makes the solute diffusion approximately steady.

In order to study the complex heat flux effect on the dendrite structure forming process, we simulated the dendrite growing process with heat flux at all boundaries. The morphology, concentration, and temperature distribution maps at $t = 2$ ms are shown in Figure 5. With heat input from the south and north boundaries coupling with heat extraction from east and west boundaries, the dendrite primary arms growing in

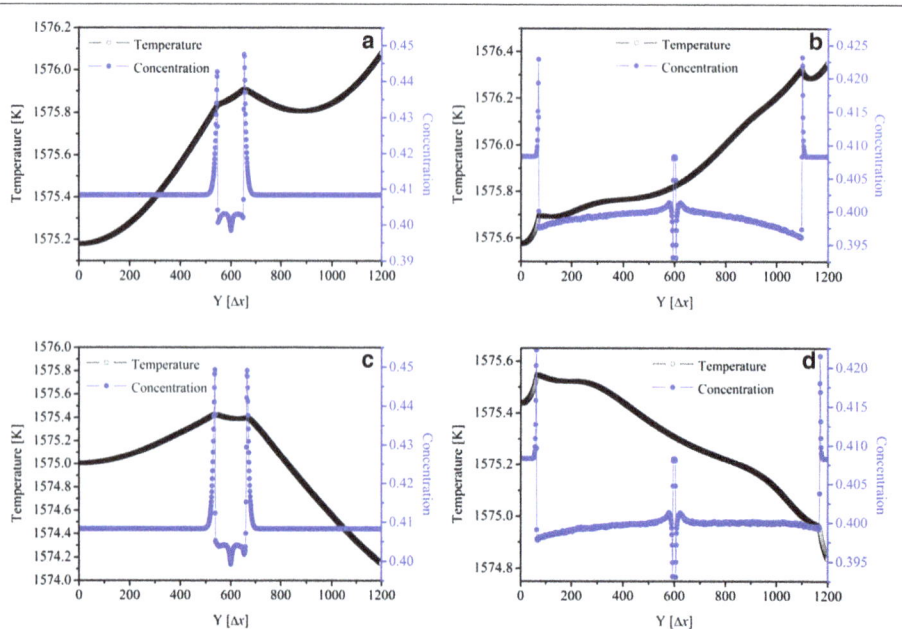

Figure 4 The concentration distribution and corresponding temperature distribution profiles associated with heat flux input/extraction of $|\lambda i| = 10 \times 10^{-3}$ W/m^2 along the vertical lines labeled in Figure 1b. **(a, c)** The vertical $X = 200$ and **(b, d)** $X = 600$.

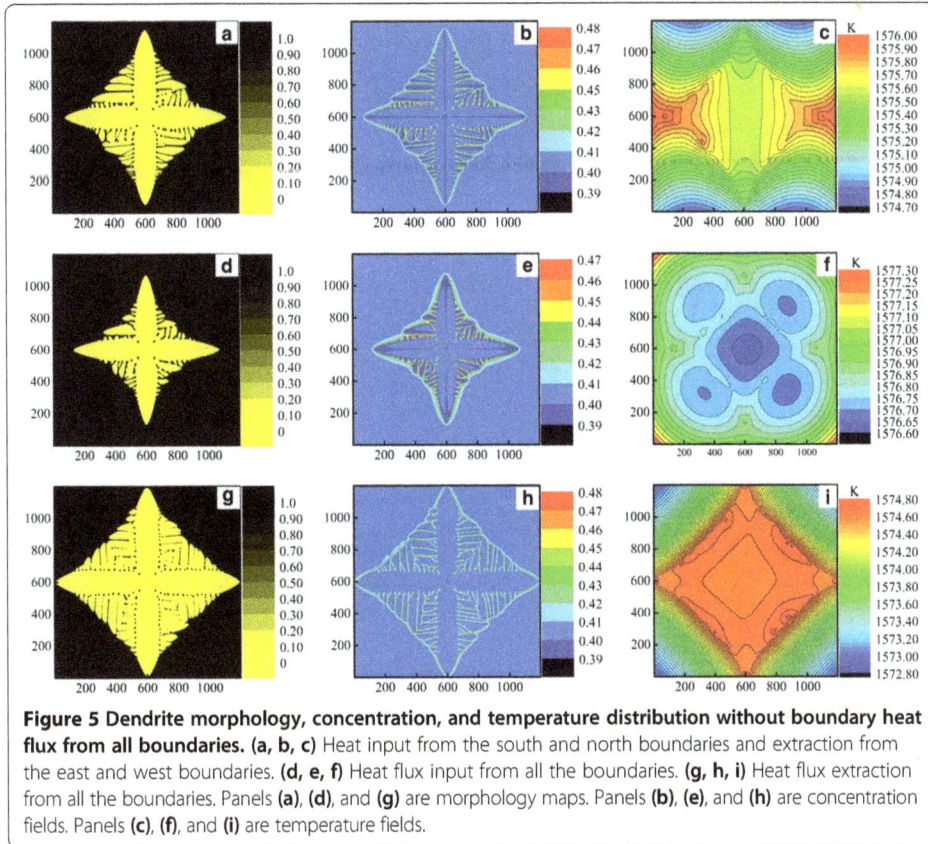

Figure 5 Dendrite morphology, concentration, and temperature distribution without boundary heat flux from all boundaries. (a, b, c) Heat input from the south and north boundaries and extraction from the east and west boundaries. **(d, e, f)** Heat flux input from all the boundaries. **(g, h, i)** Heat flux extraction from all the boundaries. Panels **(a)**, **(d)**, and **(g)** are morphology maps. Panels **(b)**, **(e)**, and **(h)** are concentration fields. Panels **(c)**, **(f)**, and **(i)** are temperature fields.

the vertical direction grow fast with much more side branching, while the side branching on the primary arms growing in the horizontal direction is prevented. The length of the primary dendrite arms growing in the horizontal direction is 2.5% less than that growing in the vertical direction. A large temperature gradient formed between the heat extraction boundaries and solid front, and the temperature has an approximately symmetric distribution both in the vertical and horizontal orientations. Comparing the dendrite pattern formed with heat input/extraction from all boundaries, full dendrite structure can be achieved with heat extraction, and the secondary arms are almost full size and the distance between these arms is small. While with heat input from all boundaries, dendrite growth is suppressed and the number and size of the secondary dendrite arms both decrease. The length of the primary arm growing with heat flux input is 10% less than that with heat flux extraction. But with high temperature, the solute diffusion is enhanced, and thus, the solute level inside the dendrite is low, especially at the tip of the dendrite arms. But with the large undercooling caused by heat extraction from all boundaries, the solute level inside the solid almost keeps steady with the dendrite arm growth, and the solute diffusion layers is thinner than that with high temperature.

Figure 6 shows the concentration distributions and the corresponding temperature profiles along different vertical lines. The solute segregation beside the primary arms is enlarged by high temperature, and the solute distribution along the center of the primary arm in the vertical direction shows a large difference with different boundary heat conditions. Heat input from the boundary can enhance the solute diffusion and

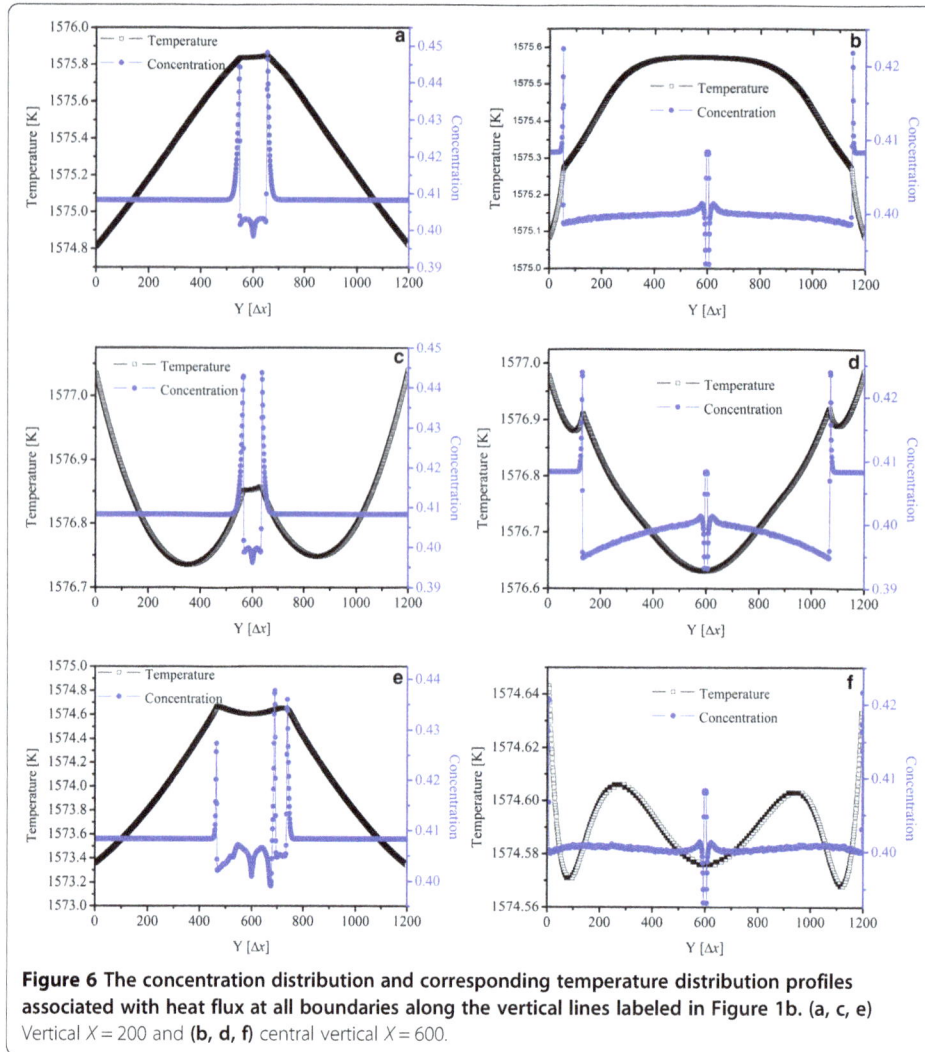

Figure 6 The concentration distribution and corresponding temperature distribution profiles associated with heat flux at all boundaries along the vertical lines labeled in Figure 1b. (a, c, e) Vertical X = 200 and (b, d, f) central vertical X = 600.

can steadily decrease the solute level with the primary dendrite arm growth while the heat extraction from the boundaries keeps the solute level steady inside the primary arms. High temperature caused by the heat input led to an unsteady solute level, while heat extraction from the boundary makes the solute diffusion steady at the solid-liquid interface region, and thus leads to the uniform solute distribution along the primary arms, which can be considered as a well-distributed phase.

We plotted the temperature distribution profiles along the primary arm growth in the vertical direction at different time steps in Figure 7, and we can see that with heat input, the temperature in front of the dendrite tip keeps increasing. Thus, the dendrite forms at different solute diffusion levels caused by the temperature change, which results in the decreasing solute distribution inside the dendrite arms. While with heat extraction from the boundary, the temperature gradient change with time is small, though the temperature distribution shows an irregular change with time evolution. But at the solid-liquid interface, which is pointed out by arrows in Figure 7b, the temperature difference is very small; thus, the phase transition almost takes place at the same temperature. So, the solute diffusion keeps a steady level, and this leads to the uniform solute dendrite microstructure.

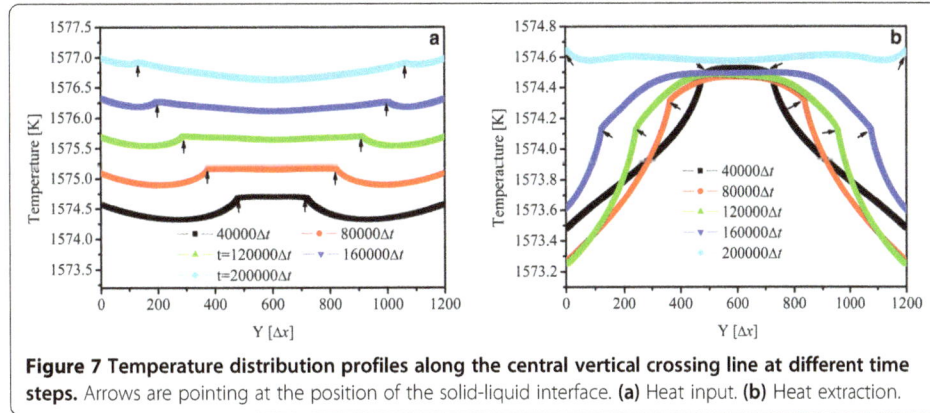

Figure 7 Temperature distribution profiles along the central vertical crossing line at different time steps. Arrows are pointing at the position of the solid-liquid interface. **(a)** Heat input. **(b)** Heat extraction.

Effect of different magnitudes of heat flux

From the 'Dendrite growth with boundary heat flux' section, we can see that the heat flux induced from all boundaries has a significant effect on the heat and solute diffusion as well as the microstructure forming during dendrite growth in undercooled melts. In this section the effect of boundary heat fluxes of different values is investigated. Figure 8 gives the concentration distribution maps with different heat fluxes at the same time. It is clear that with heat extraction, the dendritic structure formation is enhanced, and the more heat extracted, the much more enhanced the structure was. The solute diffusion was suppressed with large undercooling caused by heat extraction, so the solute inside dendritic structures showed a relatively uniform distribution. At the same time, with heat input from all boundaries, the temperature is raised in melts, and high temperature makes the dendrite growth constrained with enhanced solute diffusion. The solute diffusion layer is enlarged with large heat flux at boundaries, which leads to solute diffusion change in the dendritic structure. Especially with large heat input, the dendrite forms with weak side branching, and the primary arms have small growing velocity and tip radius. Figure 9a shows the velocity of the primary dendrite tip at different times; we can see that at the early stage of solidification, all tip growths decrease because of the latent heat release at

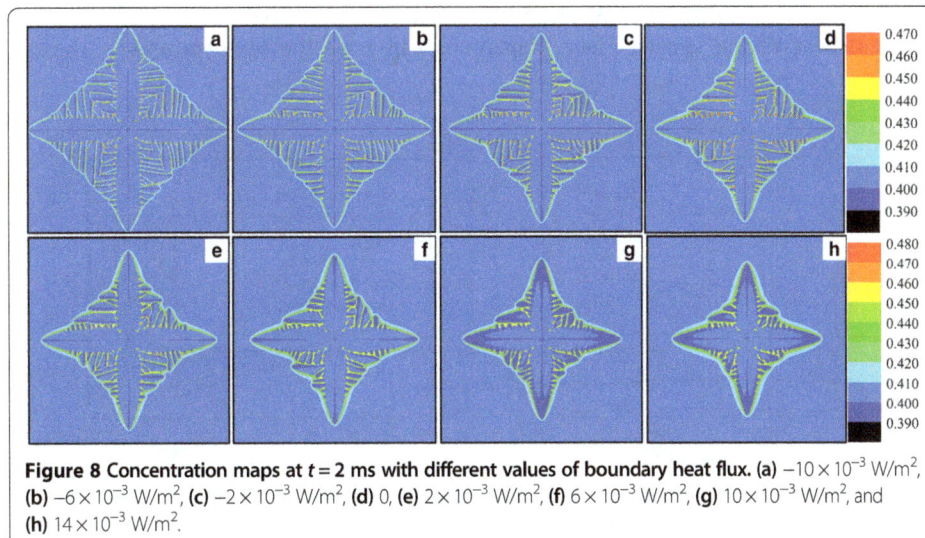

Figure 8 Concentration maps at $t = 2$ ms with different values of boundary heat flux. (a) -10×10^{-3} W/m^2, **(b)** -6×10^{-3} W/m^2, **(c)** -2×10^{-3} W/m^2, **(d)** 0, **(e)** 2×10^{-3} W/m^2, **(f)** 6×10^{-3} W/m^2, **(g)** 10×10^{-3} W/m^2, and **(h)** 14×10^{-3} W/m^2.

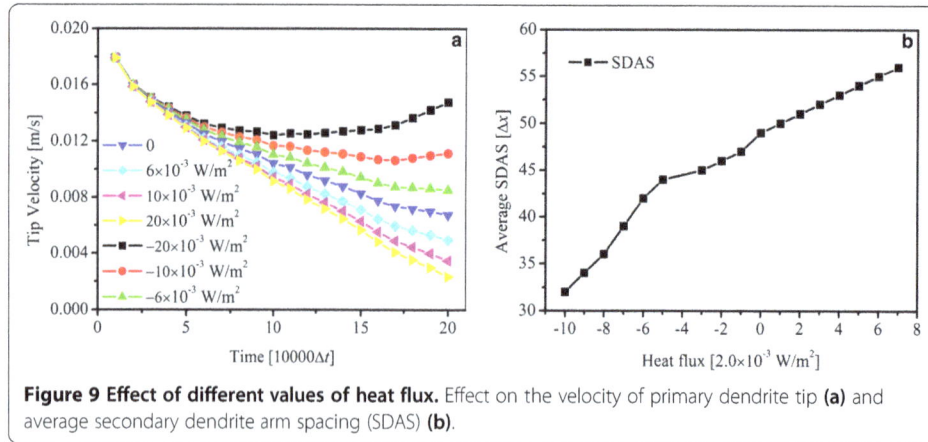

Figure 9 Effect of different values of heat flux. Effect on the velocity of primary dendrite tip **(a)** and average secondary dendrite arm spacing (SDAS) **(b)**.

the liquid-solid interface. With time evolution, boundary heat flux begins to show its influence on the tip growth: a proper heat extraction from the boundaries can make the tip growth reach a steady speed, and large heat extraction from the boundaries increases the tip growth speed, while heat flux input slows down the tip growth compared with that without heat flux at the boundaries. The average secondary dendrite arm spacing (SDAS) changes with different heat fluxes are plotted in Figure 9b, and the average SDAS keeps increasing with heat flux, changing from large heat extraction to large heat input. A similar effect has been reported during the secondary dendrite arm coarsening in Sn-Bi alloy by means of synchrotron microradiography [29].

In order to understand the relations between temperature and solute segregation inside the interface during phase transition, we plot the temperature and concentration distributions along the central vertical lines at $t = 2$ ms with different values of boundary heat flux in Figure 10. The convex in each temperature line indicates the position of the interface since the latent heat released during the phase transition can directly increase the temperature at the interface. The concentration distribution lines plotted in Figure 10b indicate that the solute segregation at the interface is non-equilibrium due to the rapid solidification caused by undercooling. Since the heat diffusion almost takes place in liquid melt in these simulations, the corresponding temperature distribution lines recorded the temperatures at which the phase transition occurred. As we know, latent heat release during phase transitions can raise the temperatures at the

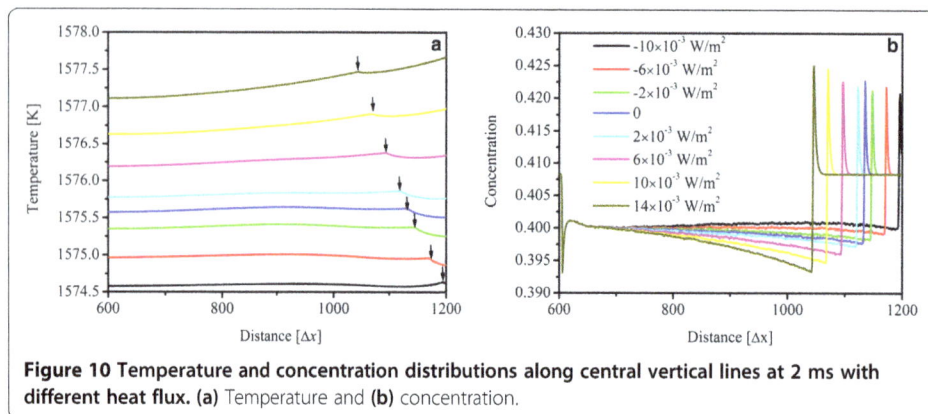

Figure 10 Temperature and concentration distributions along central vertical lines at 2 ms with different heat flux. (a) Temperature and **(b)** concentration.

interface, and this effect is enlarged with time evolution; so, the temperature at which the solidification takes place increases with time, which leads to solute diffusion changes with time. Thus, the concentration distribution inside the dendrite primary arms keeps changing with time, too. With heat flux extracted from the boundaries, the temperature in the melt decreases with time at a certain rate, and this can partly counterbalance the latent heat release, which makes the phase transition at a steady velocity (as shown in Figure 9a) with the unchanged temperature, and this leads to uniform concentration distributions inside dendrite tips, which can be seen with $\lambda_i = -10 \times 10^{-3}$ W/m^2 in Figure 10b. But at the same time, with heat flux input, the temperature is raised in the melt. Coupling with latent heat release, the temperature at which solidification takes place is significantly increased, and solute segregation is kept enhanced with time evolution, which decreases the concentration with time. And this effect could be enlarged with the increasing of heat flux input. Figure 11 gives the maximum and minimum concentration inside the solid-liquid interface near the tip of the primary dendrite arm, and a solute diffusion coefficient was defined as $kc = c_{min}/c_{max}$, where c_{min}/c_{max} is the minimum/maximum value of the concentration near the interface of the primary tip at the same time. The minimum concentration keeps decreasing with larger heat flux, while the maximum concentration keeps increasing; thus, the solute diffusion coefficient shows an increasing trend with increasing heat flux. The solute diffusion coefficient k_c is equal to the solute segregation coefficient $k_v = c_{solid}/c_{liquid}$ in the solid-liquid interface near the dendrite tip, where c_{solid}/c_{liquid} represents the equilibrium concentration at the interface. As shown in Figure 11, the large undercooling resulting from the heat extraction drives k_v toward 1 during rapid solidification which is confirmed with the approximate analytic solution about the rapid dendrite growth in undercooled melts [30].

Therefore, heat flux induced from boundaries can significantly affect the dendrite growth in the undercooled melt. As shown in Figure 12, the undercooling of the simulation zone is decreasing with enhanced heat flux input from the boundaries, which leads to the solid rate decreasing with large heat flux. Large heat extraction from the boundaries can enhance the dendrite microstructure formation with strong side branches, while large heat input at the boundaries prevents the dendrite growth as well as secondary arm formations. Similar results have been observed and reported in many experimental studies in recent years [31-35].

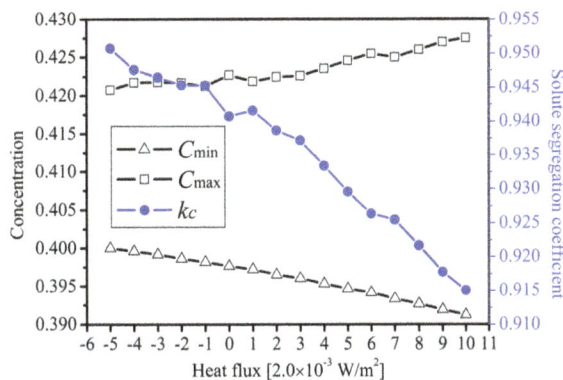

Figure 11 Effect of different heat fluxes on the solute diffusion coefficient.

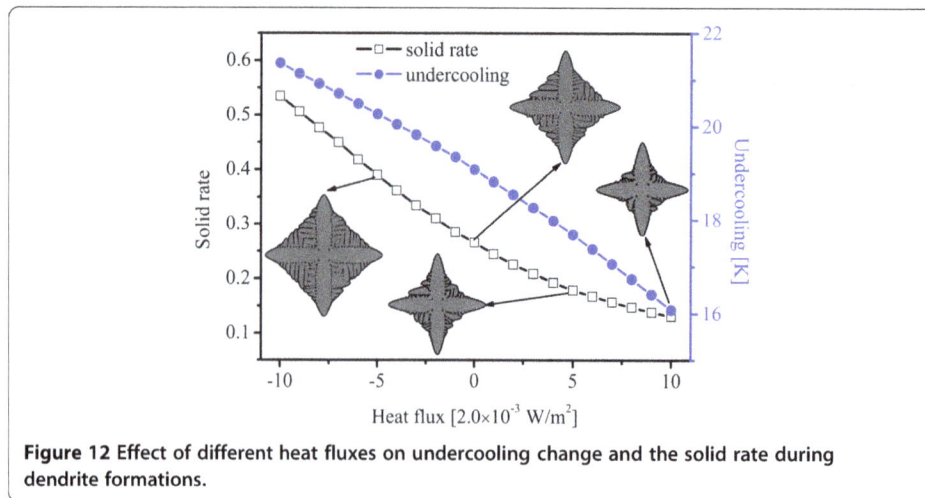

Figure 12 Effect of different heat fluxes on undercooling change and the solid rate during dendrite formations.

Conclusions

A phase field model for simulating solidification in binary alloys under non-isothermal conditions is implemented to study the effect of heat flux input/extraction from boundaries on the dendrite structure forming process in a Ni-Cu alloy. In order to suppress the solute trapping due to the larger interface width used in simulations, the anti-trapping current is introduced into the diffusion equation. Simulations were carried out to study the dendrite structure formation, concentration, and temperature changes with different boundary heat fluxes. It is obvious that heat flux input can raise the temperature near the boundary and results in the prevention of dendrite arm growth and side branching, while heat flux extraction from the boundaries can enhance the dendrite arm growth and secondary arm formations. The heat flux can also lead to the concentration distribution changes by changing the temperature distribution. High temperature can enlarge the solute segregation and decrease the solute level inside the dendrite structure decrease. The effect of different heat fluxes on solute segregation and concentration distributions inside dendrite structures is studied. With heat flux changing from large extraction to large input, the dendrite tip speed of the primary arm shows different changes, and the average secondary dendrite arm spacing decreases. Heat extraction from the boundaries could partly counterbalance latent heat release from solidification and lead to uniform concentration distributions in solid, while heat input, coupling with latent heat, would significantly raise temperatures at interfaces and decreases concentration with time evolution. These effects could be enlarged with increasing heat flux. Therefore, the heat flux input/extraction from the boundary can significantly change the morphology, concentration, and temperature distribution of the dendrite growth in undercooled melts, and these morphology and concentration changes will affect the properties of the alloys.

Availaiblity of supporting data

The computer code to reproduce all of the case studies presented in this manuscript is available upon request.

Competing interests

The authors declare that they have no competing interests.

Authors' contributions

LD did all calculations and prepared the first version of the article. RZ discussed the simulation results and corrected the article. Both authors read and approved the final manuscript.

Authors' information

LD is a Ph.D. Candidate in the Department of Applied Physics in Northwestern Polytechnical University, whose research interest is phase-field simulation of solidification in metal alloys under different conditions. RZ is a Professor in the Department of Applied Physics in Northwestern Polytechnical University, whose research interest is solidifying process under mutiphysics fields.

Acknowledgements

The authors would like to thank the financial support from the NPU Foundation of Fundamental Research, China (No. JC201272).

References

1. Asta M, Beckermann C, Karma A, Kurz W, Napolitano R, Plapp M, Purdy G, Rappaz M, Trivedi R (2009) Solidification microstructures and solid-state parallels: recent developments, future directions. Acta Mater 57:941–971
2. Chen LQ (2002) Phase-field models for microstructure evolution. Annu Rev Mater Res 32:113–140
3. Steinbach I (2009) Phase-field models in materials science. Model Simul Mater Sc 17:073001
4. Singer-Loginova I, Singer HM (2008) The phase field technique for modeling multiphase materials. Rep Prog Phys 71:106501
5. Boettinger WJ, Warren JA, Beckermann C, Karma A (2002) Phase-field simulation of solidification. Annu Rev Mater Res 32:163–194
6. Kobayashi R (1993) Modeling and numerical simulations of dendritic crystal-growth. Physica D 63:410–423
7. Wheeler AA, Boettinger WJ, Mcfadden GB (1992) Phase-field model for isothermal phase-transitions in binary-alloys. Phys Rev A 45:7424–7439
8. Wheeler AA, Boettinger WJ, Mcfadden GB (1993) Phase-field model of solute trapping during solidification. Phys Rev E 47:1893–1909
9. Kim SG, Kim WT, Suzuki T (1999) Phase-field model for binary alloys. Phys Rev E 60:7186–7197
10. Kim SG, Kim WT, Suzuki T (1998) Interfacial compositions of solid and liquid in a phase-field model with finite interface thickness for isothermal solidification in binary alloys. Phys Rev E 58:3316–3323
11. Karma A (2001) Phase-field formulation for quantitative modeling of alloy solidification. Phys Rev Lett 87:045501
12. Karma A, Rappel WJ (1998) Quantitative phase-field modeling of dendritic growth in two and three dimensions. Phys Rev E 57:4323–4349
13. Tong X, Beckermann C, Karma A, Li Q (2001) Phase-field simulations of dendritic crystal growth in a forced flow. Phys Rev E 63:061601
14. Jeong JH, Goldenfeld N, Dantzig JA (2001) Phase field model for three-dimensional dendritic growth with fluid flow. Phys Rev E 64:041602
15. Tsai YL, Chen CC, Lan CW (2010) Three-dimensional adaptive phase field modeling of directional solidification of a binary alloy: 2D-3D transitions. Int J Heat Mass Tran 53:2272–2283
16. Du LF, Zhang R, Zhang LM (2013) Phase-field simulation of dendritic growth in a forced liquid metal flow coupling with boundary heat flux. SCIENCE CHINA Technological Sciences 56:2586–2593
17. Nestler B, Choudhury A (2011) Phase-field modeling of multi-component systems. Curr Opin Solid St M 15:93–105
18. Nestler B, Garcke H, Stinner B (2005) Multicomponent alloy solidification: phase-field modeling and simulations. Phys Rev E 71:041609
19. Zhang RJ, Li M, Allison J (2010) Phase-field study for the influence of solute interactions on solidification process in multicomponent alloys. Comp Mater Sci 47:832–838
20. Feest EA, Doherty RD (1973) Dendritic solidification of Cu-Ni alloys. 2. Influence of initial dendrite growth temperature on microsegregation. Metall Trans 4:125–136
21. Viskanta R (1988) Heat-transfer during melting and solidification of metals. J Heat Trans-T Asme 110:1205–1219
22. Kumar TSP, Prabhu KN (1991) Heat-flux transients at the casting chill interface during solidification of aluminum base alloys. Metall Trans B 22:717–727
23. Juric D, Tryggvason G (1996) A front-tracking method for dendritic solidification. J Comput Phys 123:127–148
24. Amberg G, Tonhardt R, Winkler C (1999) Finite element simulations using symbolic computing. Math Comput Simulat 49:257–274
25. Tang JJ, Xue XA (2009) Phase-field simulation of directional solidification of a binary alloy under different boundary heat flux conditions. J Mater Sci 44:745–753
26. Loginova I, Amberg G, Agren J (2001) Phase-field simulations of non-isothermal binary alloy solidification. Acta Mater 49:573–581
27. Warren JA, Boettinger WJ (1995) Prediction of dendritic growth and microsegregation patterns in a binary alloy using the phase-field method. Acta Metall Mater 43:689–703

28. Lan CW, Shih CJ (2004) Phase field simulation of non-isothermal free dendritic growth of a binary alloy in a forced flow. J Cryst Growth 264:472–482
29. Xu JJ, Wang TM, Zhu J, Xie HL, Xiao TQ, Li TJ (2011) In situ study on secondary dendrite arm coarsening of Sn-Bi alloy by synchrotron microradiography. Mater Res Innov 15:156–159
30. Kurz WFWJ (1998) Fundamentals of Solidification. Trans Tech Publications, Switzerland
31. Herlach D (2011) Crystal nucleation and dendrite growth of metastable phases in undercooled melts. J Alloy Compd 509:S13–S17
32. Chang J, Wang HP, Zhou K, Wei B (2012) Rapid dendritic growth and solute trapping within undercooled ternary Ni-5%Cu-5%Mo alloy. Appl Phys A-Mater 109:139–143
33. Cao CD, Wang F, Duan LB, Bai XJ (2011) Effect of solidification temperature range on the dendritic growth mode. Sci China Phys Mech 54:89–94
34. Song RB, Dai FP, Wei BB (2011) Dendritic growth and solute trapping in rapidly solidified Cu-based alloys. Sci China Phys Mech 54:901–908
35. Yang XB, Fujiwara K, Maeda K, Nozawa J, Koizumi H, Uda S (2011) Dependence of Si faceted dendrite growth velocity on undercooling. Appl Phys Lett 98:012113

3

Workflow for integrating mesoscale heterogeneities in materials structure with process simulation of titanium alloys

Ayman A Salem[1*], Joshua B Shaffer[1], Daniel P Satko[1], S Lee Semiatin[2] and Surya R Kalidindi[3]

* Correspondence:
ayman.salem@icmrl.net
[1]Materials Resources LLC, Dayton, OH 45402, USA
Full list of author information is available at the end of the article

Abstract

In this paper, a generalized workflow is outlined for the necessary integration of multimodal measurements and multiphysics models at multiple hierarchical length scales demanded by an Integrated Computational Materials Engineering (ICME) approach to accelerated materials development. Recognizing that multiple choices or techniques are typically available in each of the main steps, several exemplary analyses are detailed utilizing mainly the alpha/beta titanium alloys as an illustrative case. It is anticipated that the use and further refinement of these workflows will promote transparency and engender intimate collaborations between materials experts and manufacturing/design specialists by providing an understanding of the various mesoscale heterogeneities that develop naturally in the workpiece as a direct consequence of the inherent heterogeneity imposed by the manufacturing history (i.e., different thermomechanical histories at different locations in the sample). More specifically, this article focuses on three main areas: (i) data science protocols for efficient analysis of large microstructure datasets (e.g., cluster analysis), (ii) protocols for extracting reduced descriptions of salient microstructure features for insertion into simulations (e.g., regions of homogeneity), and (iii) protocols for direct and efficient linking of materials models/databases into process/performance simulation codes (e.g., crystal plasticity finite element method).

Keywords: ICME; Microstructure informatics; Higher-order statistics; Materials big data; Macrozones; Region of homogeneity; Representative orientation distribution; Alpha/beta titanium alloys

Review

Introduction

Predicting mechanical response and texture evolution during thermomechanical processing (TMP) of alpha/beta titanium alloys is challenging due to the complexity of the underlying microstructure in the alloys and the highly anisotropic response exhibited by the microscale constituents. Not only do these materials exhibit two phases with drastically different deformation mechanisms, but also the morphology, crystallography, and relative ratio of the phases changes with temperature and deformation path. Consequently, there have been many research efforts focusing on various sources of anisotropy in these materials [1,2]. With the recent push of the Materials Genome Initiative (MGI) and Integrated Computational Materials Engineering (ICME) towards a coherent integration between various stages and scales of modeling the materials behavior and the corresponding measurements, new workflows (i.e., protocols) are critically needed. In this article, an example workflow is presented that integrates materials characterization (both microstructure and mechanical response) with suitable multiscale simulations of processing conditions (Figure 1). This workflow comprises the following general operations: (i) assembly of information gathered about the local state of the materials from experimental or modeling data into a feature vector; (ii) application of data analytics techniques to identify particular features of interest; (iii) creation of representative descriptors of microstructure features that provide an optimized reduced description; and (iv) insertion of the representative descriptors into materials models that evolve the microstructure and properties during TMP, making use of a particular solver architecture (e.g., finite element method (FEM)) with various degrees of feedback. The path of the workflow can also be closed through post-processing operations whereby the results of materials models are used to update the local state information, enriching the information contained in the feature vector.

In order to provide for a greater understanding of the various operations and the flow of information throughout the workflow, detailed descriptions of exemplary illustrations are provided in the subsequent sections. In particular, discussion is focused on examples illustrating how a series of choices can be made to advance the incorporation of titanium microstructures into numerical simulations of a part production. The workflow and selected techniques are directed towards the emerging 'Big-Data' materials innovation ecosystem that utilizes modern data science techniques such as machine learning and computer vision [3,4]. While the Ti-6Al-4V microstructure data presented here were generated with standard

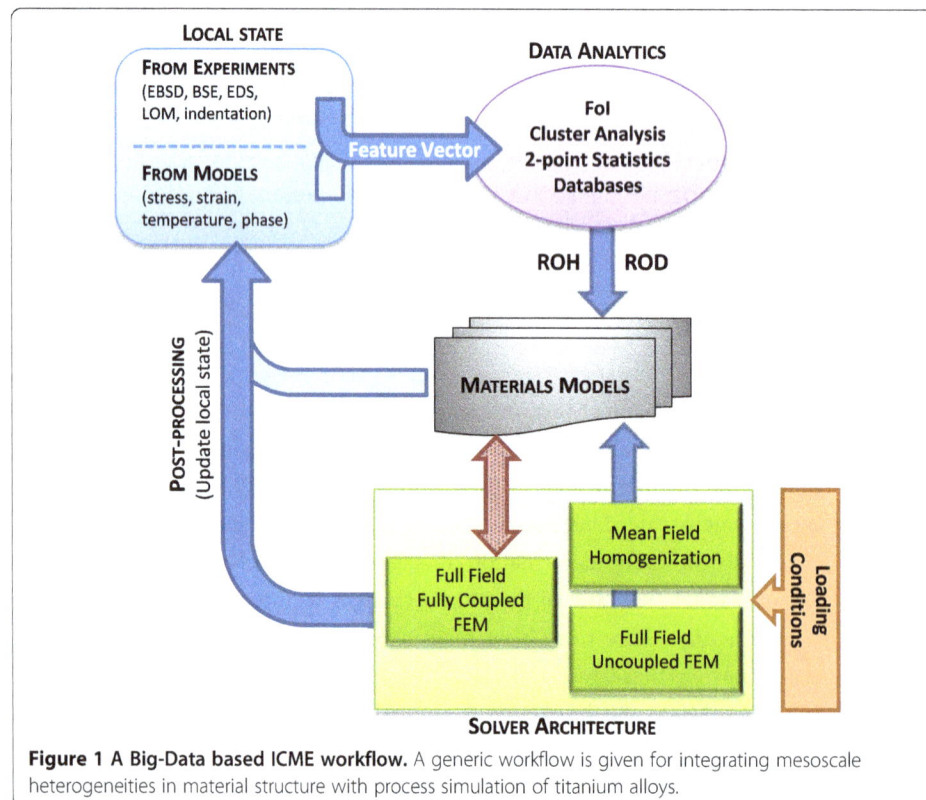

Figure 1 A Big-Data based ICME workflow. A generic workflow is given for integrating mesoscale heterogeneities in material structure with process simulation of titanium alloys.

methods (e.g., electron backscatter diffraction (EBSD) and backscattered electron (BSE) imaging), they were recorded from large scan areas resulting in 100,000,000s of EBSD data points and 10,000s of high-resolution BSE images. Consequently, new tools have had to be developed for texture analysis, image segmentation, and quantification of microstructure metrics. In addition, salient microstructure descriptors (e.g., regions of homogeneity (ROH), representative orientation distributions (ROD), and microtextured regions or macrozones (MTR)) have been generated using a new generation of data analytics techniques.

Microstructure of α-β titanium alloys

Due to the strong effect of microstructure on the mechanical behavior of materials, it is important to have a clear understanding of the various constituents of the microstructure and their influence on the mechanical behavior and texture evolution under a broad range of loading conditions. Ti-6Al-4V was selected as the primary model material in the following sections to describe our workflow. This alloy exhibits a dual-phase microstructure with a hexagonal close-packed (HCP) alpha phase and a body-centered cubic (BCC) beta phase that coexist in various volume fractions based on the temperature history of the sample or component. Thermomechanical processing of this alloy produces a range of microstructures that can be categorized into three major types: fully lamellar, fully equiaxed, and bimodal (Figure 2). For the same chemistry, the volume fractions, morphology, and texture of the constituents are known to have a major effect on the mechanical behavior of the alloy [5].

Lamellar and equiaxed microstructures have constituents with distinct morphologies that can be extracted using standard image segmentation techniques applied on BSE or optical images. Consequently, extracting alpha laths in the lamellar microstructure or alpha particles (not grains) in the equiaxed microstructure (Figure 2a,b) can be done using contrast thresholding in a commercial software package (e.g., ImageJ and Photoshop) due to the image contrast resulting from etching (optical) or atomic number difference (BSE) of the two phases. However, the bimodal microstructure is a challenge because the alpha phase has two morphological constituents, namely, primary alpha particles and secondary alpha colonies with alternating alpha laths and beta layers (Figure 2c). Because both constituents have the same HCP crystal structure, regular EBSD maps cannot automatically distinguish between them (Figure 3a). The same challenge lies while using the BSE images (Figure 2c) alone. On smaller areas, the use of multimodal signals and traditional thresholding techniques has succeeded to segment (α_p) particles and (α_s) colonies based

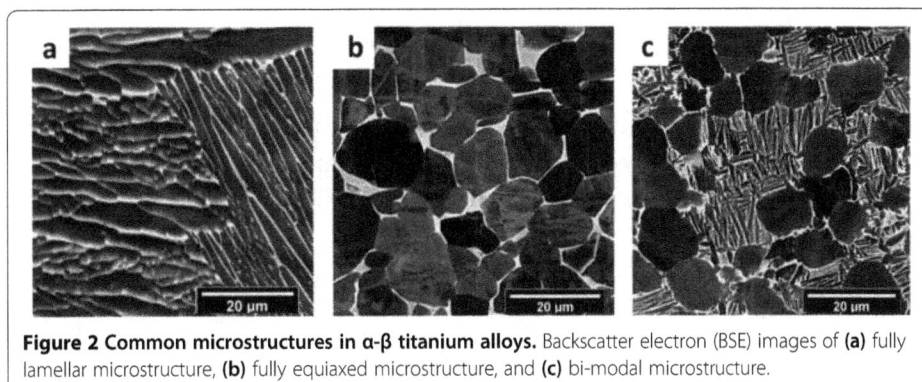

Figure 2 Common microstructures in α-β titanium alloys. Backscatter electron (BSE) images of **(a)** fully lamellar microstructure, **(b)** fully equiaxed microstructure, and **(c)** bi-modal microstructure.

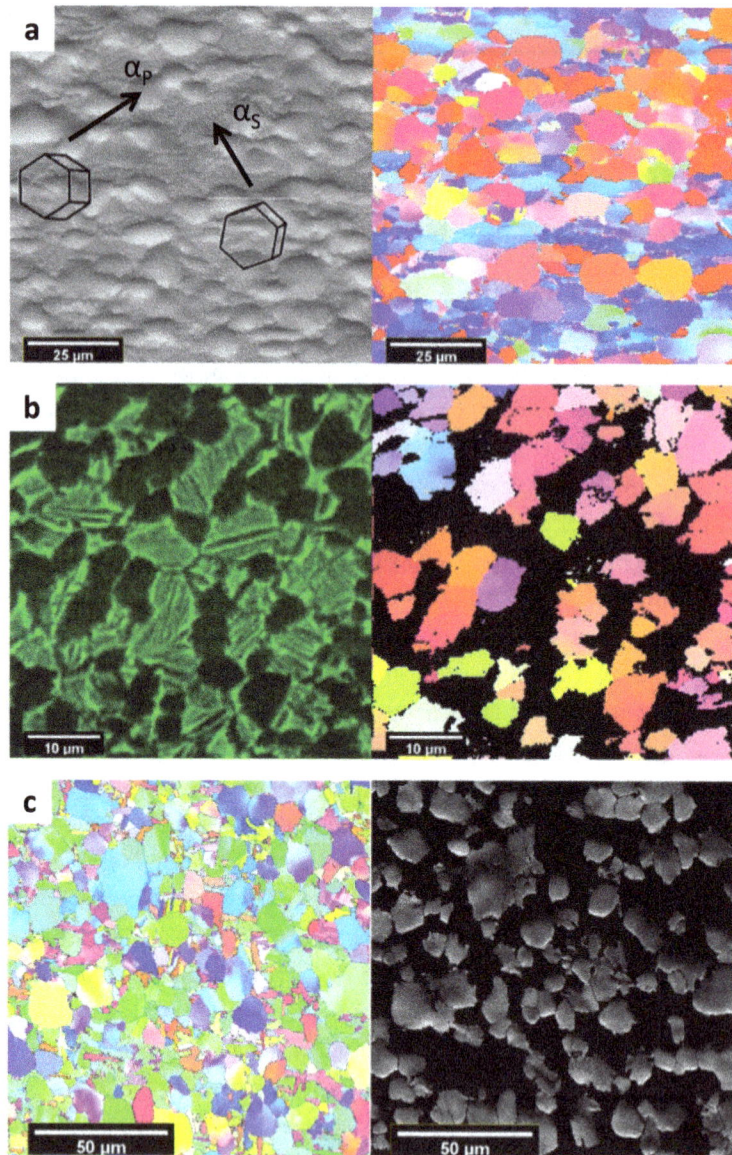

Figure 3 Segmenting HCP constituents (α_p and α_s) in Ti-6Al-4V. **(a)** Secondary electron image of a bimodal Ti-6Al-4V alloy with primary particles (α_p) and secondary alpha colonies (α_s) and ND-IPF map of Ti-6Al-4V with indistinguishable HCP constituents (α_p grains and α_s colonies). **(b)** Using a map of vanadium concentration generated by EDS showing higher concentration in secondary alpha (α_s) colonies to segment the α_p particles. **(c)** Alpha-phase inverse pole figure map and resultant EBSD pattern quality map of auto-segmented α_p grains using TiSeg™.

on vanadium partitioning (Figure 3b) [6]. However, applying these methods to large areas for practical applications is expensive and time-consuming. The use of data science approaches (as described in detail in subsequent sections) has enabled automated segmentation of 10,000s of EBSD and BSE datasets (Figure 3c).

Data analytics for large microstructure datasets

The bimodal microstructure contains many microstructure features (α_p particles, α_p grains, α_s colonies, α_s laths, layers of alpha on beta grain boundaries, prior β grains, microtextured regions, etc.) that can affect the response of the material under loading. However, the data captured by typical characterization techniques (EBSD, BSE, spectroscopy, etc.) do not directly identify these microstructure features automatically. Rather, at each probed location, the data include signals from the materials that reflect the internal local state of the materials. These internal variables change with time under externally applied variables (e.g., temperature, cooling rate, strain). As such, one can assign the local state at each voxel location in the studied area using a list of internal variables that are related to the crystal structure and alloying elements of the underlying materials. The microstructure features mentioned earlier are then defined by certain morphological characteristics displayed by the local states. Consequently, it should be possible to use various cluster analysis techniques [7] to identify the features of interest (FoI) from large datasets (e.g., EBSD scan of 10×10 mm at 1 µm step size giving 10^8 measurements of the local state). An example is the familiar spatial clustering of measurement points sharing a common orientation that is understood as a 'grain'. Similarly, the spatial clustering of a group of grains indicates a microtextured region (macrozone), while the spatial clustering of similar chemical elements is indicative of micro/macrosegregation or precipitates. As such, taking advantage of recent advances [8] in data science has enabled the development of multiple algorithms and tools for quantifying the microstructural features of interest in computationally efficient ways [9]. In the next few sections, the main terminologies and methodologies used in these workflows are introduced.

Feature vector

Each voxel in a materials dataset is assigned an n-dimensional feature vector [10] of variables needed to obtain a concise mathematical representation of all the distinct local states in the dataset. Such representation facilitates image processing, statistical analysis, and utilization of numerous algorithms from the pattern recognition and machine learning communities [11] to extract salient information about the material. The list of variables used in the feature vectors can range from scalar variables (e.g., chemical composition) to tensorial variables (e.g., crystal lattice orientation). These feature vectors play an important role in identifying proper materials models for process simulations (e.g., thermomechanical processing of Ti alloys).

While the main goal of materials Big-Data efforts is to improve the accuracy of predictive modeling, the implementation of such efforts is often faced with the doubt that 'bigger is not always better' [12]. This unfortunate stigma has surfaced in fields other than materials science where low signal-to-noise ratio is the norm [13] and the main question often is exactly what data needs to be collected. Furthermore, the high cost of collecting materials data often results in the production of very sparse datasets relative to the high degree of heterogeneity possible in large manufactured structures. Similar challenges have faced predictive modeling of human behavior, though the large economic gains that have been realized from the success of these models have in turn helped to improve the popularity and the accuracy of such efforts [12,13]. Recently, Fortuny et al. [12] provided empirical support that predictive performance from sparse fine-grained (behavior) data was increased by using Big-Data. For example, using KDD Cup 2010 student evaluation data, Fortuny et al. [12] showed that the predictive power (area under receiver operating characteristic, ROC) for the multinomial Naïve Bayes variants increased with the addition of more data (Figure 4). Such correlation is likely also applicable to predictive modeling with materials Big-Data with an expected improvement in accuracy of predictions due to the absence of noise inherent in the often emotion-driven behavior of human populations.

Identification of features of interest

As noted earlier, the local states do not necessarily identify the microstructure features of interest directly, necessitating additional analysis. As such, prior knowledge in the field will help identify the specific microstructural FoI (presumably these features control the performance characteristics of the final finished product) that need to be tracked in the process simulations. Automated protocols for the extraction of selected features of interest are needed to obtain consistency and reproducibility independent of operator bias and with minimum computational cost. Due to the importance of the FoI as the main building block for further analysis and modeling efforts, computer vision algorithms built on supervised and unsupervised machine learning techniques [11] should be employed for automated feature identification and extraction.

To extract specific FoI, some or all components of a feature vector can be used in the cluster analysis [7]. For example, for identifying only the beta phase, a single variable in the feature vector (e.g., BCC phase in EBSD dataset) is adequate. However, segmenting primary alpha grains in bimodal microstructures requires clustering data regarding orientation, chemistry, and/or imaging contrast at each voxel in the microstructure dataset. Distinguishing between primary alpha grains and primary alpha particles is important for accurate predictions of materials responses that depend on the slip distance (e.g., crystal plasticity). An α_p particle (Figure 5a) may engulf multiple α_p grains (Figure 5b), though in the case of a microstructure that has been fully spherodized during the ingot breakdown process, a particle may include only one grain. As such, while identification of particles can be accomplished directly from BSE images (Figure 2c) by finding the remnant beta

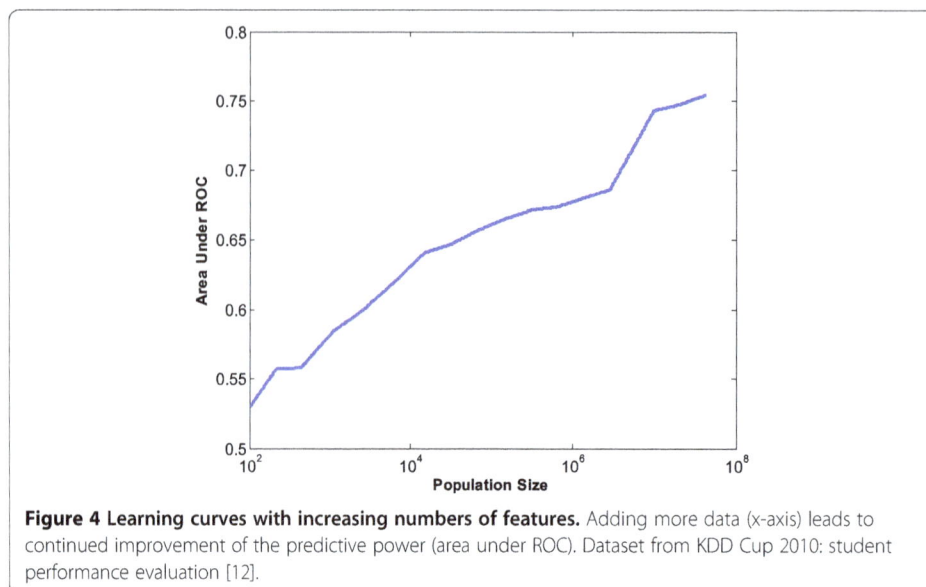

Figure 4 Learning curves with increasing numbers of features. Adding more data (x-axis) leads to continued improvement of the predictive power (area under ROC). Dataset from KDD Cup 2010: student performance evaluation [12].

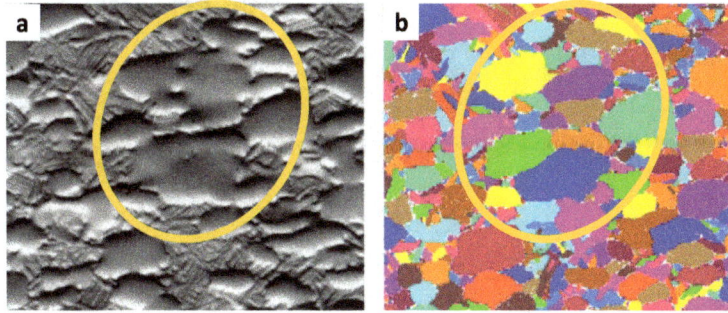

Figure 5 Distinguishing between particles and grains in Ti-6Al-4V. (a) Secondary electron image with primary alpha particles shown as continuous islands in a sea of secondary alpha laths and **(b)** EBSD data of the same location with individual primary alpha grains colored differently. Almost all primary alpha particles in **(a)** contain multiple primary alpha grains as shown in **(b)** with an example highlighted in the circle.

layer separating particles, further differentiation of α_p grains within a particle requires orientation information in addition to BSE information (Figure 3c, Figure 5b).

Independent of the length scale, a feature of interest may be defined as *a region in physical space (3D) that depicts similar characteristic feature vectors as some other regions in the microstructure*. This definition enables setting up automated extraction tools that are mainly dependent on the components of the feature vector. In the case mentioned above, the feature vector for alpha 'particle' identification needs only one element (BSE gray level), while a larger feature vector is necessary to define primary alpha 'grains' (BSE gray level + orientation). To achieve automation and to increase the speed of extraction, the identification is conducted in a two-step process in two separate domains. The first step is conducted in the feature vector domain (i.e., feature space) in which volumes with similar feature vectors are classified using cluster analysis [7]. The second step is conducted in physical space by mapping the classified clusters from the first step to corresponding spatial locations in real space. To demonstrate the FoI concept practically on Ti-6Al-4V, EBSD, and BSE data within a feature vector are used to extract information about MTRs which can be defined as 3D regions in physical space that contain similarly oriented primary alpha grains [14].

Prior attempts to identify MTRs were based on traditional EBSD data analysis methods available in commercial EBSD data analysis tools [15]. However, these methods do not scale well for large datasets ($>10^7$ EBSD data points covering hundreds of square millimeters and $>10^5$ BSE images). However, the two-step process [14] described above has enabled the rapid identification of MTR clusters, each distributed about a common texture component with a defined misorientation range ($<10°$ in this case) within each cluster. The cluster analysis was conducted with a feature vector of 551 dimensions in domain of the generalized spherical harmonics (GSH), a mathematical construct commonly used to analytically describe the distribution of crystallographic orientations [16-21]. The normal direction (ND) inverse pole figure (IPF) maps in Figure 6a, b show the variability of the length scale from primary alpha grains in (b) to MTRs in (a). One of the identified MTR cluster families is shown in real space in Figure 6c, along with the corresponding GSH-space projection (Figure 6d), demonstrating the orientation clustering of various MTR families.

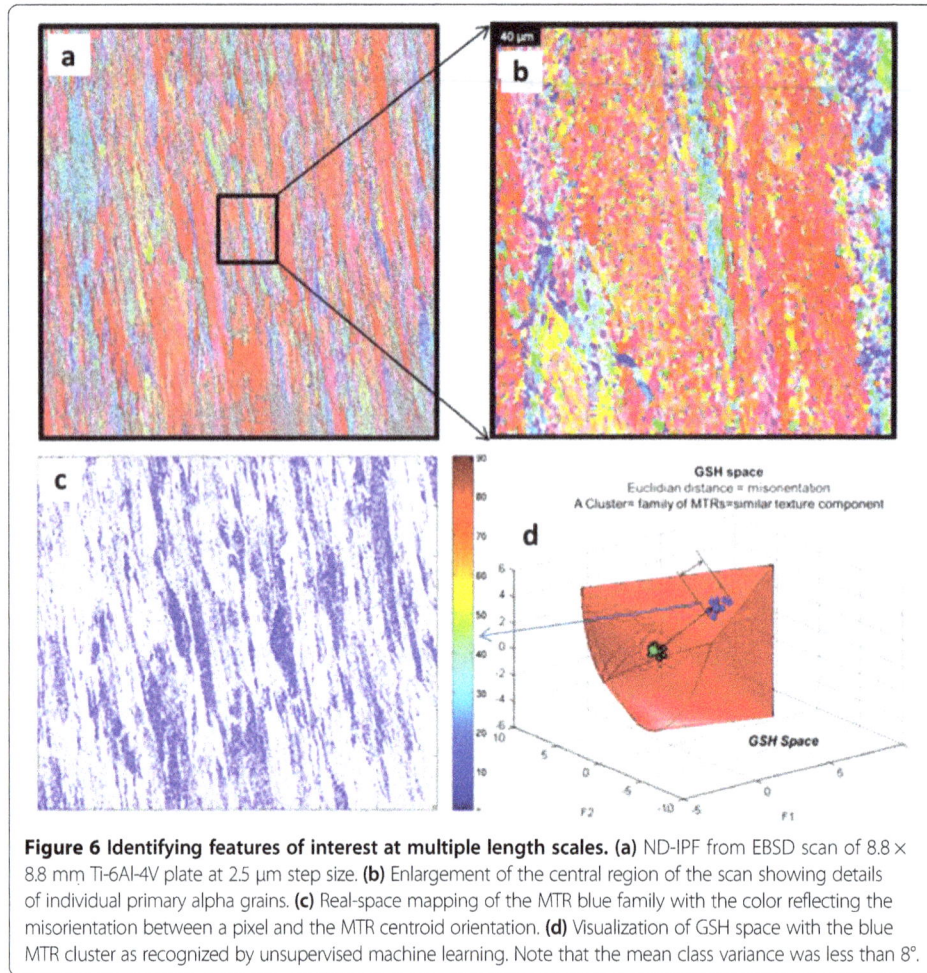

Figure 6 Identifying features of interest at multiple length scales. (a) ND-IPF from EBSD scan of 8.8 × 8.8 mm Ti-6Al-4V plate at 2.5 μm step size. **(b)** Enlargement of the central region of the scan showing details of individual primary alpha grains. **(c)** Real-space mapping of the MTR blue family with the color reflecting the misorientation between a pixel and the MTR centroid orientation. **(d)** Visualization of GSH space with the blue MTR cluster as recognized by unsupervised machine learning. Note that the mean class variance was less than 8°.

Representative descriptions of microstructure

Regions of homogeneity

Identifying specific subsets that represent the whole material has been the target of many research efforts including finding a representative volume element (RVE) [22,23] and/or the statistical volume element (SVE) [24]. One of the defining features of an RVE or SVE description is the linking of materials properties of the ensemble to the properties of the defined RVE or SVE. For example, Drugan and Willis [23] defined an RVE as 'a smallest material volume element of the composite for which the usual spatially constant "overall modulus" macroscopic constitutive representation is a sufficiently accurate model to represent mean constitutive response'. However, estimating the materials response depends on the defined RVE/SVE, so defining the RVE or SVE that is based on the response of the material to use it in models that predict the response of the material becomes a challenging task. Furthermore, many of the traditional methods used to define RVEs/SVEs are based on highly simplified metrics (e.g., average grain size and its distribution) and ignore the spatial correlations of individual FoI (e.g., complex morphologies that may not fit a standard geometrical shape such as an ellipse). While the traditional methods may work for materials with homogeneous spatial distributions of the FoI, it may not be efficient for heterogeneous spatial distributions of FoI such as MTRs (Figure 6c) or gradient microstructures. In both cases, there is a need for microstructure representation that captures the essence of spatial and morphological heterogeneities. Presented below is a new microstructure descriptor (regions of homogeneity) that has been developed based on the spatial homogeneity of the two-point statistics that were calculated for sampled microstructure regions.

The hierarchical materials systems described here exhibit salient features at different length scales. For example, in the Ti alloys described here, one scale of heterogeneity occurs at the length scale of individual α_p grains (on the order of 10 to 30 μm) and another scale of heterogeneity occurs at the scale of each MTR (up to several mm in length). In multiscale modeling, one typically identifies different length scales where one might be able to homogenize the materials response by aggregating in some way all of the inherent heterogeneity below that length scale. In materials with complex structures, one has to identify suitable hierarchical length scales for homogenization, which are called regions of homogeneity. ROH can be established objectively at different hierarchical length scales by carefully quantifying spatial correlations (e.g., using two-point spatial correlations [17,25-30]) and finding suitable window sizes in the microstructure that capture the inherent heterogeneity (FoI and their spatial distributions) to a desired accuracy. It is also desired to keep these ROH small enough to enable cost-effective modeling (the computational cost rises steeply with increases in the number of voxels needed to capture the ROH).

For example, the dataset in Figure 6 was iteratively divided into 12 subsets with 12 different window sizes. Each subset consisted of multiple fixed size windows resulting in a total of 2,912 windows. For each subset, the two-point correlations were calculated for the microstructure inside each window. Then, the difference between the median of the two-point correlations of each subset and the whole dataset was measured via a Euclidean distance (D) and plotted in Figure 7 as a function of window size within each subset. The two-point correlation of the median for subset with window size of 2.2×2.2 mm was the closest to the two-point statistics of the original larger dataset (smallest distance D in Figure 7). Consequently, the median of windows with size of 2.2×2.2 mm can represent the microstructure details regarding the features of interest (i.e., MTRs) including size, morphology, and spatial distributions for dataset in Figure 6.

It is worth noting that in the above example, the FoI (specifically MTRs) were large relative to the scanned area. As such, a window size smaller than 2.2×2.2 mm did not

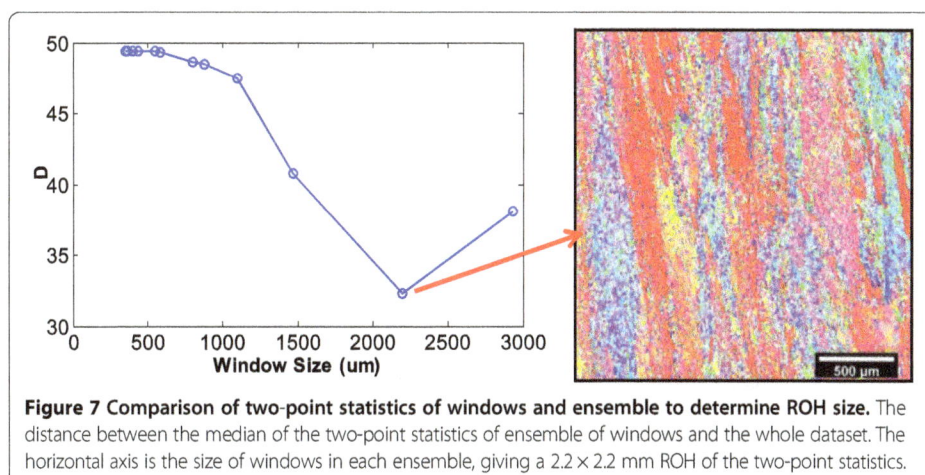

Figure 7 Comparison of two-point statistics of windows and ensemble to determine ROH size. The distance between the median of the two-point statistics of ensemble of windows and the whole dataset. The horizontal axis is the size of windows in each ensemble, giving a 2.2 × 2.2 mm ROH of the two-point statistics.

provide an adequate representation of the microstructure in the original scanned area. However, the same technique to find ROH can be applied to FoI with a size much smaller than the scanned area. For example, individual primary alpha particles in a Ti-6Al-4V plate hot rolled at 955°C [31] have a typical size of about 12 μm, covering an area of <0.05% relative to total area of the captured image (490 × 745 μm, Figure 8a). Following the method developed by Niezgoda et al. [28], the selection of ROH is demonstrated using weighted two-point statistics of subsets of the microstructure. Niezgoda et al. [28] showed that an RVE generated with this method can result in minimum deviation in the effective modulus compared to the original microstructure modulus. Similar results have also been presented by Qidwai et al. [32] and Wargo et al. [33] in very different application domains. Here, the same technique is applied without a reference to the material properties, and to distinguish between both approaches, the ROH approach is used to reflect the weighted representative microstructure without calculating any

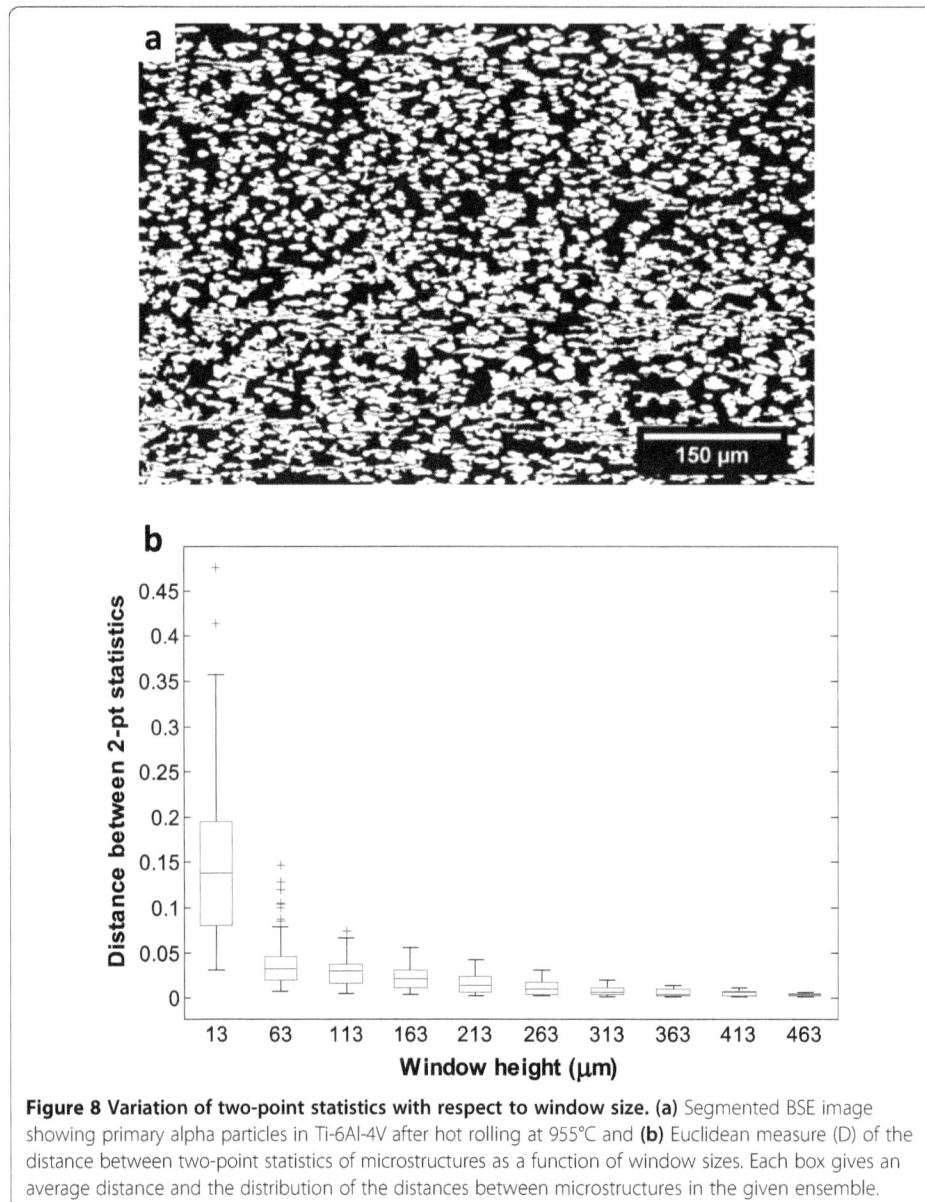

Figure 8 Variation of two-point statistics with respect to window size. (a) Segmented BSE image showing primary alpha particles in Ti-6Al-4V after hot rolling at 955°C and **(b)** Euclidean measure (D) of the distance between two-point statistics of microstructures as a function of window sizes. Each box gives an average distance and the distribution of the distances between microstructures in the given ensemble.

associated properties. For the microstructure in Figure 8a, 100 windows were sampled from the image and the median and variance within each sample were quantitatively measured in the two-point statistics domain in which the differences can be measured via a Euclidean distance (D) in Figure 8b.

A window size of 247×163 μm was observed to show saturation in the mean separation (D) of the two-point statistics within an ensemble. A comparison of the two-point statistics of one of these selected windows and the statistics of the full window (Figure 9a,b) shows a close similarity in the short-range order, with only minor differences at longer range. The example window and its position relative to the total image are shown in Figure 9c,d, respectively. The inclusion of additional weighted windows results in a smaller error between the ensemble statistics and the ROH from the weighted windows providing a trade-off between increased accuracy of the ROH description and computational cost. Window selection is accomplished by nonnegative least squares regression after decomposition of the two-point statistics via principal component analysis (PCA; Figure 10a) [8]. The weighted combination of the three windows given in Figure 10b gives the ROH with the smallest error in PCA space (Figure 11a) compared to the total image (Figure 11b, also shown in red in Figure 10a).

The next step after generating the two-point correlations of the ROH is to generate the microstructure in the ROH to be used in materials models. This inverse reconstruction process is meant to create statistically similar microstructures that account for morphological and spatial heterogeneities that were captured within the ROH [30,34,35]. The process of reconstruction from the two-point correlation has been proven accurate for

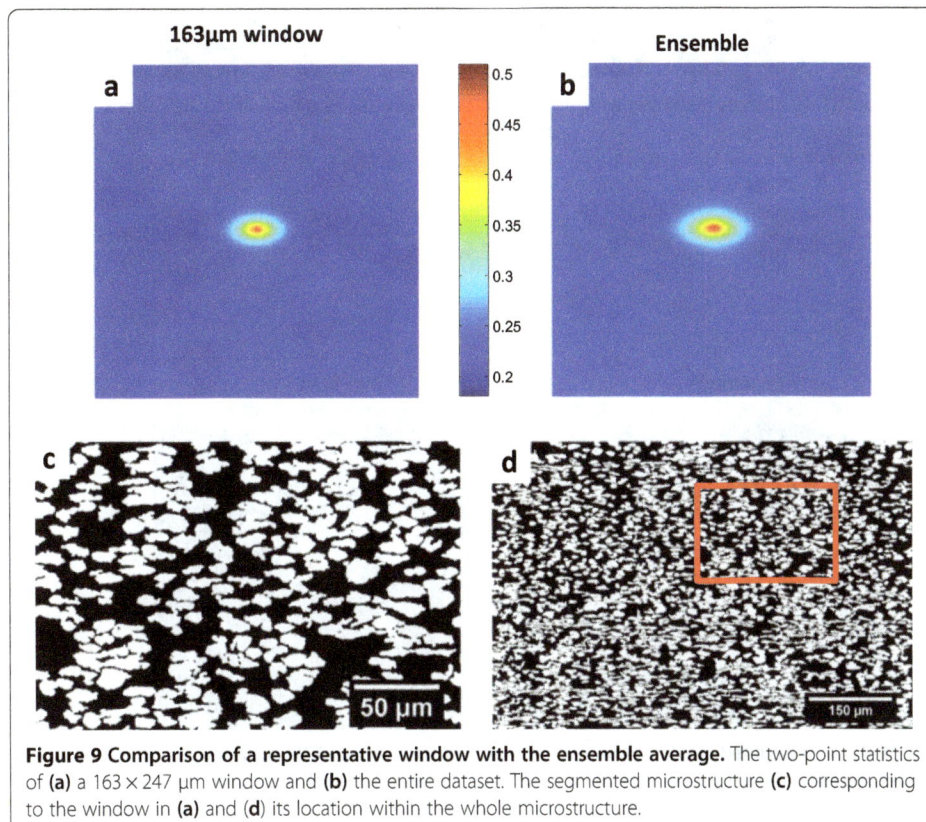

Figure 9 Comparison of a representative window with the ensemble average. The two-point statistics of **(a)** a 163×247 μm window and **(b)** the entire dataset. The segmented microstructure **(c)** corresponding to the window in **(a)** and **(d)** its location within the whole microstructure.

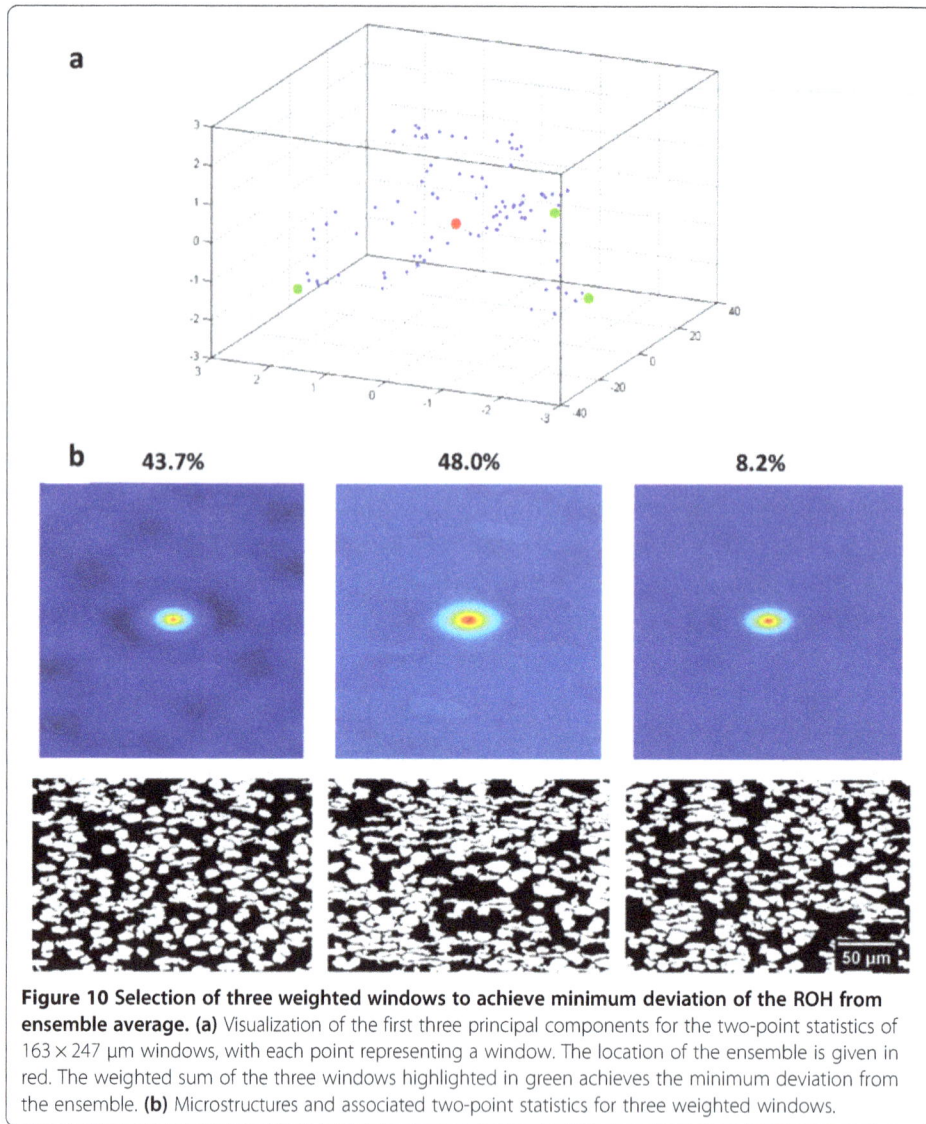

Figure 10 Selection of three weighted windows to achieve minimum deviation of the ROH from ensemble average. (a) Visualization of the first three principal components for the two-point statistics of 163×247 µm windows, with each point representing a window. The location of the ensemble is given in red. The weighted sum of the three windows highlighted in green achieves the minimum deviation from the ensemble. **(b)** Microstructures and associated two-point statistics for three weighted windows.

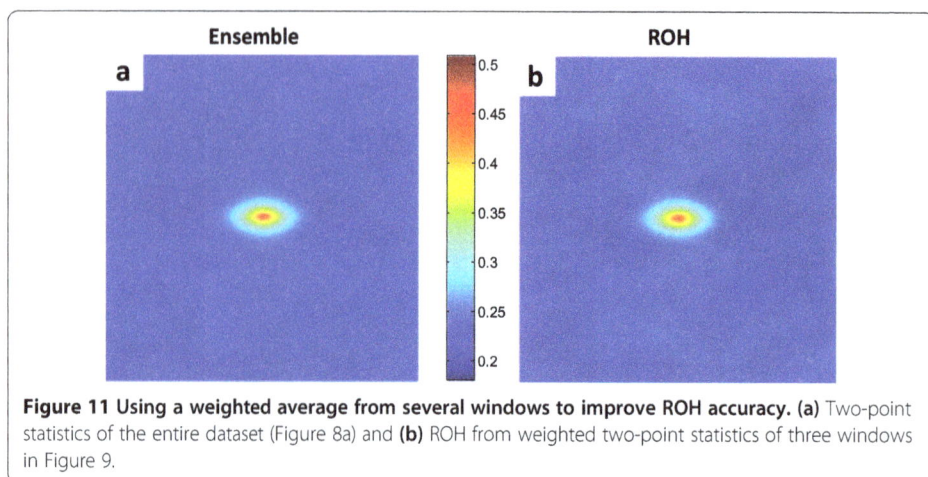

Figure 11 Using a weighted average from several windows to improve ROH accuracy. (a) Two-point statistics of the entire dataset (Figure 8a) and **(b)** ROH from weighted two-point statistics of three windows in Figure 9.

images with two phases within an inversion and translation of the original image, while prescribing periodic boundary conditions (Figure 12) [30].

Representative orientation distribution

The exceedingly large number of crystallographic orientations recorded during data collection for MTR identification represents a challenge for predicting texture evolution using any crystal plasticity model. Consequently, it is crucial to develop a procedure that finds a representative orientation distribution consisting of a reduced number of orientations giving the same texture as the measured large dataset. Recently, in an example case study, it has been demonstrated that a set of weighted 551 orientations is able to reproduce exactly (with an error of 10^{-17} in the orientation distribution function (ODF)) the same texture as the 12,467,961 (Figure 13) orientations measured in the Ti-6Al-4V sample given in Figure 6. To give an estimate of the time savings, a Taylor-type crystal plasticity modeling of simple compression to strain of -1.0 was executed in 1.6 mins on a standard desktop computer (quad-core 3.0 GHz) using the ROD. Under an approximation of linear time complexity with the number of orientations evolved, the identical simulation is estimated to take more than 25 days to run with the original dataset on the same computer.

Materials models: crystal plasticity

Titanium exhibits highly anisotropic properties at the single-crystal level, which can be attributed to the operation of different deformation mechanisms under different external stimuli (temperature, strain rate, etc.). In addition to the numerous competing deformation mechanisms [36], there also exist additional challenges due to the fact that an allotropic transformation occurs from alpha-HCP to beta-BCC at high temperatures (beta transus is dependent on alloying elements). The addition of alloying elements causes the alpha and beta phases to coexist with varying ratios and morphologies based on the temperature and the amount of alpha or beta stabilizers. This consequently alters the activity of various deformation mechanisms depending on the resultant microstructure. For example, the Ti-6Al-4V with bimodal microstructure shown previously (Figure 2) accommodates plastic deformation through slip in both alpha and beta phases, with the latter contribution increasing as the deformation temperature increases. As such, accurate

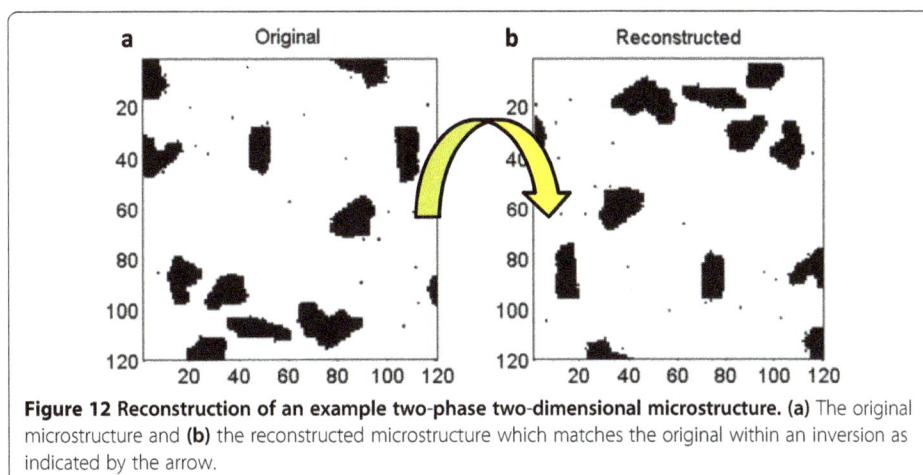

Figure 12 Reconstruction of an example two-phase two-dimensional microstructure. (a) The original microstructure and **(b)** the reconstructed microstructure which matches the original within an inversion as indicated by the arrow.

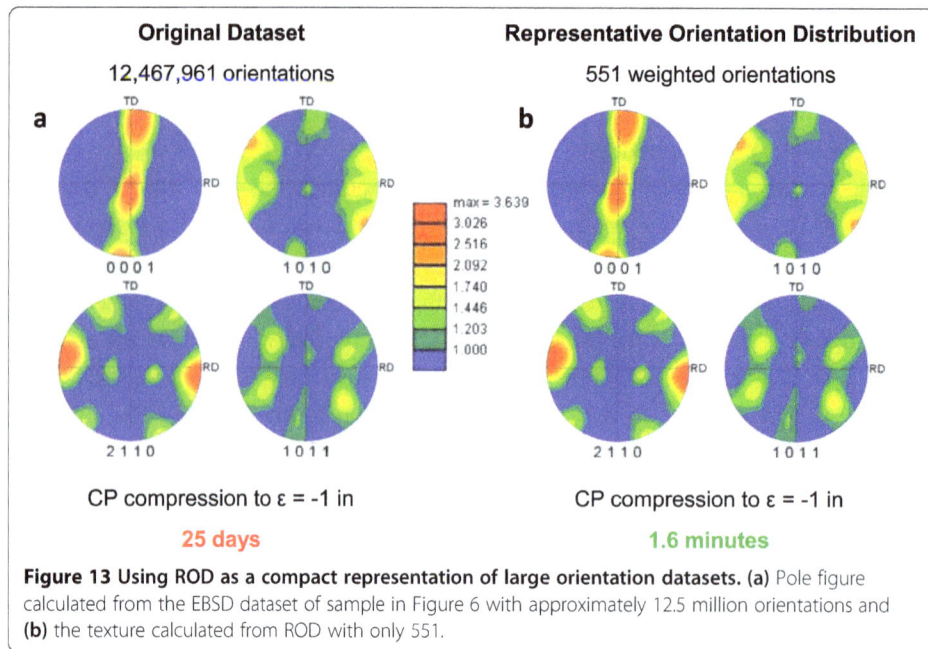

Figure 13 Using ROD as a compact representation of large orientation datasets. (a) Pole figure calculated from the EBSD dataset of sample in Figure 6 with approximately 12.5 million orientations and **(b)** the texture calculated from ROD with only 551.

modeling of the mechanical behavior and texture evolution of this material requires an understanding on the crystal level of the various deformation mechanisms of each phase, the interaction between both phases, and their evolution as a function of temperature.

Single-crystal deformation behavior

Deformation mechanisms in the alpha phase (HCP) of unalloyed titanium and the two-phase materials are limited to a finite number of slip systems and/or deformation twinning which results in pronounced anisotropic behavior and strong deformation texture [36]. In particular, deformation can be accommodated by prism < a > and basal < a > slip which result in only four independent slip systems [36]. Hence, extension or contraction of the HCP c-axis requires activation of pyramidal < c + a > slip and/or deformation twinning, both of which exhibit high resistance to activation. The activity of any slip or twin system occurs once the resolved shear stress exceeds the critical resolved shear stress (CRSS) on that system. The values of the CRSS can be calibrated using experimental stress-strain data [37,38]. Once these values are estimated for the materials under a selected deformation path, the validated values can be subsequently used to predict the mechanical behavior and texture evolution under any applied deformation using crystal plasticity modeling [39].

On the other hand, the beta phase (BCC) is known to accommodate plastic deformation exclusively by slip on various reported slip systems. Some simulations have used pencil glide on any slip plane containing <111 > slip systems [40,41]. Others have assumed slip on {110}, {112}, and {123} planes [42,43]. The abovementioned options should be available in any practical crystal plasticity model.

In two-phase Ti-6Al-4V with a colony microstructure, the alpha and beta phases are known to maintain a Burgers orientation relationship [44], creating an easy pathway for dislocation transmission across the interface for aligned slip systems, with resistance increasing with increased misalignment [45]. Consequently, the presence of such interfaces and limitations imposed by slip transmission across the interface results in anisotropy within individual slip systems. For example, compression experiments conducted on large single-colony samples demonstrated a significant difference between yield and hardening on the prism $< a_1 >$ and prism $< a_3 >$ slip systems (Figure 14) [45].

A single crystal under external applied strain deforms from a reference configuration to a deformed configuration, with this change being described by a deformation gradient tensor, $\mathbf{F}$. To separate elastic and plastic deformations, Kröner [46] suggested the multiplicative decomposition of the deformation gradient into elastic ($\mathbf{F}^*$) and plastic ($\mathbf{F}^p$) components (Figure 15). The plastic component $\mathbf{F}^p$ causes the permanent deformation of the materials, and it is applied in an intermediate configuration which maintains a perfect lattice. As such, estimating how a single crystal accommodates plastic deformation by crystallographic slip or twinning (pseudo slip [47]) can be conducted in the intermediate configuration using the starting orientation of the single crystal [37]. Further details of the crystal plasticity theory are presented in Figure 15 and have been explained elsewhere [39].

Polycrystalline deformation behavior

Most metallic parts are made of polycrystalline and/or multiphase alloys which require homogenization methods to predict their response starting with the single-crystal constitutive equations. Furthermore, calibrating single-crystal CRSSs and its evolutions (strain hardening) require comparisons with experimental measurements which are mostly conducted on polycrystalline materials.

In addition to homogenization methods, strain partitioning between coexisting phases is also used to capture the overall mechanical response and texture evolution of two phase materials (e.g., Ti-6Al-4V) [48].

Homogenization method: Taylor approach

To relate the deformation of an aggregate to the deformation of a single crystal, homogenization (i.e., mean field) methods have been used based on iso-strain, iso-

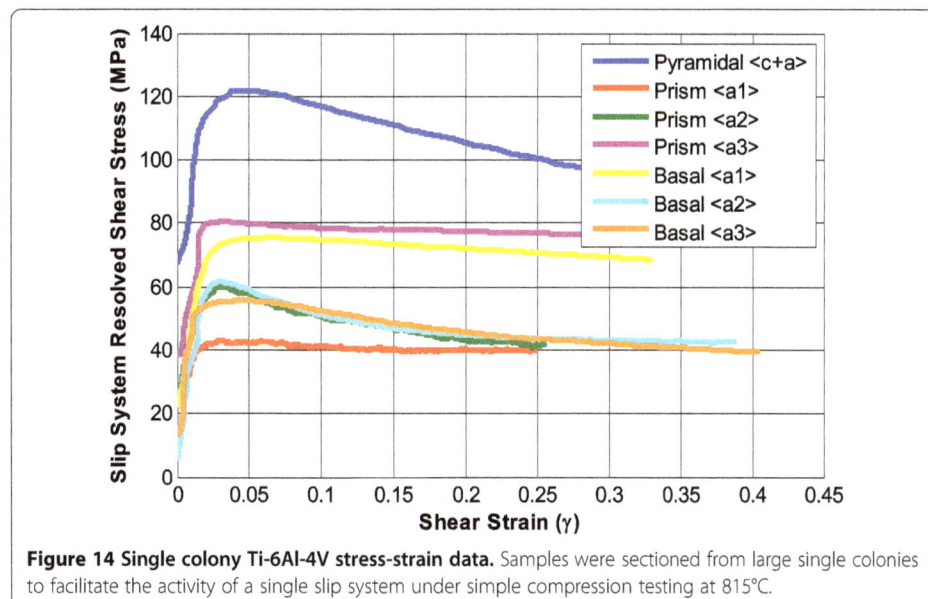

Figure 14 Single colony Ti-6Al-4V stress-strain data. Samples were sectioned from large single colonies to facilitate the activity of a single slip system under simple compression testing at 815°C.

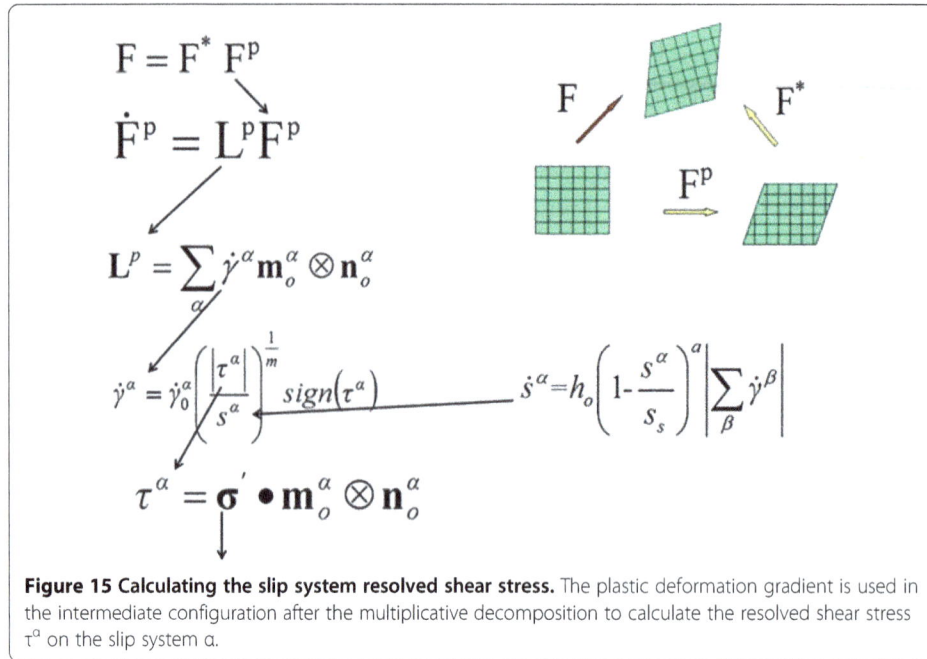

Figure 15 Calculating the slip system resolved shear stress. The plastic deformation gradient is used in the intermediate configuration after the multiplicative decomposition to calculate the resolved shear stress τ^α on the slip system α.

stress, or viscoplastic self-consistent assumptions [49]. The simplest approach is the iso-strain approach based on Taylor's model [50], which assumes that the deformation gradient experienced by individual crystallites is equal to the applied deformation gradient for the aggregate. While the method violates stress compatibility, it has been popularly applied as an upper-bound technique and has been used in simulating texture evolution of the beta phase during hot working breakdown operations above the beta transus temperature for production-scale ingots of Ti-6Al-4V [42] and during hot rolling in the beta field [51], with deformation accommodated by slip. At room temperatures, the alpha phase deforms by a combination of slip and twinning which was successfully simulated (Figure 16) using the Taylor assumption in both high purity Ti [37] and commercial purity Ti [52]. Due to the wide use of the Taylor approach, attention is focused in subsequent

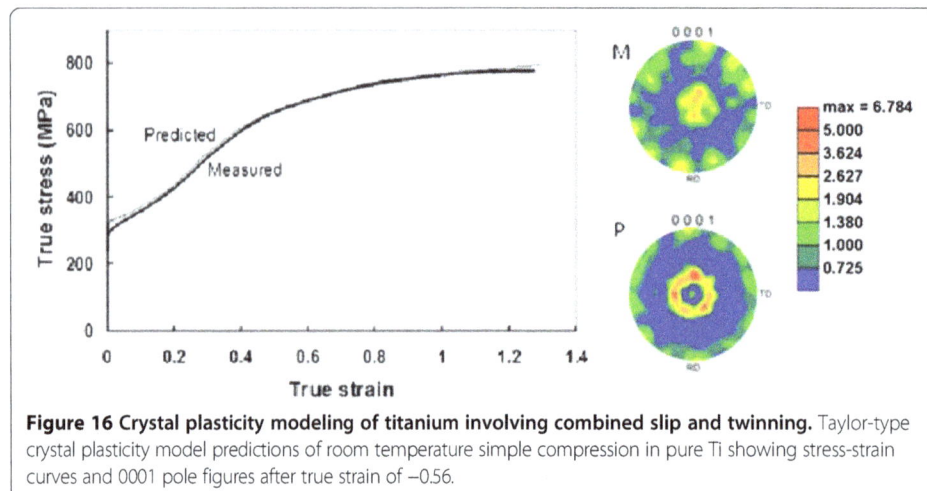

Figure 16 Crystal plasticity modeling of titanium involving combined slip and twinning. Taylor-type crystal plasticity model predictions of room temperature simple compression in pure Ti showing stress-strain curves and 0001 pole figures after true strain of −0.56.

sections on this method and to examining various methods to increase the speed of integration into FEM codes.

Spectral crystal plasticity

The use of Taylor-type models requires repeating the single-crystal calculations for each orientation within the input texture. However, when simulating real parts with varying textures at different locations in the part (as identified by the ROH), traditional Taylor calculations take a very long time to execute, and full-field simulations, such as crystal plasticity finite element method (CPFEM) (described in more detail below), can require several orders of magnitude more computational time. However, the use of spectral crystal plasticity based on fast Fourier transforms (FFTs) [53,54] has resulted in major time savings (Figure 17). The use of discrete Fourier transform (DFT) databases enables efficient calculations using spectral crystal plasticity [55]. In particular, this database is constructed for all possible orientations under all possible deformation gradients. The building of such a database constitutes a onetime effort investment. Consequently, subsequent computations of crystal plasticity are carried out by using the database. Figure 17 shows about 45× increase in the speed of calculations for interstitial-free (IF) steel (BCC) [56]. Similar databases for HCP are under development.

Full-field crystal plasticity finite element simulations

In the previously discussed Taylor-type approach, the local deformation gradient is assumed to be the same as the global deformation gradient an assumption that needs to be relaxed for most commercially used alloys with complex dual-phase microstructures. Finite element methods offer a practical solution to calculating the grain-scale heterogeneity present in the deformation field. Consequently, crystal plasticity models have been coupled with commercial finite element codes to explore highly sophisticated homogenization schemes. Some commercial FEM codes enable the implementation of crystal plasticity materials constitutive behavior as a user-subroutine (e.g., UMAT/

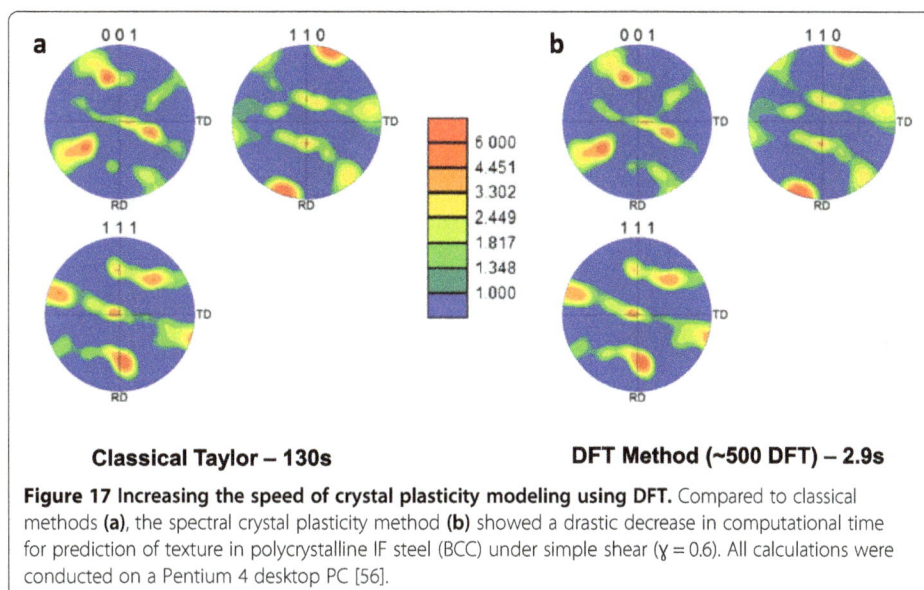

Classical Taylor – 130s   **DFT Method (~500 DFT) – 2.9s**

Figure 17 Increasing the speed of crystal plasticity modeling using DFT. Compared to classical methods **(a)**, the spectral crystal plasticity method **(b)** showed a drastic decrease in computational time for prediction of texture in polycrystalline IF steel (BCC) under simple shear ($\gamma = 0.6$). All calculations were conducted on a Pentium 4 desktop PC [56].

VUMAT in ABAQUS [57] and Hypela2 in Marc [58] and Simufact [59]. For implicit codes, the crystal plasticity user-subroutine is expected to (1) calculate the stress based on the imposed local deformation history and (2) calculate the local Jacobian matrix (defined as $d\Delta\sigma/d\Delta\varepsilon$). As such, once the appropriate user subroutine is verified to work for a crystal structure, further validation and calibration can be done by users for various materials within the same crystal structure. Adopting this method results in full-field simulations, also known as CPFEM simulations. While they provide higher accuracy in predicting the materials behavior, they are known to be slow for large simulations. The use of spectral crystal plasticity [60] has demonstrated an improvement of the calculation speed more than 30× compared to the conventional approaches (Figure 18). Further increase in the speed of calculation can be achieved by using the ROH and the ROD concepts described above.

In certain situations, it becomes necessary to simulate multiscale coupled phenomena at two well-separated length scales. As an example, consider the simulation of a complex processing operation where different macroscale spatial locations in the workpiece experience different thermomechanical histories (often an unavoidable consequence of the boundary conditions imposed at the macroscale). Consequently, strong variations in the microstructure should be expected at different locations in the workpiece. In other words, it is not enough to track the evolution of a single representative microstructure for the entire workpiece. The development of such microstructure heterogeneities can be expected to influence the macroscale simulation by altering the local effective properties at different locations in the workpiece. In such a situation, it is necessary to track independently microstructures at multiple macroscale locations in the workpiece and pass high value information in both directions (between the microscale and the macroscale). This is extremely difficult, if not impossible, using any of the currently employed computational strategies.

The challenge described above can be addressed with modest computational resources using a data science approach called materials knowledge systems (MKS) [61-66]. In the MKS framework, the focus is on localization (i.e., opposite of homogenization) relationships that capture the spatial distribution of the response field of interest (e.g., stress or

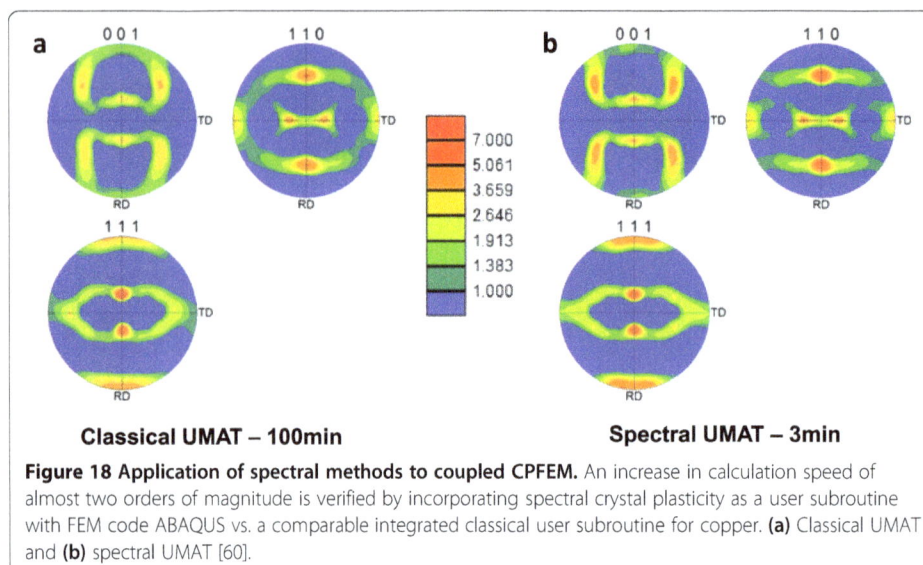

Classical UMAT – 100min   **Spectral UMAT – 3min**

Figure 18 Application of spectral methods to coupled CPFEM. An increase in calculation speed of almost two orders of magnitude is verified by incorporating spectral crystal plasticity as a user subroutine with FEM code ABAQUS vs. a comparable integrated classical user subroutine for copper. **(a)** Classical UMAT and **(b)** spectral UMAT [60].

strain rate fields) at the microscale (on an RVE) for an imposed loading condition at the macroscale. In this approach, the localization relationships are expressed as calibrated metamodels that take the form of a simple algebraic series whose terms capture the individual contributions from a hierarchy of local microstructure descriptors. Each term in this series expansion is expressed as a convolution of the appropriate local microstructure descriptor and its corresponding influence at the microscale. The series expansion in the MKS framework is in complete accord with the series expansion obtained in the physics-based statistical continuum theories [30,67-70]. The MKS approach dramatically improves the accuracy of these expressions by calibrating the convolution kernels using results from previously validated physics-based models. The most impressive benefit of the MKS approach lies in the dramatic reduction of the computational cost, often by several orders of magnitude compared to numerical approaches typically employed in microstructure design problems. In various preliminary demonstrations, the MKS methodology has been successfully applied to capturing thermoelastic stress (or strain) distributions in composite RVEs [64,65,71] and multiscale structures[61], rigid-viscoplastic deformation fields in composite RVEs [72], and the evolution of the composition fields in spinodal decomposition of binary alloys [62]. Efforts are currently underway to extend the MKS to address multiscale plastic deformation of complex alloys such as Ti alloys.

On the other hand, the use of uncoupled CPFEM simulations may also result in some computational cost savings by eliminating the feedback loop between the crystal plasticity calculations and the FEM and using only the crystal plasticity after the FEM simulation is finished [73]. In this approach, the FEM exports post-simulation strain tensor and rigid-body rotations at each integration point which is then input to the crystal plasticity code to estimate texture evolution. The main drawback of this method is the lack of coupling.

In the cases where commercial FEM codes use only empirical constitutive laws (e.g., Hill yield surface), a virtual testing laboratory method [74] using a RVE approach can be used to overcome some limitations of empirical laws by fitting the yield surface to series of simulated tests. Such a numerical test protocol predicts the shape of a yield locus and then uses it to calibrate empirical constitutive models. The accuracy of this method is dependent on the ability to find best fit for the limited number of preset yield surface variables. In addition, a new yield locus needs to be generated for each new texture (starting or evolving).

Conclusions

A workflow to incorporate microstructure morphology and crystallography into modeling of TMP of titanium alloys has been summarized based on the application of data analytics for generating representative descriptors of large microstructure datasets coupled with novel techniques that may increase throughput of multiscale materials models. Representing texture using weighted orientations (based on GSH representations of ODF) enables fast data mining of orientation information and feature of interest extraction. Using the concept of a region of homogeneity defined based on the spatial distribution of the two-point correlations enables efficient insertion of a detailed quantitative microstructure description into materials models without the need to know a priori the properties of the ROH. Reducing the size of crystallographic orientation datasets using the representative orientation distribution allows for significant increases in the speed of crystal plasticity calculations while maintaining the accuracy of predictions.

A number of options exist for solution of the crystal plasticity constitutive equations during deformation of a sample volume. Uncoupled Taylor-type simulations provide a simple methodology for estimation of final texture. For fully coupled integration of crystal plasticity into FEM simulations, the FE codes need to allow user developed subroutines that allow specification of a broad range of materials constitutive descriptions. Once again, smartly constructed databases might prove to be very practical in integrating these sophisticated materials descriptions with typically used process simulation codes.

Competing interests
The authors declare that they have no competing interests.

Authors' contributions
AAS conceived the workflow concept and created the initial draft. JBS and DPS helped with manuscript writing and analysis of exemplary datasets. SLS provided materials science guidance and expertise. SRK contributed to the overall development of the main concepts presented in this paper. All authors read and approved the final manuscript.

Acknowledgements
Support from the Air Force Research Laboratory and Air Force Office of Scientific Research is gratefully acknowledged. In particular, AAS, JBS, and DPS were partially supported by contract no. FA865009D5600 (Dr. J. Calcaterra, program manager). SRK was supported by the Air Force Office of Scientific Research, MURI contract no. FA9550-12-1-0458.

Author details
[1]Materials Resources LLC, Dayton, OH 45402, USA. [2]Air Force Research Laboratory, Materials and Manufacturing Directorate, Wright-Patterson AFB, OH 45433, USA. [3]Georgia Institute of Technology, Atlanta, GA 30332, USA.

References
1. Semiatin SL, Glavicic MG, Shevchenko SV, Ivasishin OM, Chun YB, Hwang SK (2009) Modeling and simulation of texture evolution during the thermomechanical processing of titanium alloys. In: Semiatin SL, Furrer DU (eds) ASM Handbook, Vol 22A: fundamentals of modeling for metals processing. ASM International, Materials Park, pp 536–552
2. Semiatin SL, Furrer DU (2009) Modeling of microstructure evolution during the thermomechanical processing of titanium alloys. In: Semiatin SL, Furrer DU (eds) ASM Handbook, Volume 22A: fundamentals of modeling for metals processing. ASM International, Materials Park, pp 522–535
3. Agrawal A, Deshpande PD, Cecen A, Basavarsu GP, Choudhary AN, Kalidindi SK (2014) Exploration of data science techniques to predict fatigue strength of steel from composition and processing parameters. Integr Mater Manuf Innov 3:8
4. Gibbs JW, Voorhees P (2014) Segmentation of four-dimensional, X-ray computed tomography data. Integr Mater Manuf Innov 3:6
5. Lütjering G, Williams JC (2007) Titanium. Springer, New York
6. Salem A, Glavicic M, Semiatin S (2008) A coupled EBSD/EDS method to determine the primary-and secondary-alpha textures in titanium alloys with duplex microstructures. Mater Sci Eng A 494(1):350–359
7. Kaufman L, Rousseeuw PJ (2009) Finding groups in data: an introduction to cluster analysis. John Wiley & Sons, Hoboken, NJ, USA
8. Kalidindi SR, Niezgoda SR, Salem AA (2011) Microstructure informatics using higher-order statistics and efficient data-mining protocols. JOM 63(4):34–41
9. Niezgoda SR, Kalidindi SR (2009) Applications of the phase-coded generalized hough transform to feature detection, analysis, and segmentation of digital microstructures. CMC: Comput Mater Cont 14(2):79–89
10. Bunke H, Wang PS (1997) Handbook of character recognition and document image analysis. World Scientific, New Jersey
11. Mohri M, Rostamizadeh A, Talwalkar A (2012) Foundations of machine learning. MIT Press, Cambridge, MA, USA
12. Junqué de Fortuny E, Martens D, Provost F (2013) Predictive modeling with big data: is bigger really better? J Big Data 1(4):215–226
13. Silver N (2012) The signal and the noise: Why so many predictions fail—but some don't. Penguin Press, New York
14. Salem AA, Shaffer JB (2013) Identification and quantification of microtextured regions in materials with ordered crystal structure., US Patent 13/761,612
15. Germain L, Gey N, Humbert M, Bocher P, Jahazi M (2005) Analysis of sharp microtexture heterogeneities in a bimodal IMI 834 billet. Acta Mater 53(13):3535–3543
16. Bunge H (1982) Texture analysis in materials science. Butterworths, London
17. Adams BL, Kalidindi SR, Fullwood D (2012) Microstructure sensitive design for performance optimization. Butterworth-Heinemann, Newton, MA, USA
18. Fullwood DT, Niezgoda SR, Adams BL, Kalidindi SR (2010) Microstructure sensitive design for performance optimization. Prog Mater Sci 55(6):477–562

19. Houskamp JR, Proust G, Kalidindi SR (2007) Integration of microstructure-sensitive design with finite element methods: elastic-plastic case studies in FCC polycrystals. Int J Multiscale Com 5(3–4):261–272

20. Knezevic M, Kalidindi SR (2007) Fast computation of first-order elastic-plastic closures for polycrystalline cubic-orthorhombic microstructures. Comput Mater Sci 39(3):643–648

21. Proust G, Kalidindi SR (2006) Procedures for construction of anisotropic elastic-plastic property closures for face-centered cubic polycrystals using first-order bounding relations. J Mech Phys Solids 54(8):1744–1762

22. Hill R (1963) Elastic properties of reinforced solids: some theoretical principles. J Mech Phys Solids 11(5):357–372

23. Drugan W, Willis J (1996) A micromechanics-based nonlocal constitutive equation and estimates of representative volume element size for elastic composites. J Mech Phys Solids 44(4):497–524

24. Ostoja-Starzewski M (1998) Random field models of heterogeneous materials. Int J Solids Struct 35(19):2429–2455

25. Torquato S (2002) Random heterogeneous materials. Springer-Verlag, New York

26. Niezgoda SR, Kanjarla AK, Kalidindi SR (2013) Novel microstructure quantification framework for databasing, visualization, and analysis of microstructure data. Integr Mater Manuf Innov 2:3

27. Niezgoda SR, Yabansu YC, Kalidindi SR (2011) Understanding and visualizing microstructure and microstructure variance as a stochastic process. Acta Mater 59(16):6387–6400

28. Niezgoda SR, Turner DM, Fullwood DT, Kalidindi SR (2010) Optimized structure based representative volume element sets reflecting the ensemble-averaged 2-point statistics. Acta Mater 58(13):4432–4445

29. Niezgoda SR, Fullwood DT, Kalidindi SR (2008) Delineation of the space of 2-point correlations in a composite material system. Acta Mater 56(18):5285–5292

30. Fullwood DT, Niezgoda SR, Kalidindi SR (2008) Microstructure reconstructions from 2-point statistics using phase-recovery algorithms. Acta Mater 56(5):942–948

31. Salem AA, Glavicic M, Semiatin S (2008) The effect of preheat temperature and inter-pass reheating on microstructure and texture evolution during hot rolling of Ti-6Al-4 V. Mater Sci Eng A 496(1):169–176

32. Qidwai SM, Turner DM, Niezgoda SR, Lewis AC, Geltmacher AB, Rowenhorst DJ, Kalidindi SR (2012) Estimating response of polycrystalline materials using sets of weighted statistical volume elements (WSVEs). Acta Mater 60:5284–5299

33. Wargo EA, Hanna AC, Cecen A, Kalidindi SR, Kumbur EC (2012) Selection of representative volume elements for pore-scale analysis of transport in fuel cell materials. J Power Sources 197:168–179

34. Torquato S (2009) Inverse optimization techniques for targeted self-assembly. Soft Matter 5(6):1157–1173

35. Torquato S (2010) Optimal design of heterogeneous materials. Annu Rev Mater Res 40:101–129

36. Salem A, Kalidindi SR, Doherty RD (2003) Strain hardening of titanium: role of deformation twinning. Acta Mater 51(14):4225–4237

37. Salem A, Kalidindi S, Semiatin S (2005) Strain hardening due to deformation twinning in α-titanium: constitutive relations and crystal-plasticity modeling. Acta Mater 53(12):3495–3502

38. Li H, Mason D, Bieler T, Boehlert C, Crimp M (2013) Methodology for estimating the critical resolved shear stress ratios of α-phase Ti using EBSD-based trace analysis. Acta Mater 61(20):7555–7567

39. Kalidindi SR, Bronkhorst CA, Anand L (1992) Crystallographic texture evolution in bulk deformation processing of FCC metals. J Mech Phys Solids 40(3):537–569

40. Morris PR, Semiatin SL (1979) The prediction of plastic properties of polycrystalline aggregates of BCC metals deforming by <111 > pencil glide. Texture of Crystalline Solids 3(2):113–126

41. Piehler H, Backofen W (1971) A theoretical examination of the plastic properties of bcc crystals deforming by <111 > pencil glide. Metall Trans 2(1):249–255

42. Glavicic M, Kobryn P, Goetz R, Yu K, Semiatin S (2004) Texture evolution during primary processing of production-scale vacuum arc remelted ingots of Ti–6Al–4V. In: Proc. 10th world conf. on titanium. Wiley-VCH, Weinheim, Germany, pp 1299–1306

43. Chin G, Mammel W (1967) Computer solutions of Taylor analysis for axisymmetric flow. Trans Metall Soc AIME 239(9):1400–1405

44. Burgers W (1934) On the process of transition of the cubic-body-centered modification into the hexagonal-close-packed modification of zirconium. Physica 1(7):561–586

45. Salem A, Semiatin S (2009) Anisotropy of the hot plastic deformation of Ti–6Al–4 V single-colony samples. Mater Sci Eng A 508(1):114–120

46. Kröner E (1959) Allgemeine kontinuumstheorie der versetzungen und eigenspannungen. Arch Rational Mech Anal 4(1):273–334

47. Kalidindi SR (1998) Incorporation of deformation twinning in crystal plasticity models. J Mech Phys Solids 46(2):267–290

48. Glavicic M, Goetz R, Barker D, Shen G, Furrer D, Woodfield A, Semiatin S (2008) Modeling of texture evolution during hot forging of alpha/beta titanium alloys. Metall Mater Trans A 39(4):887–896

49. Lebensohn R, Tomé C (1993) A self-consistent anisotropic approach for the simulation of plastic deformation and texture development of polycrystals: application to zirconium alloys. Acta Metall Mater 41(9):2611–2624

50. Taylor GI (1938) Plastic strain in metals. J Inst Metals 62:307–324

51. Gey N, Humbert M, Philippe MJ, Combres Y (1996) Investigation of the α- and β- texture evolution of hot rolled Ti-64 products. Mater Sci Eng A 219(1–2):80–88

52. Wu X, Kalidindi SR, Necker C, Salem AA (2007) Prediction of crystallographic texture evolution and anisotropic stress–strain curves during large plastic strains in high purity α-titanium using a Taylor-type crystal plasticity model. Acta Mater 55(2):423–432

53. Kalidindi SR, Duvvuru HK, Knezevic M (2006) Spectral calibration of crystal plasticity models. Acta Mater 54(7):1795–1804

54. Lebensohn RA, Rollett AD, Suquet P (2011) Fast Fourier transform-based modeling for the determination of micromechanical fields in polycrystals. JOM 63(3):13–18

55. Knezevic M, Al-Harbi HF, Kalidindi SR (2009) Crystal plasticity simulations using discrete Fourier transforms. Acta Mater 57(6):1777–1784

56. Al-Harbi HF, Knezevic M, Kalidindi SR (2010) Spectral approaches for the fast computation of yield surfaces and first-order plastic property closures for polycrystalline materials with cubic-triclinic textures. CMC: Comput Mater Cont 15(2):153–172

57. ABAQUS (2014) 6.13. Dassault Systèmes, Providence, RI, USA
58. Marc (2013) 2013.1. MSC Software, Newport Beach, CA, USA
59. Simufact.forming (2014) Simufact engineering GmbH. Hamburg, Germany
60. Al-Harbi HF, Kalidindi SR (2014) Crystal plasticity finite element simulations using a database of discrete Fourier transforms. Int J Plast doi:10.1016/j.ijplas.2014.04.006
61. Al-Harbi HF, Landi G, Kalidindi SR (2012) Multi-scale modeling of the elastic response of a structural component made from a composite material using the materials knowledge system. Modell Simul Mater Sci Eng 20:055001
62. Fast T, Niezgoda SR, Kalidindi SR (2011) A new framework for computationally efficient structure–structure evolution linkages to facilitate high-fidelity scale bridging in multi-scale materials models. Acta Mater 59(2):699–707
63. Kalidindi SR, Niezgoda SR, Landi G, Vachhani S, Fast T (2010) A novel framework for building materials knowledge systems. CMC: Comput Mater Cont 17(2):103–125
64. Landi G, Niezgoda SR, Kalidindi SR (2010) Multi-scale modeling of elastic response of three-dimensional voxel-based microstructure datasets using novel DFT-based knowledge systems. Acta Mater 58(7):2716–2725
65. Landi G, Kalidindi SR (2010) Thermo-elastic localization relationships for multi-phase composites. CMC: Comput Mater Cont 16(3):273–293
66. Adams BL, Kalidindi SR, Fullwood DT (2012) Microstructure sensitive design for performance optimization. Science, Elsevier
67. Kroner E (1986) Statistical modelling. In: Gittus J, Zarka J (eds) Modelling small deformations of polycrystals. Elsevier Science Publishers, London, pp 229–291
68. Kroner E (1977) Bounds for effective elastic moduli of disordered materials. J Mech Phys Solids 25(2):137–155
69. Binci M, Fullwood D, Kalidindi SR (2008) A new spectral framework for establishing localization relationships for elastic behavior of composites and their calibration to finite-element models. Acta Mater 56(10):2272–2282
70. Kalidindi SR, Binci M, Fullwood D, Adams BL (2006) Elastic properties closures using second-order homogenization theories: case studies in composites of two isotropic constituents. Acta Mater 54(11):3117–3126
71. Fast T, Kalidindi SR (2011) Formulation and calibration of higher-order elastic localization relationships using the MKS approach. Acta Mater 59(11):4595–4605
72. Kalidindi SR (2012) Computationally-efficient fully-coupled multi-scale modeling of materials phenomena using calibrated localization linkages. ISRN Materials Science doi:10.5402/2012/305692
73. Kalidindi S, Anand L (1992) An approximate procedure for predicting the evolution of crystallographic texture in bulk deformation processing of FCC metals. Int J Mech Sci 34(4):309–329
74. Kraska M, Doig M, Tikhomirov D, Raabe D, Roters F (2009) Virtual material testing for stamping simulations based on polycrystal plasticity. Comput Mater Sci 46(2):383–392

4

Advancing quantitative description of porosity in autogenous laser-welds of 304L stainless steel

Jonathan D Madison[1*], Larry K Aagesen[2], Victor WL Chan[2] and Katsuyo Thornton[2]

* Correspondence:
jdmadis@sandia.gov
[1]Computational Materials & Data Science, Sandia National Laboratories, 87185 Albuquerque, NM, USA
Full list of author information is available at the end of the article

Abstract

Porosity in linear autogenous laser welds of 304L stainless steel has been investigated using micro-computed tomography to reveal defect content in fifty-four welds made with varying delivered power, travel speed and focal lens. Trends associated with porosity size and frequencies are shown and interfacial measures are employed to provide quantitative descriptors of pore shape, directionality, interspacing and solid linear fraction. Lastly, the coefficient of variation associated with equivalent pore radii is reported toward a discussion of microstructural variability and the influence of process-parameters on such variability.

Keywords: Micro-computed tomography; Porosity; Stainless steel; Interfacial shape distribution; Interfacial normal distribution

Background

Among joining and metal processing techniques used in industrial and scientific capacities, laser welding is relatively new. Due to its ability to supply high densities of power to very controlled areas with minimal peripheral excess heat input, it has become a rapidly growing and highly attractive joining process for metals [1,2]. Common interrogation practice for welds are often performed via post-mortem failure analysis [3], post-process radiography [4], or ultrasonic scan [5]. Typically, these evaluations provide an opportunity to identify the most probable cause of failure, or produce a qualitative understanding of the internal structure of the weld.

For most engineering metals, there exists a fairly clear inverse correlation between pore volume and mechanical properties such as strength or modulus with varying degrees of sensitivity. As a specific example, defects such as pores, occurring naturally or imposed artificially, have been shown to serve as preferred sites for the initiation or propagation of failure in creep in both conventional and high cycle fatigue of aluminum [6,7], a material system having high formability and broad applications. 304L stainless steel is unique in this regard as the effects of porosity on some material properties challenge intuition. Two examples in the literature which illustrate this phenomena can be found in the work of Boyce et al. [3] and Kuo and Jeng [8]. In the work of Boyce et al., autogenous continuous-wave and pulsed-wave laser welds were made across the gauge section of 304L stainless-steel tensile bars, which were subsequently strained to failure. While one weld schedule was noted to produce higher amounts of porosity than the other, no decrease in mechanical strength was observed. In the work of Kuo and Jeng, a variety of weld schedules were created for 304L stainless steel, where increasing porosity levels coincided with decreases in hardness and relatively small variations in yield strength. Additionally, the continuous-wave-laser weld sample, which contained higher amounts of porosity than any pulsed-wave-laser weld sample, demonstrated significantly higher tensile strength than all pulsed-wave-laser weld samples [8]. These findings suggest that the interplay of processing parameters may affect laser-welded microstructure in ways that complicate the individual effect of porosity, particularly in 304L. Furthermore, both examples illustrate that the effects of laser-welding induced porosity in 304L on certain mechanical properties is not clearly understood. We suggest that advancing the quantitative description of porosity in 304L laser weldments and relating them directly to carefully controlled weld parameters can assist in better understanding the concomitant effects of porosity in this ubiquitous and highly damage-tolerant material system.

Fortunately, for nearly all metallic systems, the parameters used to form the laser-weld are among the most pivotal factors that determine the local microstructure. Typical processing parameters may include; shielding gas, laser power, power profile, filler material, travel speed and focal distance between the laser source and weld surface. The combination of these factors is often referred to as the 'weld schedule'. In this study, parameters of the weld schedule investigated have been limited to weld power, travel speed and focal length. Fortunately, recent advances in characterization and microstructure visualization have provided a rich set of tools being increasingly brought to bear on laser-weld induced porosity in a variety of metals [9-13]. The work presented here builds upon such investigations and utilizes micro-computed tomography and other emerging state-of-the-art three-dimensional (3D) characterization techniques to quantitatively relate porosity in autogenous laser-welds of 304L stainless-steel to specific processing parameters [9,10,14].

Methods

Using a ROFIN-Sinar, Inc. CW 0.15 HQ, fiber-optic delivered Nd:YAG laser, over fifty unique weld-schedules were used to produce autogenous standing-edge seam welds in 304L stainless steel having an elemental composition of Fe–0.04C–18.12Cr–1.21Mn–8.09Ni–0.028 N–0.022P–0.001S–0.34Si (wt.%). In all subsequent depictions, the x-direction denotes the weld width, the z-direction denotes the weld length and the y-direction denotes the weld depth, which is also the axis of incidence of the laser relative to the sample. For each sample, a standing-edge weld was formed by affixing two 2.54 cm × 10.16 cm × 0.1 cm flat plates together face-to-face by base-clamp, then laser-welding their upper seam at one of five constant travel speeds (254 mm × min^{-1}, 508 mm × min^{-1}, 1016 mm × min^{-1}, 1524 mm × min^{-1} or 2032 mm × min^{-1}) and at one of six delivered powers ranging from 200 W to 1200 W as measured with a Macken P2000Y laser power-probe. This parameter set included a total of 27 separate weld schedules and was performed for two separate focal lengths, 80 mm and 120 mm, bringing the total weld schedule count to 54. For a complete matrix of the weld schedules investigated, see Table 1.

Following welding, micro-computed tomography (μCT) was performed on each sample to reveal the size, location and morphology of the internal porosity. A Kevex PSX 10-65 W x-ray tube was employed using a 250 μA current and 130 kV operating

Table 1 Maximum pore volume and total pores observed per case

	80 mm lens				
	252 mm × min^{-1}	510 mm × min^{-1}	1016 mm × min^{-1}	1524 mm × min^{-1}	2032 mm × min^{-1}
1200 W			0.49 mm^3 (373)	0.05 mm^3 (550)	0.03 mm^3 (425)
1000 W		0.76 mm^3 (160)	0.30 mm^3 (337)	0.03 mm^3 (603)	0.02 mm^3 (403)
800 W	0.28 mm^3 (60)	0.79 mm^3 (190)	0.1 mm^3 (612)	0.03 mm^3 (835)	0.016 mm^3 (349)
600 W	0.17 mm^3 (145)	0.38 mm^3 (247)	0.045 mm^3 (652)	0.013 mm^3 (406)	0.008 mm^3 (192)
400 W	0.08 mm^3 (394)	0.02 mm^3 (343)	0.009 mm^3 (116)	0.0005 mm^3 (10)	0.0007 mm^3 (20)
200 W	– (0)	– (0)	– (0)	– (0)	– (0)
	120 mm lens				
	252 mm × min^{-1}	510 mm × min^{-1}	1016 mm × min^{-1}	1524 mm × min^{-1}	2032 mm × min^{-1}
1200 W			0.95 mm^3 (431)	0.10 mm^3 (391)	0.03 mm^3 (736)
1000 W		1.50 mm^3 (130)	0.51 mm^3 (190)	0.57 mm^3 (381)	0.01 mm^3 (263)
800 W	0.17 mm^3 (77)	0.59 mm^3 (129)	0.09 mm^3 (302)	0.01 mm^3 (290)	0.006 mm^3 (264)
600 W	0.24 mm^3 (120)	0.26 mm^3 (284)	0.01 mm^3 (267)	0.007 mm^3 (132)	0.009 mm^3 (91)
400 W	0.07 mm^3 (81)	0.01 mm^3 (6)	0.001 mm^3 (1)	– (0)	– (0)
200 W	– (0)	– (0)	– (0)	– (0)	– (0)

voltage. Samples were rotated for one full rotation at a speed of approximately 0.12°/s. In an effort to identify optimal trade-offs in scanning time and resolution, multiple scanning resolutions ranging from 9–30 μm per voxel edge were employed with most scans being performed at 15 μm per voxel edge, except the six weld schedules produced at 1200 W. A minimum of ten contiguous voxels was imposed for identification as a pore across all datasets. To verify changes in tomographic resolution did not bias results, one weld produced at 1524 mm × min^{-1} was re-scanned at higher resolution. No significant differences in average pore size, maximum pore or total number of pores were observed. To provide a basis for comparison with more conventional methods, standard metallographic preparation and imaging via optical microscopy were performed on weld-sample cross-sections. Measurements of weld depth, width, crown height and pore volume fraction were also performed [9,10]. In these studies, pore volume fraction was shown to vary from 1-8% and result largely from key-hole collapse of the weld pool [13,15-18]. Transverse and longitudinal micrographs illustrating pore structures are shown in Figure 1 for three separate weld schedules made at 1200 W.

Following tomography, three-dimensional reconstructions of pore populations were performed by two separate methods. A cone-beam reconstruction algorithm operating in tandem with the μCT experiment was carried out using VG StudioMax® and two-dimensional images derived from the μCT experiments were independently segmented in Adobe Photoshop® and reconstructed using IDL®. This was done to transfer the data into a form amenable to higher-level morphological analyses such as interfacial and topological characterization, which will be discussed later. A series of images from IDL®-generated reconstructions of weld porosity at 600 W and the previously identified five travel speeds are shown in Figure 2. The color designations for each pore set in Figure 2 have been utilized to denote a specific travel speed and will be employed throughout the paper to aid the reader in making associations between weld parameters in subsequent figures and plots. For two full color-coded matrices of the welds investigated, the reader may see the Additional files 1 and 2.

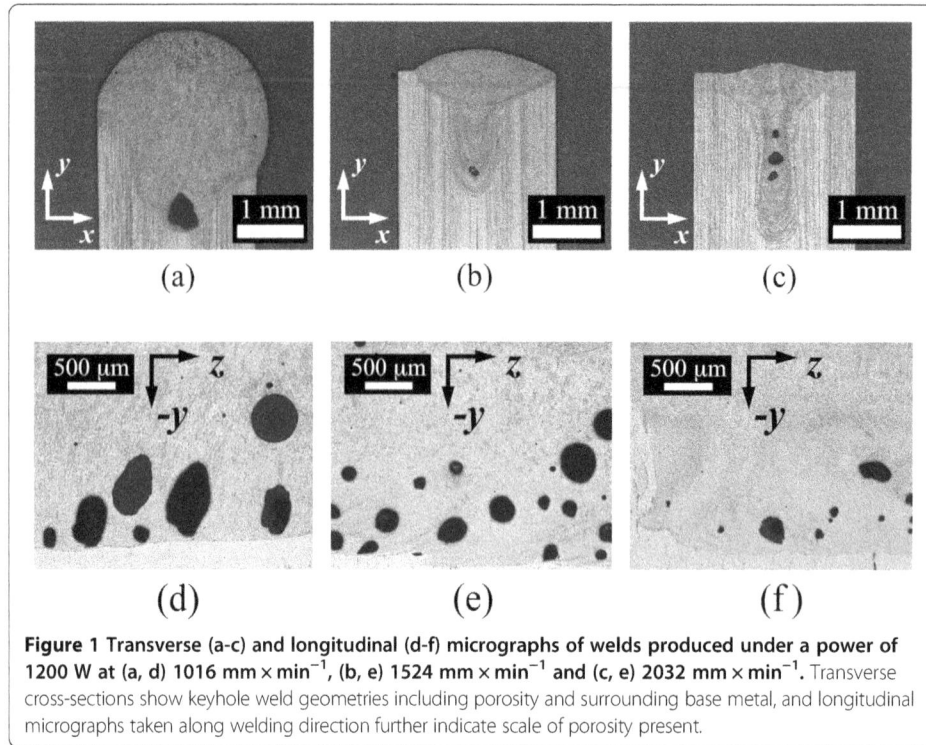

Figure 1 Transverse (a-c) and longitudinal (d-f) micrographs of welds produced under a power of 1200 W at (a, d) 1016 mm × min^{-1}, (b, e) 1524 mm × min^{-1} and (c, e) 2032 mm × min^{-1}. Transverse cross-sections show keyhole weld geometries including porosity and surrounding base metal, and longitudinal micrographs taken along welding direction further indicate scale of porosity present.

Characterization

Pore characterization

Utilizing the reconstructions obtained and the known voxel resolutions for each weld sample, physical measures of pore size, population and frequency were calculated for pores constituting ninety-percent or more of the voided space within each sample. These values serve as a baseline and comparison for readily employed measures of pore presence.

Interfacial morphology

Using tools developed by Voorhees et al. [19-24], the interfacial morphology of porosity was also examined. The primary tools used to illustrate these quantitative descriptors of shape are the interfacial shape [20,22] and interfacial normal [24] distributions. Given a triangulation representing the interface between two phases of a discretized microstructure, the mean (H) and Gaussian (K) curvatures can be calculated for each patch of interface according to the method of Meyer et al. [25]. These in turn, can be used to calculate the minimum and maximum principal curvatures, κ_1, κ_2 respectively, for each patch of interfacial area as follows.

$$\kappa_1 = H - \sqrt{H^2 - K} \tag{1}$$

$$\kappa_2 = H + \sqrt{H^2 - K} \tag{2}$$

This analysis allows for categorization of all patches into a specific (κ_1, κ_2) pairing. The method used to visualize this probability of principal curvature pairings is a two-dimensional color contour plot in which the horizontal and vertical axes are κ_1 and κ_2, respectively, and the color indicates the probability associated with each pair of

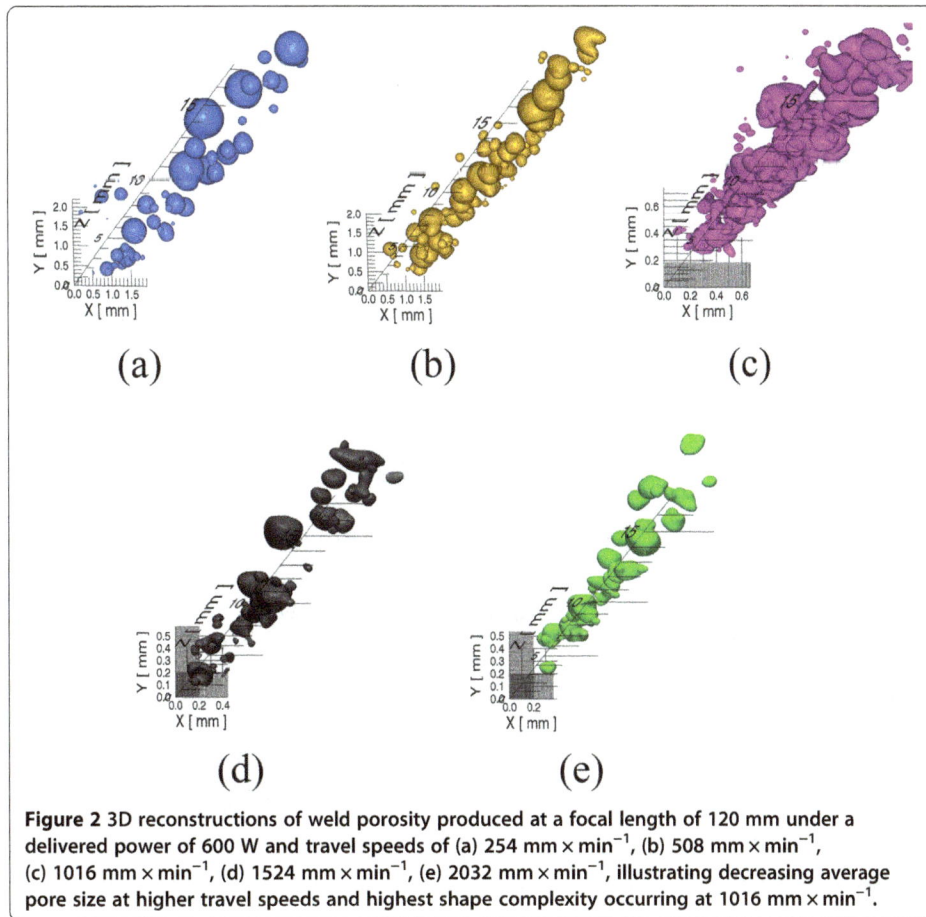

Figure 2 3D reconstructions of weld porosity produced at a focal length of 120 mm under a delivered power of 600 W and travel speeds of (a) 254 mm × min^{-1}, (b) 508 mm × min^{-1}, (c) 1016 mm × min^{-1}, (d) 1524 mm × min^{-1}, (e) 2032 mm × min^{-1}, illustrating decreasing average pore size at higher travel speeds and highest shape complexity occurring at 1016 mm × min^{-1}.

principal curvatures. This visualization technique is called the Interfacial Shape Distribution (ISD) [20,22,23] and its descriptive legend is reproduced here for convenience, Figure 3. For the purposes of this study, the "L" phase, as illustrated in the figure, corresponds to the pores within welds (the figure was originally made for a solid–liquid system). To interpret the ISD, it is useful to arrange the curvature pairings into four major regions or categories. The uniqueness of these four categories is determined by the combination of positive or negative signs of H and K. Patches in which both H and K are positive correspond to Region 1, which are spherical or ellipsoidal with solid within. Patches in which the H is positive and K is negative correspond to region 2. Region 3 consists of patches that have negative H and K values. Regions 2 and 3 contain saddle shaped patches. Lastly, patches that have a negative H and positive K correspond to Region 4, which are spherical or ellipsoidal with the pore phase within. Based upon the population of each principal curvature within one complete three-dimensional reconstruction, which we will also refer to as a dataset, a probability can be assigned denoting the likelihood of encountering a particular pairing of principal curvatures for a given patch on the interface. It is important to note that the ISDs presented in this study have not been normalized by any characteristic length-scale and therefore illustrate a combined effect of both shape and size. As a result, the color bar associated with the ISDs presented later in the results section have units of μm^2 so that the integration of the probability function over the entire ISD is equal to 1.

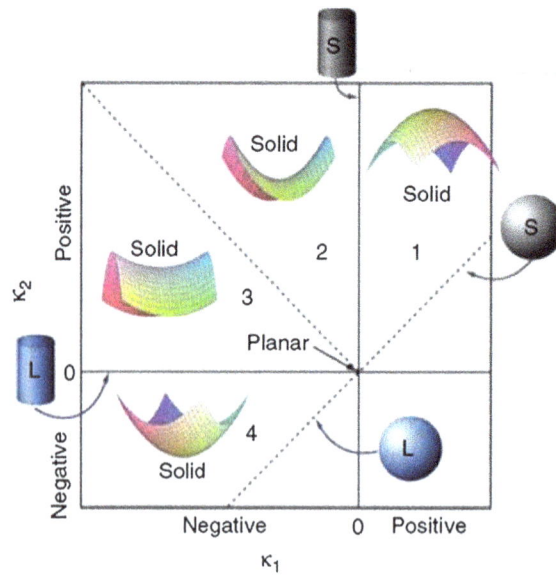

Figure 3 Interfacial Shape Distribution Legend indicating curvature types and their location within the ISD [20,22,23]. Reprinted from Acta Materialia, Vol. 54, issue 6, D. Kammer, P. Voorhees, "The Morphological Evolution of Dendritic Microstructures During Coarsening", pp. 1549–1558, (2006), with permission from Elsevier.

Interfacial orientation

The interfacial normal ($\hat{n}$) associated with each interfacial patch of a dataset is used to define a probability distribution for their orientation in three-dimensional space. The method used to visualize this probability distribution is the Interfacial Normal Distribution or IND [23,24]. In this visualization technique, the two-dimensional projection of a sphere with respect to a given axis displays the probability of occurrence of a given normal orientation. In this study, all INDs are presented as projections along the positive z-axis, which is also the direction of travel for the work-piece beneath the welding laser. Thus, the upper and lower hemispheres correspond to the direction toward and away from the laser, respectively. The color values at each location in the IND indicate the probability of encountering a particular normal based on the population of normals within the dataset. The color bar associated with each IND presented later in the section on results represents non-dimensional probability.

Spatial analysis

To better understand the spatial distribution of pore interfaces, we develop a method to calculate interfacial-distance distributions (IDD) that provide the probability distribution of the inter-pore distances, as measured between the pore surfaces nearest to one another. In this paper, we refer to the IDD of the pores as "pore-interspacing distribution" (PID). The method is based on the scheme for calculating the channel-size distributions of complex three-dimensional microstructures [26]. Further details regarding this methodology will be available in a forthcoming publication. Briefly, to calculate the interface-distance distribution, the image data of the 3D microstructure of the weld is first converted to a signed distance function, the magnitude of which represents the distance from the nearest pore-solid interface. The sign of the value indicates phase within the dataset. In this study, points inside the pores are represented as positive distances and points in the solid are represented as negative distances. Each isodistance of a microstructure is an isosurface defined by thresholding the distance function associated with the microstructure at a particular distance value. We refer to the regions with distance function values greater than the threshold as "voids" to distinguish them from the original pores as the sizes and topological characteristics of the voids differ from those of the pores. (Note that the interface of the microstructure is described by the isosurface with a threshold value of zero.) Topological changes in the isodistance structure from one threshold to another denote the feature sizes of the structure. The differences between isodistance structures resulting from thresholding in the positive and negative directions are illustrated in the two-dimensional schematics of Figure 4, where the value of the distance function is represented by grayscale and the isocontour (analogous to isosurface in three dimensions) of each threshold value is marked by a black line.

When the isodistance structures join together from thresholding with a negative threshold value, a change in the number of voids arises. This occurs at a distance value corresponding to half of the pore interspacing (the distance between interfaces at the narrowest point). Since we are examining systems that contain a variety of spatial distributions of pores, PIDs are calculated by measuring the rate at which pores are joining as a function of the distance threshold. Specifically, the PID is calculated by taking the negative derivative (-1 times the derivative) of the number of voids as a function of twice the distance threshold. Numerically, a central differencing method is used to calculate the derivatives. Each point in the PID represents the probability of finding a pair of pores with the pore interspacing at the corresponding distance threshold value.

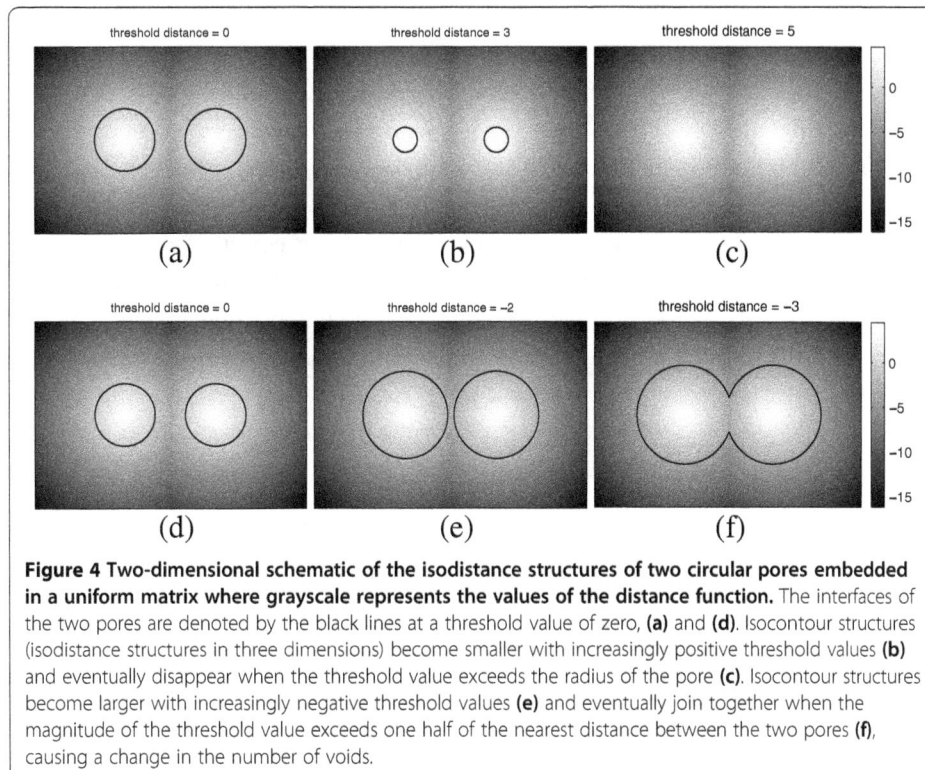

Figure 4 Two-dimensional schematic of the isodistance structures of two circular pores embedded in a uniform matrix where grayscale represents the values of the distance function. The interfaces of the two pores are denoted by the black lines at a threshold value of zero, **(a)** and **(d)**. Isocontour structures (isodistance structures in three dimensions) become smaller with increasingly positive threshold values **(b)** and eventually disappear when the threshold value exceeds the radius of the pore **(c)**. Isocontour structures become larger with increasingly negative threshold values **(e)** and eventually join together when the magnitude of the threshold value exceeds one half of the nearest distance between the two pores **(f)**, causing a change in the number of voids.

Furthermore, a characteristic pore interspacing is calculated by taking the weighted mean of the pore interspacing.

Similarly, when the isodistance structures are created for positive threshold values, a change in the number of voids also arises when the threshold becomes larger than the largest distance function value within a pore. The corresponding distance value is half the smallest dimension (the radius for a sphere) of the pore. Once again, measuring the rate at which pores disappear as a function of the distance threshold gives a pore-size distribution. Lastly, the characteristic pore interspacing is scaled to yield solid linear fraction (SLF):

$$SLF = \frac{R}{r + R} \tag{3}$$

where R is one half of the characteristic pore interspacing and r is the characteristic pore radius. The SLF provides a measure of local linear fraction of solid along the path connecting the center of the particles and passing through the narrowest matrix region. Unlike pore volume fraction, another commonly used measure of density, the SLF does not depend on the volume used for the calculation. This is of particular note for each weld schedule studied here, as laser weld porosity is generally a localized phenomena often occurring at the centerline of the weld and not distributed homogeneously throughout the weld. Furthermore, the SLF is useful as it yields a quantitative metric of solid material between regions of densely populated pores relative to the size of pores present. It is expected that this type of spacing sensitivity metric would have a strong influence on the mechanical properties of the weld.

Results and discussion

Population statistics

As mentioned previously, a variety of population metrics were extracted from each weld schedule including but not limited to average and maximum pore volume, linear frequency, and total number of pores observed per weld case. Average pore volume for both focal lengths are shown in Figure 5 where each data series corresponds to a specific travel speed and data are presented as a function of delivered power. The trend is consistent in that for a given travel speed, increases in welding power produce larger pores. Slower travel speeds also generally produce larger pores with some exceptions to this trend. Linear frequency was obtained by taking the total number of observed pores for a given weld and dividing it by the weld length. Figure 6 illustrates a more complex dependence of linear frequency on power and travel speed. With the exception of the 2032 mm × min^{-1} data series, an inflection point for the maximum frequency of pores was observed across all series of travel speeds, indicating that fewer pores can be obtained by increasing power delivered while maintaining the same travel speed. Furthermore, increases in travel speed shift the inflection point for diminishing pore presence to higher powers. An inflection point may be observed for the 2032 mm × min^{-1} travel speed at higher powers as the data suggests an inflection point for this travel speed may lie just beyond the bounds of this study. While the aforementioned trends are consistent across both focal lenses, it is interesting to note that the 120 mm lens produces higher average porosity volume (Figure 5) but generally lower frequencies (Figure 6) at all parameter sets. In Table 1, maximum pore volume observed and total pore

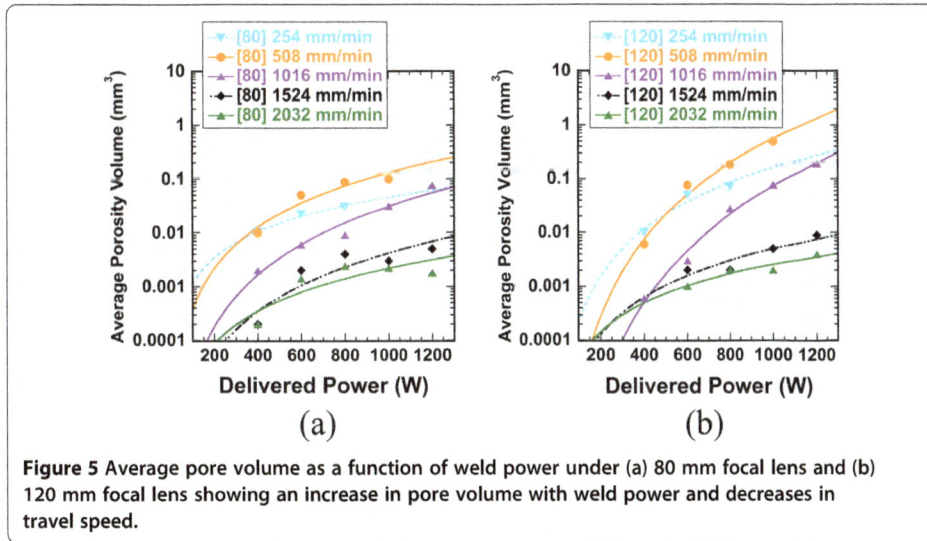

Figure 5 Average pore volume as a function of weld power under (a) 80 mm focal lens and (b) 120 mm focal lens showing an increase in pore volume with weld power and decreases in travel speed.

populations within each weld schedule are shown, with population counts denoted within parentheses. No porosity was observed in any weld produced at 200 W nor at higher travel speeds at 400 W under the 120 mm focal length.

Interfacial shape distributions

Interfacial shape distributions for porosity observed at 600 W across all five travel speeds are shown for a focal length of 120 mm in Figure 7. All ISDs are plotted on a uniform color scale in units of square microns to aid in comparison. In the discussion in this section and those following, welds are grouped by travel speed into three categories: low (254 and 508 mm × min^{-1}), moderate (1016 mm × min^{-1} only), and high (1524 and 2032 mm × min^{-1}). At low travel speeds, the majority of interfacial patches lie in Region 4 of the ISD, which largely corresponds to elliptical shapes with the edge of the shape distributions reaching the $\kappa_1 = \kappa_2$ line, which corresponds to completely

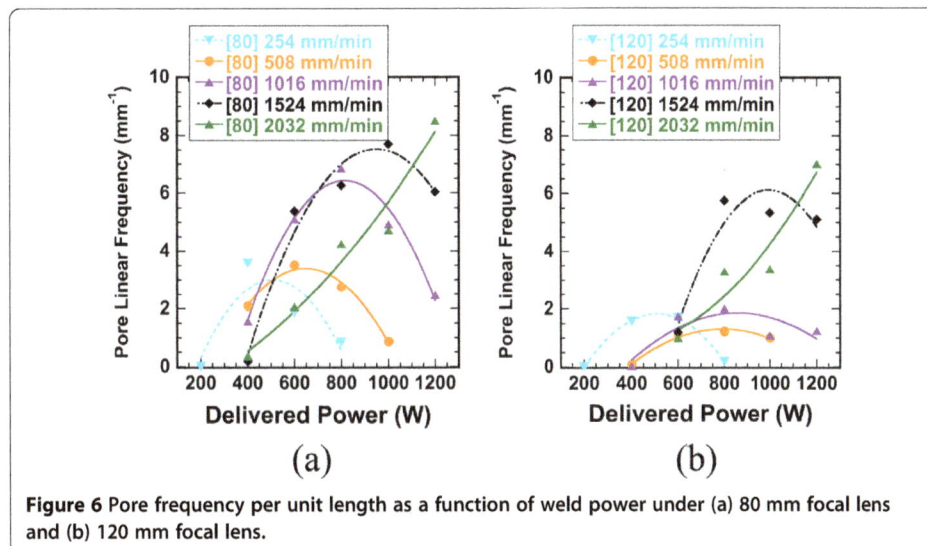

Figure 6 Pore frequency per unit length as a function of weld power under (a) 80 mm focal lens and (b) 120 mm focal lens.

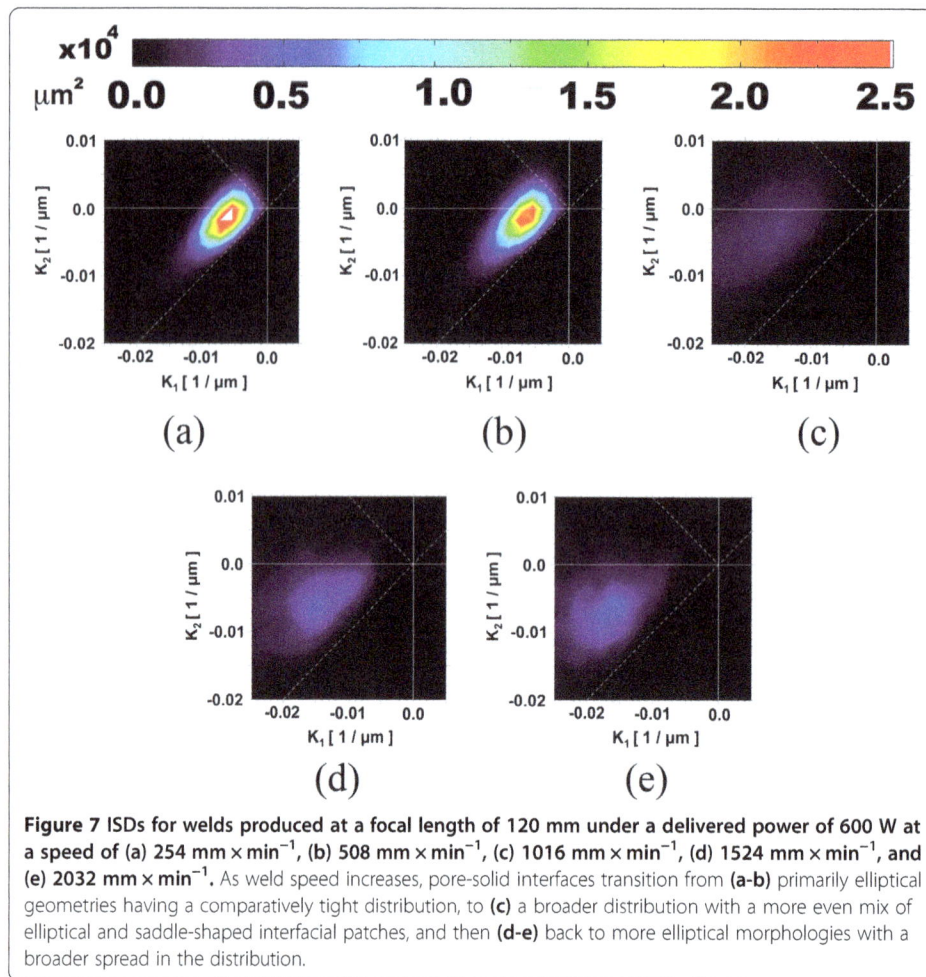

Figure 7 ISDs for welds produced at a focal length of 120 mm under a delivered power of 600 W at a speed of (a) 254 mm × min^{-1}, (b) 508 mm × min^{-1}, (c) 1016 mm × min^{-1}, (d) 1524 mm × min^{-1}, and (e) 2032 mm × min^{-1}. As weld speed increases, pore-solid interfaces transition from **(a-b)** primarily elliptical geometries having a comparatively tight distribution, to **(c)** a broader distribution with a more even mix of elliptical and saddle-shaped interfacial patches, and then **(d-e)** back to more elliptical morphologies with a broader spread in the distribution.

spherical shapes. For low travel speeds, the shape distributions also extend into Region 3 due to the presence of some saddle-shaped patches, as can be seen in Figure 2b in the "kidney bean"-shaped void at the upper right corner of the 3D reconstruction. At moderate travel speed, curvature pair maxima tend to lie upon the $\kappa_1=0$ axis with a broad spread into regions 3 and 4. At this moderate speed, the distribution does not approach the $\kappa_1=\kappa_2$ line closely, providing indication that voids are less spherical and generally more complex in nature. At high travel speeds, shape distributions are more diffuse than at low speeds and the overwhelming majority of the curvature distributions reside in Region 4. Additionally, the peaks of these shape distributions are again very close to the $\kappa_1=\kappa_2$ line, indicating voids are rather spherical at these higher travel speeds with very few saddle-shaped patches. The ISD peaks shift toward lower magnitudes of curvatures with increasing travel speed as observed in Figure 7, which is consistent with the fact that average pore volume decreases at higher travel speeds as shown in Figure 5.

The curvature distributions for a given travel speed are rather consistent across all power levels. The primary difference in ISDs relating to power variation, see Additional file 3, 4 is that the peak of the curvature distributions exist at increasingly negative values of κ_2 with decreases in power. This change corresponds to more spherical pore morphologies being formed with decreases in weld power. These trends were observed consistently across both focal length welds. A full set of calculated ISDs for all weld cases in this study having more than twenty pores each are included in the Additional file 3, 4 to this article.

Interfacial normal distributions

Interfacial normal distributions with respect to the positive z-axis are shown in Figure 8 for porosity in weld samples discussed in the previous section, again at 600 W and under a focal length of 120 mm. Each IND is plotted with a uniform color scale with a maximum of 3.0×10^{-3} (units of dimensionless probability) to enable direct comparison. At moderate and high travel speeds, local anisotropy in directionality appears such that pronounced clusters of normal orientations populate the lower and upper hemispheres of each projection in Figure 8c – e. Taken together with the ISDs, the net change in pore morphology with travel speed can be summarized as follows. At low travel speeds, pores are on average larger and are mainly near spherical or ellipsoidal (Figure 7a – b). At the same time, the orientation of the normals are more isotropic, as observed from the more uniform INDs (Figure 8a – b). Thus, we can conclude that ellipsoidal pores are randomly oriented. At the moderate travel speed, pore sizes decrease, and pore morphologies become less ellipsoidal and more irregular (Figure 7c), and the IND shows preferential normal orientation toward and away from the incident laser (see the enhanced probability in the upper and lower hemispheres in Figure 8c). At high travel speeds, pore morphologies once again become closer to spherical, with some ellipticity still present, as evidenced by the fact that the peaks of the ISDs do not lie directly on the $\kappa_1 = \kappa_2$ lines (Figure 8d – e). Also, as in the moderate-speed case, interfacial normals are aligned with the direction of laser incidence, resulting in the

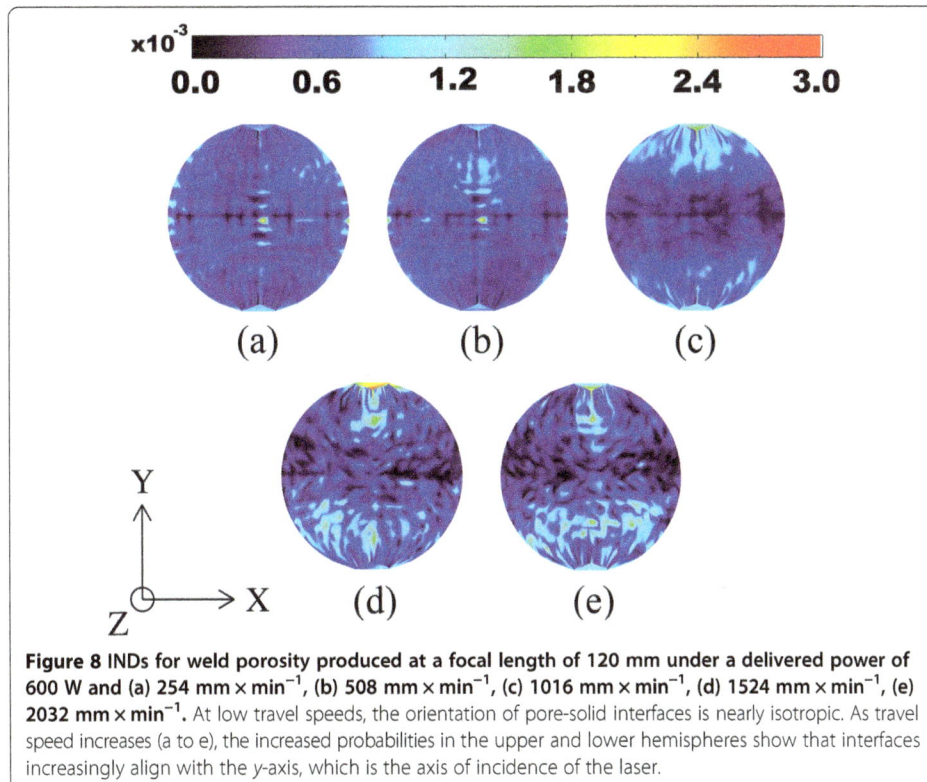

Figure 8 INDs for weld porosity produced at a focal length of 120 mm under a delivered power of 600 W and (a) 254 mm × min^{-1}, (b) 508 mm × min^{-1}, (c) 1016 mm × min^{-1}, (d) 1524 mm × min^{-1}, (e) 2032 mm × min^{-1}. At low travel speeds, the orientation of pore-solid interfaces is nearly isotropic. As travel speed increases (a to e), the increased probabilities in the upper and lower hemispheres show that interfaces increasingly align with the y-axis, which is the axis of incidence of the laser.

peaks of the INDs occurring in the upper and lower hemispheres at these high speeds. The same trends were observed in the negative z-axis projections as well. Succinctly stated, increasing weld speed results in decreased pore sizes and in interfacial normals preferentially orienting in the direction of laser incidence (both toward and away from). These trends in ISDs and INDs were observed across all powers and across both focal lengths. For completeness, calculated INDs for all weld cases in this study are also included in the Additional file 5, 6, 7 and 8 to more fully demonstrate the trend.

Pore interspacing

The pore-interspacing and pore-size distributions were calculated for a total of 18 weld schedules to more closely examine the interspacing of pores in three dimensions as a function of weld power and weld speed. These cases span both focal lenses and include welds for which power was varied while travel speed was held constant at 1016 mm × min^{-1} and for which the travel speed was varied for a constant power of 600 W; see Table 2. These selections traverse the central column and row of both sample matrices where porosity was observed, shown in Table 1, encompassing median values for mm-scale laser-welds. The probability distributions in Figures 9 and 10 were calculated as described in the section on spatial analysis. To alleviate redundancy, PIDs for only the 120 mm focal length welds are plotted in Figure 9, as the trends illustrated here are consistent for the 80 mm focal length results as well. In Figure 9, each data series represents a speed of 1016 mm × min^{-1} under a different power. Overall, the pore-interspacing probability distributions shift to higher values of distances as the weld power increases. Stated simply, pores are spaced farther apart at higher power.

Pore interspacing was also calculated for various weld speeds, as shown in Figure 10. Again, to reduce redundancy and to make the trend clear, only select results are shown for welds made at multiple speeds in conjunction with the 120 mm focal length at 600 W. While the probability of finding pores at interspacing distances below 250 microns is relatively high across all cases, the distributions appear to be broader for low and high travel speeds, with the high travel-speed case potentially exhibiting a bimodal

Table 2 Pore interspacing, radius and SLF as functions of weld power and speed

Weld power (W)	80 mm lens			120 mm lens		
	Pore interspacing (μm)	Pore radius (μm)	SLF	Pore interspacing (μm)	Pore radius (μm)	SLF
400	300 (14.8)	51 (7.4)	0.75 (0.05)	–	–	–
600	124 (9.1)	52 (4.6)	0.54 (0.05)	78 (9)	41 (4.5)	0.49 (0.07)
800	144 (9.2)	69 (4.6)	0.51 (0.04)	107 (9)	88 (4.5)	0.38 (0.04)
1000	170 (14.2)	82 (7.1)	0.51 (0.05)	110 (14.6)	104 (7.8)	0.35 (0.05)
1200	240 (15.5)	89 (7.8)	0.58 (0.05)	210 (20)	120 (10)	0.47 (0.05)
Weld speed (mm × min^{-1})	Pore interspacing (μm)	Pore radius (μm)	SLF	Pore interspacing (μm)	Pore radius (μm)	SLF
252	270 (14.8)	108 (7.4)	0.55 (0.04)	340 (14.6)	129 (7.8)	0.57 (0.04)
510	170 (14.8)	142 (7.4)	0.37 (0.04)	160 (14.6)	124 (7.8)	0.40 (0.04)
1016	124 (9.1)	52 (4.5)	0.54 (0.05)	78 (9)	41 (4.5)	0.49 (0.07)
1524	110 (14.3)	63 (7.1)	0.47 (0.07)	230 (14.6)	51 (7.8)	0.70 (0.06)
2032	190 (14.3)	57 (7.1)	0.62 (0.06)	310 (14.6)	58 (7.8)	0.72 (0.05)

Variable weld powers shown are for a constant travel speed of 1016 mm × min^{-1} and variable weld speeds shown are for a constant power of 600 W.

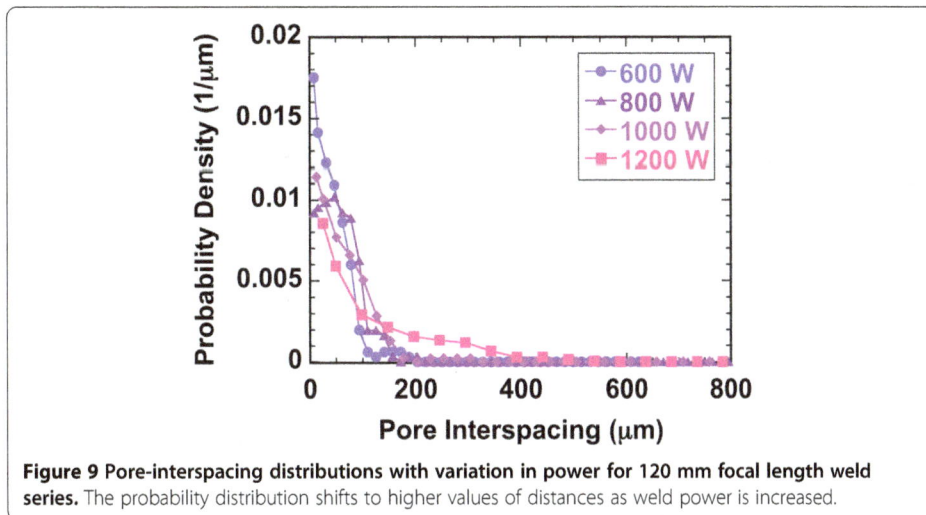

Figure 9 Pore-interspacing distributions with variation in power for 120 mm focal length weld series. The probability distribution shifts to higher values of distances as weld power is increased.

distribution. However, the statistics are insufficient to conclusively determine whether a bimodal distribution exists; further examination of larger weld samples or a larger number of samples under the same processing parameters are required to do so.

The characteristic pore interspacing, characteristic pore radius, and SLF are listed in Table 2 for all pore structures considered in Figures 9 and 10. To better illustrate the variation with process parameters, calculated SLF are plotted as functions of delivered power and weld speed across all process pairs examined in Figure 11. The method for calculating pore interspacing and radius is accurate up to half a voxel [26]. Therefore, the accuracy of each calculation is dependent on the resolution of the measurements; the uncertainty is listed within parenthesis in Table 2 and shown by error bars in Figure 11.

As described earlier, pore interspacing is a measure of the proximity of pores in the weld structure, while the SLF measures the proximity of pores relative to the distance between their centers and the characteristic pore size. For the samples where weld power is varied, the smallest pore interspacing was found at 600 W for a speed of

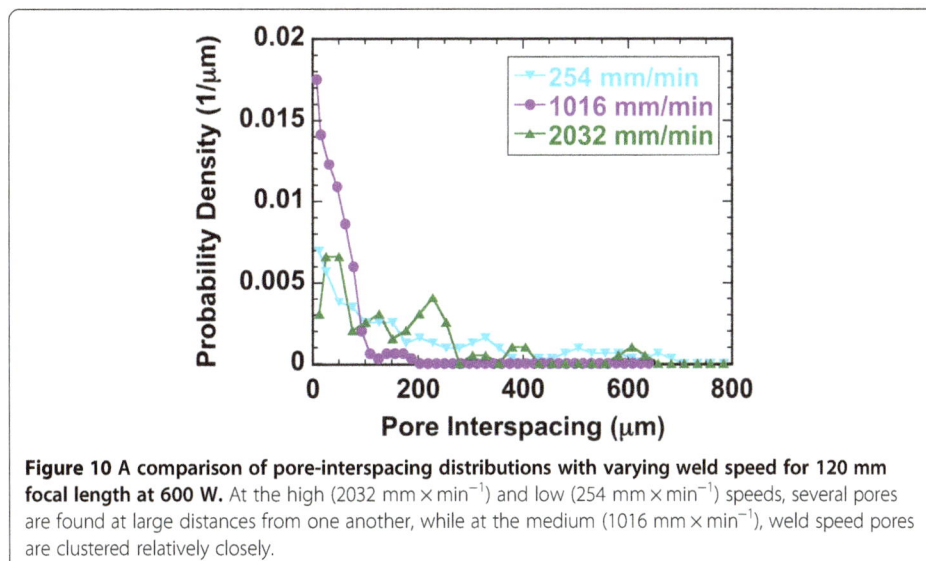

Figure 10 A comparison of pore-interspacing distributions with varying weld speed for 120 mm focal length at 600 W. At the high (2032 mm × min⁻¹) and low (254 mm × min⁻¹) speeds, several pores are found at large distances from one another, while at the medium (1016 mm × min⁻¹), weld speed pores are clustered relatively closely.

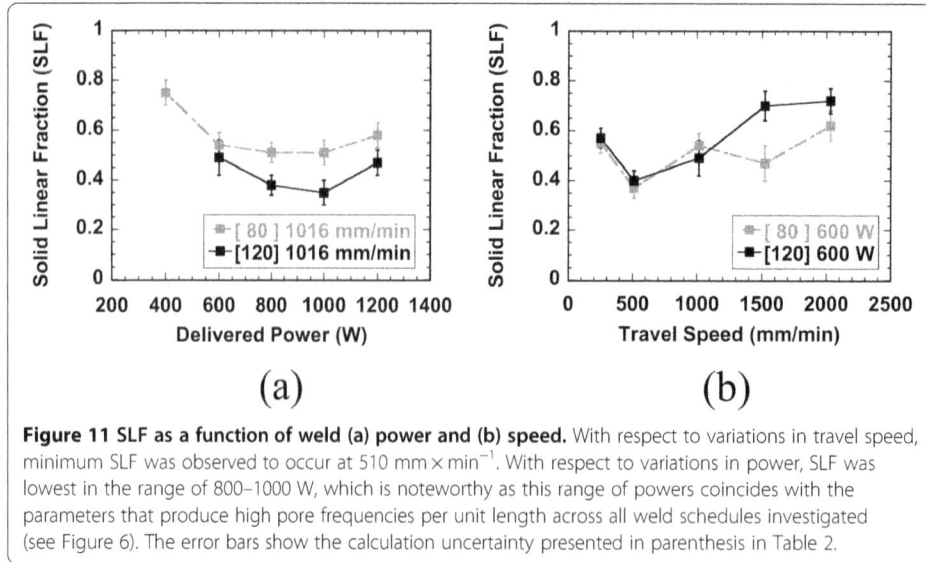

Figure 11 SLF as a function of weld (a) power and (b) speed. With respect to variations in travel speed, minimum SLF was observed to occur at 510 mm × min^{-1}. With respect to variations in power, SLF was lowest in the range of 800–1000 W, which is noteworthy as this range of powers coincides with the parameters that produce high pore frequencies per unit length across all weld schedules investigated (see Figure 6). The error bars show the calculation uncertainty presented in parenthesis in Table 2.

1016 mm × min^{-1} for both 80 and 120 mm lens welds (see Table 2), while the minimum SLF occurs at weld powers of 800 – 1000 W for the same travel speed (Figure 11a). This is consistent with the results of Figure 6, where the structure with the highest pore frequency per unit length arises at a weld power of 800 W for the 1016 mm × min^{-1} speed weld series. While these results are consistent, SLF provides a more insightful detail of the pore structures present; for example, in the case of 800 W welds formed at 1016 mm × min^{-1} with a 120 mm focus lens, the pore interfaces are separated by a distance that is 0.39 times the center-to-center distance between neighboring pores on average. Additionally, it is valuable to point out that the SLF is in the range of 0.4 to 0.6 for welds with a broad range of process parameters, which indicates that characteristic pore interspacing is approximately the same as the characteristic pore diameter in these cases. This suggests that for many weld cases, the characteristic pore interspacing can be approximated by the average pore diameter, which is generally easier to measure. However, high SLF values (> 0.6) are observed at the lowest power and the highest speed, indicating that pores may be spaced farther apart relative to their size at low delivered energy (Table 2 and Figure 11).

Pore size variability

The measures presented above are useful in quantifying the influence of specific weld parameters on resultant pore microstructures. These measures revealed various degree of variability within each weld. For example, ISDs are not sharply peaked, indicating some pore surfaces have high curvatures while others have low curvatures. Here, we examine the width of the distribution for pore sizes to better elucidate the variability in this measure. To this end, the coefficient of variation (*c.v.*) associated with equivalent pore radii is calculated. The coefficient of variation provides a quantitative measure of the variability of any population by taking the ratio of the standard deviation (σ) to the average (μ), Equation 4.

$$c.v. \quad = \frac{\sigma}{\mu} = \frac{\sqrt{\frac{1}{N}\sum_{i=1}^{N}(x_i-\mu)^2}}{\mu} \tag{4}$$

The coefficient of variation is useful as it allows for comparison across populations having very different means, varying populations and dissimilar distributions. For distributions having a standard deviation less than the average, a *c.v.* less than one is returned, indicating very low variability in the population. Alternatively, *c.v.* values greater than unity indicate notable amounts of variability where the value of *c.v.* indicates how many times greater the variability in the distribution is than the mean. To illustrate the trend is consistent across both focal lenses the *c.v.* values will be presented for both the 80 mm and 120 mm focal lens series, Figure 12. While no trend in pore size variability is apparent for changes in weld speed a minor increase in variability can be observed with an increase in delivered power. Additionally, the values of *c.v.* found for the pore size distributions, which are all well below unity, indicate that the spread in radii for all weld schedules are relatively small with respect to their mean value. Variability measures such as the one presented above will likely aid the understanding of microstructure's influence on mechanical response.

Conclusions

In this paper, quantitative characterization of porosity in laser-welds of 304L stainless steel has been performed non-destructively for 54 unique continuous-wave weld schedules via micro-computed tomography where each weld schedule represents a unique dataset. Direct correlations of pore size, shape, frequency, directionality, pore interspacing and solid linear fraction (SLF) with weld processing parameters have been made.

We find:

- Average and maximum pore volume increase with decreasing speed or increasing power.
- Pore frequency initially increases and then decreases with increasing power for a given travel speed.
- Interfacial shape distributions (ISDs) and interfacial normal distributions (INDs) illustrate that basic pore shape and directionality are similar for a given welding speed regardless of power delivered.

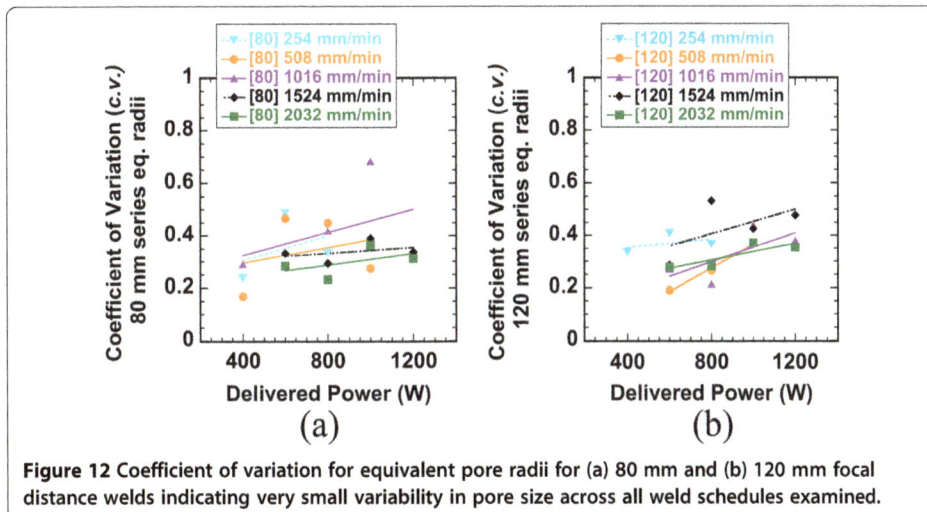

Figure 12 Coefficient of variation for equivalent pore radii for (a) 80 mm and (b) 120 mm focal distance welds indicating very small variability in pore size across all weld schedules examined.

- ISDs show that pore shapes are nearly spherical or ellipsoidal at low and high travel speeds and are far more irregular, with a mix of ellipsoidal and saddle-shape geometries at moderate travel speeds.
- INDs indicate that pore orientations become anisotropic at moderate to high travel speeds with large concentrations of pore interfacial normals pointing toward and away from the direction of laser incidence.
- Characteristic pore interspacing is nominally equivalent to characteristic pore diameter for welds with a broad range of process parameters, as reflected in the solid linear fraction (SLF) values.
- The values of *c.v.* indicate that the spread in pore radii is small with respect to their mean value for all weld schedules.
- High travel speeds and low delivered power result in the lowest pore linear frequency while increasing the amount of solid material between pores, which would likely yield improved mechanical properties.

Availability of supporting data

Animations of the five primary 3D reconstructions featured in this article for which ISDs, INDs, pore interspacing and SLF were calculated and presented have been made publicly available [27].

Additional files

Additional file 1: 3D Reconstructions of Porosity produced by laser-weld under a focal lens of 80 mm at various speeds and powers.

Additional file 2: 3D Reconstructions of Porosity produced by laser-weld under a focal lens of 120 mm at various speeds and powers.

Additional file 3: Interfacial Shape Distributions for Porosity produced by laser-weld under a focal lens of 80 mm at various speeds and powers.

Additional file 4: Interfacial Shape Distributions for Porosity produced by laser-weld under a focal lens of 120 mm at various speeds and powers.

Additional file 5: Equal Area Interfacial Normal Distributions with respect to the positive x axis for Porosity produced by laser-weld under a focal lens of 80 mm.

Additional file 6: Equal Area Interfacial Normal Distributions with respect to the negative x axis for Porosity produced by laser-weld under a focal lens of 80 mm.

Additional file 7: Equal Area Interfacial Normal Distributions with respect to the positive x axis for Porosity produced by laser-weld under a focal lens of 120 mm.

Additional file 8: Equal Area Interfacial Normal Distributions with respect to the negative x axis for Porosity produced by laser-weld under a focal lens of 120 mm.

Competing interests

The authors declare that they have no competing interests.

Authors' contributions

JM coordinated data collection, performed image segmentation, created 3d reconstructions and made basic characterization measures for all datasets. ISD and IND calculations were performed by LA and JM. Pore Interspacing and SLF calculations were performed by VC and KT. All authors read and approved the final manuscript.

Acknowledgements

Sandia National Laboratories is a multi-program laboratory managed and operated by Sandia Corporation, a wholly owned subsidiary of Lockheed Martin Corporation, for the US Department of Energy's National Nuclear Security Administration under contract DE-AC04-94AL85000. V.W.L. Chan and K. Thornton would like to acknowledge NSF DMR Grant # 0746424 "CAREER: Integrated Research and Education Program in Three-Dimensional Materials Science and Visualization." The computational resources for calculations of pore interspacing and pore sizes were provided by the Extreme Science and Engineering Discovery Environment (XSEDE), which is supported by National Science Foundation grant number OCI-1053575, under allocation No. TG-DMR110007.

Author details
[1]Computational Materials & Data Science, Sandia National Laboratories, 87185 Albuquerque, NM, USA. [2]Materials Science & Engineering, University of Michigan, 48109 Ann Arbor, MI, USA.

References
1. Mazumder J (1993) Laser-Beam Welding. In: ASM Handbook, Vol. 6. ASM International, Materials Park, OH, pp 262–269
2. Webber T, Lieb T, Mazumder J (2011) Laser Beam Welding. In: ASM Handbook, Vol. 6A. ASM International, Materials Park, OH, pp 556–569
3. Boyce BL, Reu PL, Robino CV (2006) The constitutive behavior of laser welds in 304L stainless steel determined by digital image correlation. Metall Mater Trans A 37A:2481–2492
4. Haboudou A, Peyre P, Vannes AB, Peix G (2003) Reduction of porosity content generated during Nd:YAG laser welding of A356 and AA5083 aluminum alloys. Mater Sci Eng A A363:40–52
5. Feist WD, Tillack G-R (1997) Ultrasonic Inspection of Pores in Electron Beam Welds. In: European-American Workshop Determination of Reliability and Validation Methods of NDE. BAM, Berlin, Germany. June 18–20 1997. vol 4. NDT.net
6. Zhu X, Shyam A, Jones JW, Mayer H, Lasecki JV, Allison JE (2006) Effects of microstructure and temperature on fatigue behavior of E319-T7 cast aluminum alloy in very long life cycles. Int J Fatigue 28:1566–1571
7. Shyam A, Picard YN, Jones JW, Allison JE, Yalisove SM (2004) Small fatigue crack propagation from micronotches in the cast aluminum alloy W319. Scr Mater 50:1109–1114. doi:10.1016/j.scriptamat.2004.01.031
8. Kuo TY, Jeng SL (2005) Porosity reduction in Nd-YAG laser welding of stainless steel and inconel alloy by using a pulsed wave. J Phys D Appl Phys 38:722–728. doi:10.1088/0022-3727/38/5/009
9. Madison J, Aagesen LK (2012) Porosity in Millimeter-Scale Welds of Stainless Steel: Three-Dimensional Characterization. Sandia National Laboratories, Albuquerque, NM
10. Madison J, Aagesen LK (2012) Quantitative characterization of porosity in laser welds of stainless steel. Scr Mater 67(9):783–786. doi:10.1016/j.scriptamat.2012.06.015
11. Tucker JD, Nolan TK, Martin AJ, Young GA (2012) Effect of travel speed and beam focus on porosity in alloy 690 laser welds. JOM 64(12):1409–1417. doi:10.1007/s11837-012-0481-3
12. Norris JT, Perricone MJ, Roach RA, Faraone KM, Ellison CM (2007) Evaluation of Weld Porosity in Laser Beam Seam Welds: Optimizing Continuous Wave and Square Wave Modulated Processes. Sandia National Laboratories, Albuquerque, NM
13. Norris JT, Robino CV, Hirschfeld DA, Perricone MJ (2011) Effects of laser parameters on porosity formation: investigating millimeter scale continuous wave Nd:YAG laser welds. Weld J 90:198–203
14. Madison JD, Aagesen LK, Battaile CC, Rodelas JM, Payton TKCS (2013) Coupling 3D quantitative interrogation of weld microstructure with 3D models of mechanical response. Metallography, Microstructure and Analysis 2 (6):359–363. doi:10.1007/s13632-013-0097-1
15. Matsunawa A, Kim J-D, Seto N, Mizutani M, Katayama S (1998) Dynamics of keyhole and molten pool in laser welding. J Laser Apps 10(6):247–254. doi:10.2351/1.521858
16. Pang S, Chen L, Zhou J, Yin Y, Chen T (2011) A three-dimensional sharp interface model for self-consistent keyhole and weld pool dynamics in deep penetration laser welding. J Phys D 44:1–15. doi:10.1088/0022-3727/44/2/025301
17. Rai R, Elmer JW, Palmer TA, DebRoy T (2007) Heat transfer and fluid flow during keyhole mode laser welding of tantalum, Ti-6Al-4 V, 304L stainless steel and vanadium. J Phys D Appl Phys 40:5753–5766. doi:10.1088/0022-3727/40/18/037
18. Zhou J, Tsai H-L (2007) Porosity formation and prevention in pulsed laser welding. Trans ASME 129:1014–1024. doi:10.1115/1.2724846
19. Alkemper J, Voorhees PW (2001) Three-dimensional characterization of dendritic microstructures. Acta Mater 49:897–902
20. Mendoza R, Alkemper J, Voorhees PW (2003) The morphological evolution of dendritic microstructures during coarsening. Metall Mater Trans A 34A(3):481–489
21. Mendoza R, Savin I, Thornton K, Voorhees PW (2004) Topological complexity and the dynamics of coarsening. Nat Mater 3:385–388. doi:10.1038/nmat1138
22. Kammer D, Mendoza R, Voorhees PW (2006) Cylindrical domain formation in topologically complex structures. Scr Mater 55(1):17–22. doi:10.1016/j.scriptamat.2006.02.027
23. Kammer D, Voorhees PW (2006) The morphological evolution of dendritic microstructures during coarsening. Acta Mater 54(6):1549–1558. doi:10.1016/j.actamat.2005.11.031
24. Fife JL, Voorhees PW (2009) The morphological evolution of equiaxed dendritic microstructures during coarsening. Acta Mater 57:2418–2428. doi: 10.1016/j.actamat.2009.01.036
25. Meyer N, Desbrun M, Schroder P, Barr AH (2003) Discrete Differential-Geometry Operators for Triangulated 2-Manifolds. In: Polthier K, Hege HC (ed) Visualization and Mathematics III. Springer, Berlin, Germany, pp 35–37
26. Chan VWL, Thornton K (2012) Channel size distribution of complex three-dimensional microstructures calculated from the topological characterization of isodistance structures. Acta Mater 60:2509–2517. doi:10.1016/j.actamat.2011.12.042
27. Madison JD, Aagesen LK, Chan VWL, Thornton K (2014) 3-Dimensional Reconstructions of Porosity from Laser-Welds of 304L Stainless Steel at 600W and a Variety of Travel Speeds. http://hdl.handle.net/11115/243

5

Using quality mapping to predict spatial variation in local properties and component performance in Mg alloy thin-walled high-pressure die castings: an ICME approach and case study

Joy H Forsmark[1*], Jacob W Zindel[1], Larry Godlewski[1], Jiang Zheng[2], John E Allison[2] and Mei Li[1]

* Correspondence:
jforsma5@ford.com
[1]Materials Research Department,
Ford Motor Company, Research and
Innovation Center, MD3182, P.O Box
2053, Dearborn, MI 48121, USA
Full list of author information is
available at the end of the article

Abstract

This paper explores the use of quality mapping for the prediction of the spatial variation in local properties in thin-walled high-pressure die castings (HPDC) of the magnesium alloy AM60. The work investigates the role of casting parameters on local ductility and yield strength and presents a model for predicting local ductility and yield strength in a cast component. A design of experiment (DOE) was created to examine the role of various casting parameters on local properties such as ductility and yield strength. Over 1,200 tensile samples were excised from cast parts and tested. Casting simulations were also conducted for each experimental condition. Local properties were predicted, and the local property (quality map) model was compared with a prototype production component. The results of this model were used as input to a performance simulation software code to simulate the component-level behavior under two different loading conditions. In this study, the authors bypassed the traditional Integrated Computational Materials Engineering (ICME; process-microstructure-properties) approach in favor of a semi-empirical quality mapping approach to provide estimates of manufacturing sensitive local properties for use in process and component design.

Keywords: ICME; Magnesium alloys; High-pressure die casting; Casting simulation

Background

As the need for weight savings and fuel economy has increased, so has the interest in using magnesium alloys for a wider variety of automotive applications [1-3]. The low density and favorable stiffness-to-weight ratio of Mg alloys enable considerable weight savings on components. Additionally, the excellent high-pressure die castability of many Mg alloys allows for more complex and one-piece designs compared to traditional steel stampings and even some Al castings. This enables part consolidation and potential cost savings [1]. Thus, most magnesium components are produced by high-pressure die casting (HPDC). However, HPDC Mg alloys do have a lower ductility (elongation at failure) than more commonly used wrought steel or aluminum components. In several studies of the magnesium alloys AM50 and AM60 [4,5], elongation at failure or ductility has been reported to correlate with fracture toughness in Mg alloys.

In general, higher ductility corresponds to better fracture toughness; this is a key concern in the prediction of crash response and optimal design of structures.

Castings, probably to a greater extent than wrought products, exhibit geometry and manufacturing-history-dependent spatial variation in some properties. These geometry and manufacturing-history-dependent spatial variations, also known as location-specific or local properties, should be accounted for in the design of structural applications. Studies of excised tensile samples from production instrument panel components in AM60 have shown that spatial variation in ductility can range from 1% to 14% depending on location within the component [6-9]. A key conclusion from these studies is that an understanding of the spatial variability in the mechanical properties from quasi-static tests is important because the values of ductility in a given location may influence dynamic performance of the component in service [10]. To complicate the situation further, a given property at a given location varies from casting to casting, resulting in another factor to address. One way to look at the situation is that the mean value of a given property is location dependent and each location may have its own statistical variation. Computer-aided engineering (CAE) methods used in design, however, rely on a single material model for all regions of the component to predict the component behavior without taking into account location-specific properties or their distribution about the mean. Therefore, the ability to predict spatial variation in local properties can still lead to improved accuracy in the prediction of the overall behavior of a component and thus increased confidence in and utilization of HPDC Mg components.

In a casting, the measured variation in mechanical properties (both spatial variation within a casting and statistical variation observed at a particular location across multiple castings produced under the same conditions) obtained from excised test samples can be due to several sources:

(1) Variation associated with normal foundry processing conditions (such as melt temperature or local die temperature)
(2) Variation associated with local geometry (such as melt flow paths or section thicknesses)
(3) Variation associated with the location and morphology of microstructural features (such as void size, void percentage, or phase segregation): some variation can be linked to (1) and (2) and some due to randomness of the microstructure formation
(4) Variation associated with testing (such as sample dimensions or loading forces)

A traditional Integrated Computational Materials Engineering (ICME) approach predicts variation in properties due to manufacturing history by directly linking the pertinent local microstructure to the mechanical properties of interest, such as those described by Li et al. [11] and Horstemeyer et al. [12]. Indeed, a number of researchers have investigated the influence of such microstructural features as dendrite arm spacing, externally solidified grain size, shrinkage porosity and oxide films, or cold shuts on the mechanical properties of Mg alloys [7-9,13-34]. However, there still exists a substantial amount of disagreement with regard to the level of influence that particular microstructural features have on this behavior in Mg castings, particularly with regard to local ductility. Further, models to predict the formation of many of these microstructural features are not currently available.

There has been some effort to correlate Mg alloy mechanical properties directly to variations in HPDC processing parameters [26,35-37]. Sannes et al. [26], for example, demonstrated that changing the gating configuration of a tensile bar die produced as-cast tensile specimens that had statistically significant differences in ductility. They also utilized casting simulation to qualitatively show those differences in flow characteristics. There are several commercial casting simulation codes available that can predict the filling and solidification process for a given casting and processing condition. These codes can capture variations in melt flow, air entrapment, and general solidification gradients due to processing conditions and visually map these differences onto a computer model of the part. While some of these codes have limited capability to predict certain microstructural features (e.g., macroporosity caused by unfavorable solidification gradients in certain locations and the occurrence of 'hot spots' in a casting), the capability required to predict the location and size of very fine microstructural features such as microporosity in thin-walled HPDC castings is generally not robust.

In this study, the authors bypassed the traditional ICME (process-microstructure-properties) approach in favor of a semi-empirical quality mapping approach to provide estimates of manufacturing sensitive local properties for use in process and component design. The term 'quality map' describes an empirically derived surrogate model that can be used to predict a particular property using local flow and solidification characteristics in a cast part. The equation itself is then 'mapped' onto the original cast component. In this way, the local properties can be visualized on a cast part and used in a component and engineering system performance simulation. This approach is, by its nature, a less desirable method than the development of a well-correlated model based upon physical principles. However, given that such a physical principle-based model does not exist, the development of a surrogate model using a design space approximation approach is a valid technique for understanding mechanical property behavior in HPDC castings. Other researchers have used the quality mapping approach to predict overall performance in Mg die castings. Greve et al. [38] developed a local casting quality criterion based on the flow conditions and the cooling rate in different regions in an AZ91 clutch housing. Sannes et al. [39], who first used the term 'quality map,' used a U-shaped AM60 casting with a nominal wall thickness of 2 mm in their investigation. They determined the flow and solidification characteristics in the casting which they correlated to a fracture criteria based on observed microstructure. The fracture criteria function was then used in a companion work by Dorum et al. [40-44] to predict the component behavior in quasi-static bending experiments. Weiss et al. [45] also describe a similar quality mapping approach in a study on a box structure in AM50.

The above studies all examined smaller components with box-like or bracket configurations. However, many of the Mg castings used in automotive structural applications have a frame-like geometry. For example, the Corvette engine cradle [46] and the Ford F150 grill opening radiator support [47] both have large frame-like structures. Likewise, closures such as the Aston Martin DB9 door inner [48] and, recently, the rear liftgate inner of the Ford MKT [49] are structures with window openings and have large, long sections. These frame-like structures tend to have some unique features such as the melt flow lengths that exceed 2 to 3 m and areas where melt fronts meet inside the casting.

This project was undertaken to develop a methodology to directly predict location-specific ductility and yield strength in a frame-geometry Mg HPDC casting from the outputs of a commercial casting simulation code and then use those predictions to modify the material input deck of a commercial finite element deformation simulation code to ultimately predict component performance. Figure 1 shows a flow chart of the process. The ultimate goal was to provide an industrially feasible workflow that would allow for component design iterations that takes into account manufacturing history in performance predictions prior to tooling being produced and prototypes manufactured, thus reducing design iterations for a cast component.

Methods and results

Casting of experimental components

Since production parts are complex geometries often with curvature or small changes in section thickness, they are not ideal to determine mechanical properties through excising standard test specimens. Therefore, an experimental casting called the Generic Frame Casting (GFC) was designed to contain features that would be present in a production door inner or other frame casting while still providing sufficient material to produce sub-size tensile samples and allow for ease of fixturing for component-level testing. The part is also closer in size to frame-like castings that are in production. Figure 2 shows a schematic diagram of the as-cast and machined GFC. The GFC has one large window and three smaller windows. There are two symmetrical rib sections along the sides. The GFC outer dimensions are 580 mm × 580 mm, with a 2.5-mm nominal wall thickness over the entire part. The side walls are approximately 50 mm high with a draft angle of approximately 10° for ease of casting. In the GFC, a simple, single gate is attached to the lower wall of the upper aperture and metal enters through this location. Overflows for the metal were cut at the top wall of the upper aperture and the walls of the lower apertures. Gates to the overflows could be cut or welded up, depending on the needs for a particular casting trial. The diagram also indicates the presence of a center feeder which was also used to change the metal flow pattern.

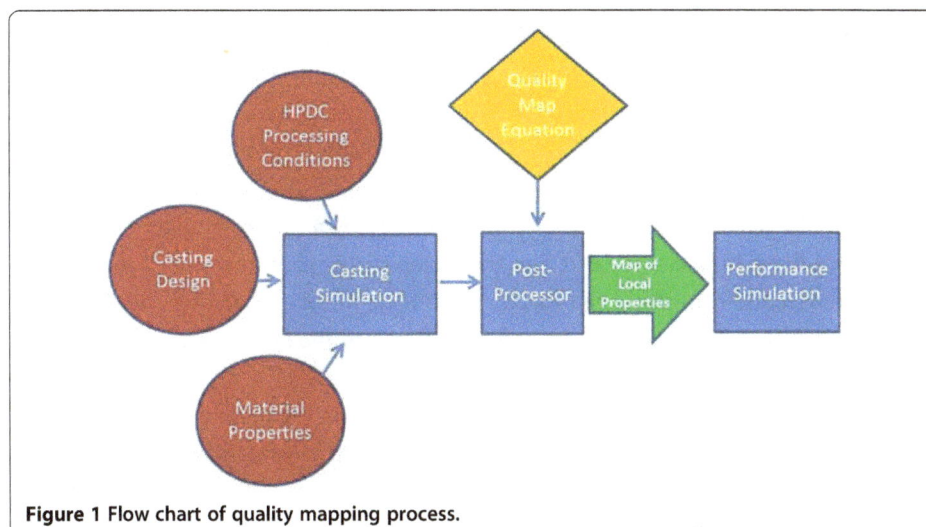

Figure 1 Flow chart of quality mapping process.

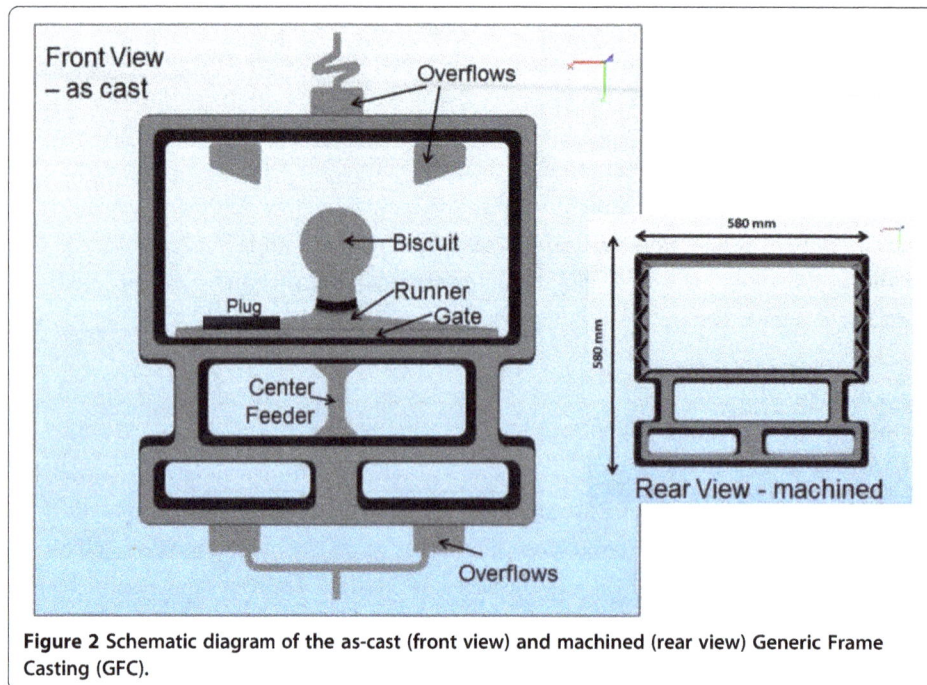

Figure 2 Schematic diagram of the as-cast (front view) and machined (rear view) Generic Frame Casting (GFC).

The gating and overflows were designed using the North American Die Casting Association (NADCA) guidelines for proper gating and casting design to optimize the quality of Mg and other castings [50,51]. They were also confirmed using a MAGMASOFT® simulation prior to the die being produced.

The GFC castings were produced by the authors at Mag-Tec Casting Corporation in Jackson, MI, USA. The parts were cast on a Model 836 Prince/Buehler cold-chamber die casting machine (BuhlerPrince, Inc., Holland, MI, USA) with a molten metal pumping system from the covered furnace. Molten Mg in a furnace must be held under a cover gas to prevent oxidation (SF_6 was used in this case). A schematic diagram of a typical cold-chamber high-pressure die cast machine is shown in Additional file 1. During casting, a specific amount of molten metal is pumped into the shot sleeve from a holding furnace. Pressure is rapidly applied to a plunger which travels the length of the shot sleeve and pushes the metal into the casting cavity in the steel die, where it is held until the metal solidifies and the die then opens, ejecting the part. The actual filling of a typical die cavity takes 20 to 40 ms, and the metal completely solidifies in less than 1 to 2 min with some surface regions solidifying within seconds.

The motion of the plunger as it fills the casting cavity can be visualized by examining the shot profile. In general, the shot profile is represented by a plot of plunger velocity as a function of distance and is divided into three steps: slow shot, transition, and fast shot. The fast shot speed controls the speed of the melt fill of the casting cavity and thus can influence the casting quality. An example of one of the shot profiles from this study can be seen in Figure 3.

Thermocouples placed in the die and furnace indicated the temperature of the melt at the beginning of the production cycle and the casting cavity throughout the production cycle of a cast part. Figure 4 shows the thermal behavior in different areas of the die and furnace during the course of a casting run to produce the component

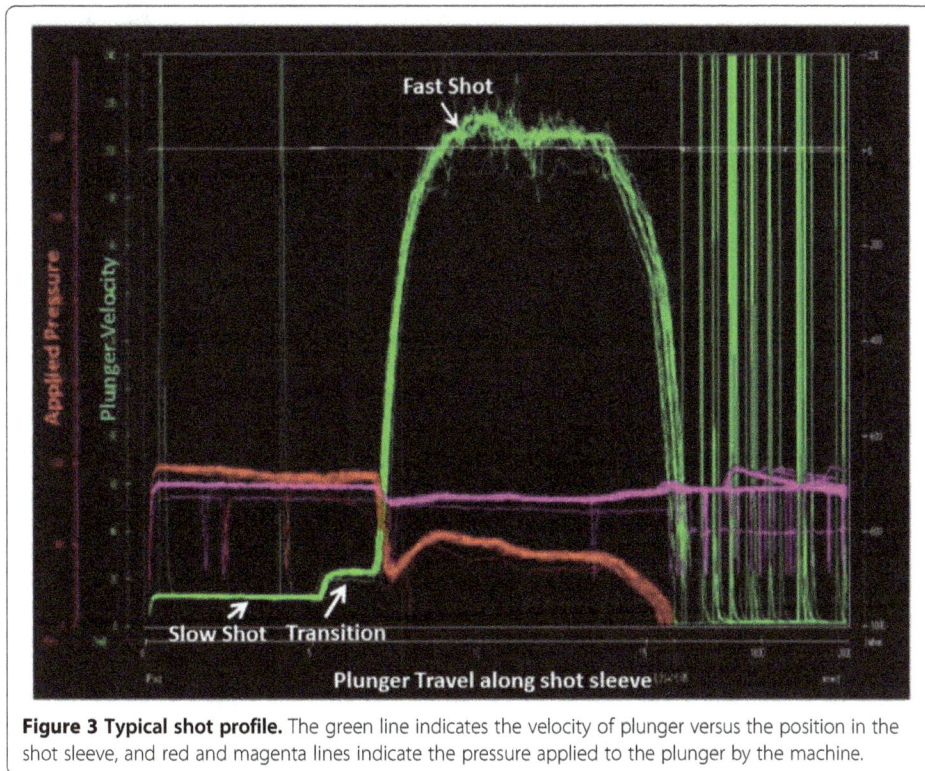

Figure 3 Typical shot profile. The green line indicates the velocity of plunger versus the position in the shot sleeve, and red and magenta lines indicate the pressure applied to the plunger by the machine.

investigated in this study. The figure indicates that the die temperature increased from the initial casting run and achieved a nominal steady state after about eight castings are produced. In many foundry practices, the first few parts cast are discarded as scrap because the die is considered too cold at this point in the trial. Cooling and heating lines were placed in the die to mitigate the thermal variation in the die or enhance a desired thermal gradient in certain locations such as near the biscuit region. The die

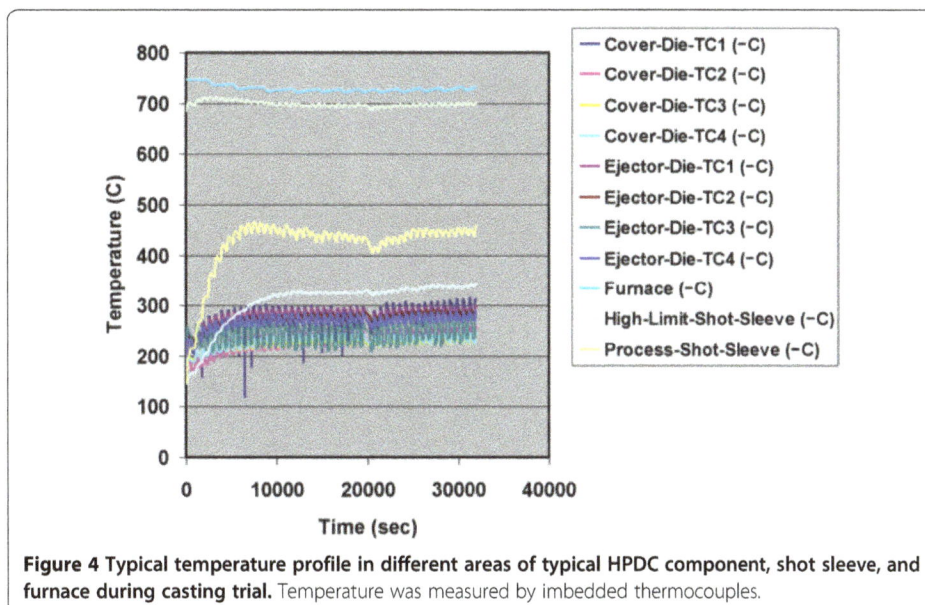

Figure 4 Typical temperature profile in different areas of typical HPDC component, shot sleeve, and furnace during casting trial. Temperature was measured by imbedded thermocouples.

temperature was partially controlled by oil lines cut into the die and run at a temperature of 240°C. Detailed processing conditions were documented for all castings.

A preliminary casting trial was run with the die to determine a process operating window for the GFC casting with the target of producing a complete cast part. Based on the results of the trial, a four-factor, two-level full factorial design of experiment (DOE) was established in order to produce testable components under a range of casting parameters within that window. A total of 16 different sets of castings were produced. The DOE approach was also used because the casting parameters (DOE factors) examined could have confounding effects on the local behavior and it was important to understand the relative weighting of each factor. Two of the casting parameters investigated were processing parameters: fast shot speed and melt temperature. The other two factors were geometrical parameters where the center feeder and a gate plug were either present or not present (see Figure 2). The plug in the gate produced an asymmetrical gating configuration that changed the flow in the part. The symmetrical gating system had a gate area of 9.7 cm^2, while the gate plug reduced the area of the gate down to 7.7 cm^2, in addition to changing the gating flow pattern. Table 1 shows the full factorial DOE with the upper and lower values for each factor.

The alloy composition was carefully controlled for all cases. Table 2 indicates the chemical composition and a comparison with the specification for AM60B. As can be seen in the table, the chemical composition of the alloy contains aluminum content at the lower end of the specification but still considered typical for production components.

In general, 15 castings were produced as die 'warm-up' shots to get the die to a quasi-steady-state temperature as determined by the embedded thermocouples. On average, the 40 to 50 castings that were subsequently produced were used for analysis. If during the course of a casting run, the machine had to be stopped for an extended

Table 1 DOE factors for GFC castings

Condition assignment	Fast shot (m/s)	Melt temperature (°C)	Gating design	Center feeder	Excised sample testing?
A	5.3	660	Symmetrical	No	Yes
B	6.6	660	Symmetrical	No	
C	5.3	700	Symmetrical	No	
D	6.6	700	Symmetrical	No	Yes
E	5.3	660	Asymmetrical	No	
F	6.6	660	Asymmetrical	No	Yes
G	5.3	700	Asymmetrical	No	Yes
H	6.6	700	Asymmetrical	No	
I	5.3	660	Symmetrical	Yes	Yes
J	6.6	660	Symmetrical	Yes	Yes
K	5.3	700	Symmetrical	Yes	Yes
L	6.6	700	Symmetrical	Yes	
M	5.3	660	Asymmetrical	Yes	Yes
N	6.6	660	Asymmetrical	Yes	
P	5.3	700	Asymmetrical	Yes	
Q	6.6	700	Asymmetrical	Yes	Yes

Table 2 Chemical composition of AM60B used in the study

	Al wt%	Zn wt%	Mn wt%	Si wt%	Fe wt%	Cu wt%
AM60	5.7	0.06	0.4	0.01	0.005	0.003
AM60B ASTM B94-07 specification	5.6 to 6.5	0.22	0.24 to 0.6	0.10	<0.005	<0.010

period of time, a small number of 'warm-up' shots would be produced and scrapped until the die returned to steady state before useable castings were made. Over 600 castings were produced during this study.

Excised tensile bar testing

After the castings were produced, tensile samples were excised from eight different locations in up to 20 castings per condition (Figure 5). A total of 8 of the 16 casting runs were selected for evaluation based on the partial factorial DOE (represented by an $L_8(2^{4-1})$ orthogonal array [52]). This experimental design was selected because it allowed for determination of the main effects while still keeping the total number of tensile tests and statistically significant number of replicates manageable; still, a total of 1,280 tensile tests were conducted to complete the experiment. As shown in Table 3, the eight conditions selected for analysis were conditions A, D, J, K, F, G, M, and Q.

The tensile sample geometry was the ASTM E8 standard [53] for sub-sized flat tensile bars with a total length of 76 mm, a gage section of 25.4 mm, and a nominal thickness of 2.5 mm. The edges of the flat, dog-bone-shaped samples were machined using a milling process that resulted in a machined surface roughness of 0.8 μm. The

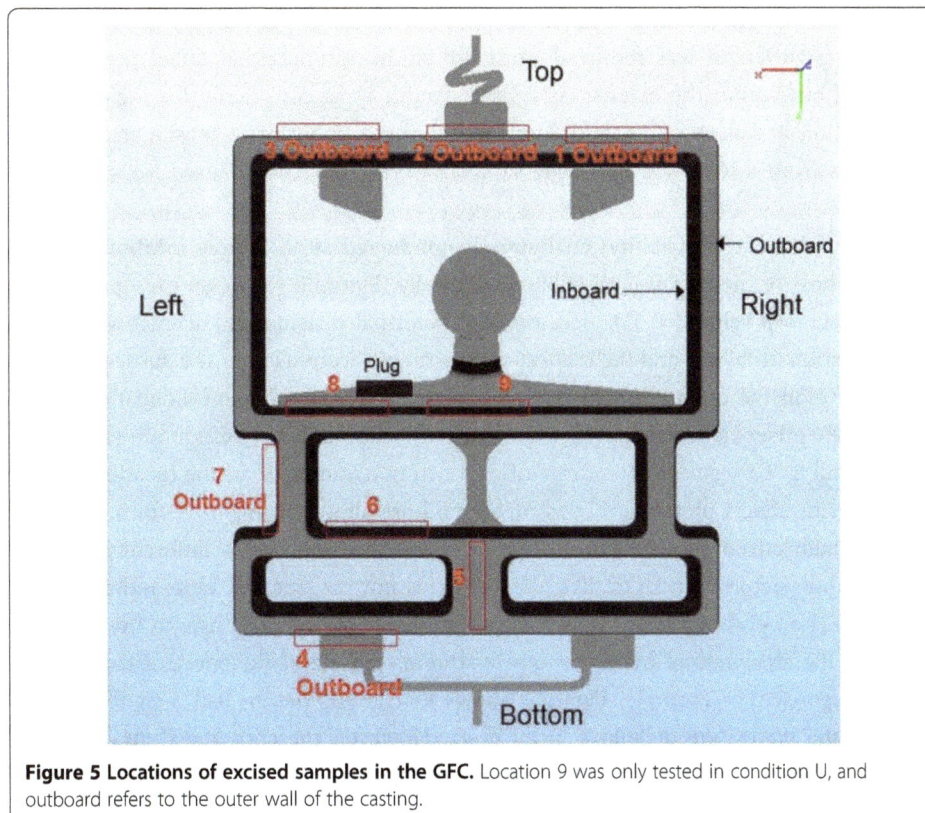

Figure 5 Locations of excised samples in the GFC. Location 9 was only tested in condition U, and outboard refers to the outer wall of the casting.

Table 3 Actual values recorded for processing conditions used for MAGMASOFT simulations

Designation	Alloy	Fast shot (m/s)	Melt temperature (°C)	Gating configuration	Center feeder
A	AM60	5.2	675	No plug	No
J	AM60	6.5	691	No plug	Yes
K	AM60	5.3	711	No plug	Yes
D	AM60	6.6	738	No plug	No
M	AM60	5.2	677	Plug	Yes
F	AM60	6.1	687	Plug	No
G	AM60	5.1	731	Plug	No
Q	AM60	6.1	727	Plug	Yes
U	AM60	5.8	720	Plug	Yes
I	AM60	5.1	685	No plug	Yes

faces of the tensile samples were as-cast (not machined). Tensile samples were tested in axial loading using a Sintech™ frame (MTS Systems Corp., Eden Prairie, MN, USA) with a 133-kN (30,000 lb) load cell and a 25.4-mm gage length extensometer. The frame was operated at a constant crosshead speed of 0.04 mm/s up to 1% strain and then transitioned to a constant cross-head speed of 0.21 mm/s thereafter. This corresponds approximately to strain rates of 1.5×10^{-3} s^{-1} and 8×10^{-3} s^{-1}, respectively.

The load and extensometer data were recorded, and the data was analyzed to give true stress and true strain tensile behavior. The total elongation at failure, also known as tensile ductility, was determined using the final extensometer measurement at failure without removing the elastic strain and was reported as 'elongation at failure' in true strain. Minimal sample necking was observed. Any sample that failed outside the extensometer gage length was removed from the study. Samples that failed prior to reaching yield but within the extensometer gage length were considered for inclusion in the elongation at failure but not for the yield strength or ultimate tensile strength values. Results from a total of 936 tensile samples were reported and used for subsequent analysis.

Statistical (cumulative probability) analysis was conducted at all sample locations indicating that most of the data was distributed normally (Figure 6 shows an example for condition A where p values > 0.05). Because of the normal distribution, average values of both elongation at failure and 0.2% offset yield strength were used in the subsequent analysis to develop the quality mapping relationships. Average values of elongation at failure and 0.2% offset yield strength as a function of location and condition are shown in Figures 7 and 8. To separate the effects of the casting parameters on the locations in the casting, main effects plots were constructed and are shown in Figure 9 for several locations. A main effects plot indicates differences in the responses to different levels used in the input factors in a DOE. If a main effect is not present, the plots indicate a horizontal line. However, if a factor does have an influence on the response (in this case the ductility), the steepness of the line indicates the magnitude of the effect of the factor on the response. For example, Figure 9 shows that gating design had a significant influence on the elongation at failure in all cases. However, the shot speed and melt temperature also had a minor influence on the ductility in locations 2 and 3, while the presence of the center feeder was a factor only for locations 4 and 5.

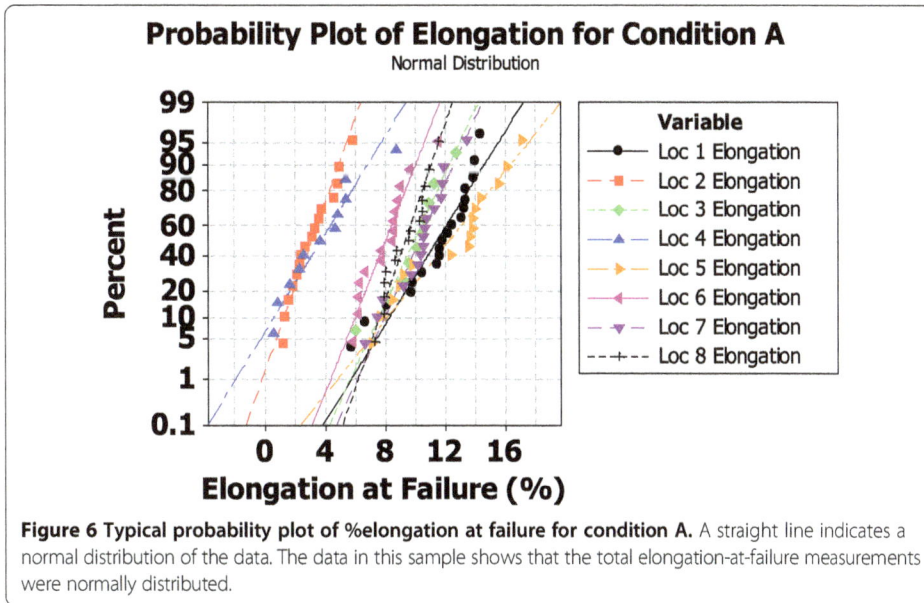

Figure 6 Typical probability plot of %elongation at failure for condition A. A straight line indicates a normal distribution of the data. The data in this sample shows that the total elongation-at-failure measurements were normally distributed.

For the DOE, care was taken to ensure that all other processing parameters outside the four factors being tested remained constant. One important parameter that remained the same throughout the DOE was the location of the overflows. Overflow position is normally optimized to account for geometry and melt flow changes in commercial production. Overflows are generally positioned in areas where melt fronts meet in order to remove defects that could result in those regions from the body of the casting. However, during processing of the DOE castings, the overflows for the asymmetrical gating configuration remained constant and thus were not optimally placed. Upon completion of the DOE experiment, the geometry of the overflows and their locations were modified for the asymmetrical gating geometry factor creating condition U. Condition U had similar processing parameters as condition Q. Figure 10 shows the differences in overflow location for the two castings. Excised sample testing indicated that overflow position did indeed make a difference in the measured ductility in several

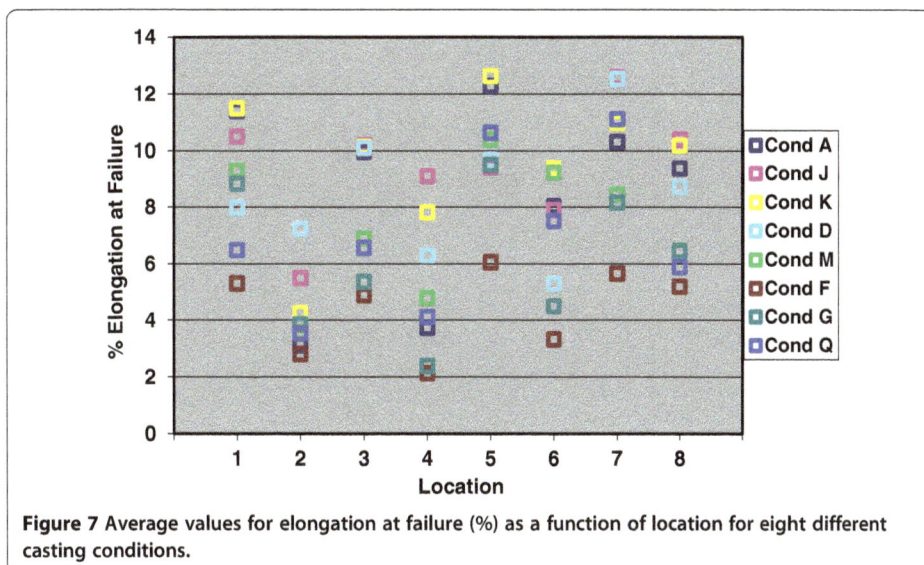

Figure 7 Average values for elongation at failure (%) as a function of location for eight different casting conditions.

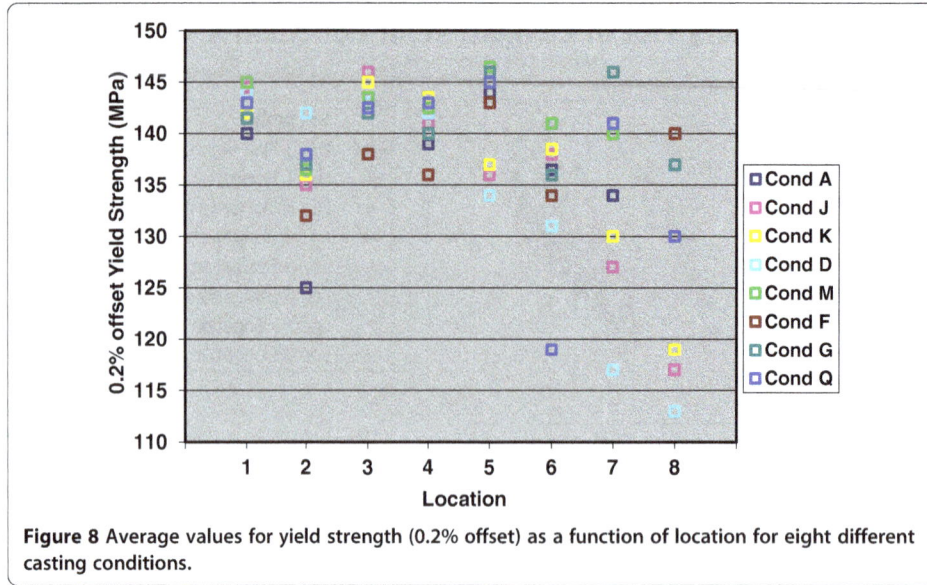

Figure 8 Average values for yield strength (0.2% offset) as a function of location for eight different casting conditions.

locations in the casting (Figure 11). Location 2, in particular, showed significant improvement in average elongation at failure.

With the process window defined, two casting conditions were selected for further study and for use in component-level experiments: condition I and condition U. An additional excised tensile bar location was also identified and tested (location 9, see Figure 5). Figure 12a,b, shows 'box and whisker plots' of the tensile results (elongation, 0.2% offset yield strength, and ultimate tensile strength, respectively) in each of the locations tested for the condition U that were generated in Minitab® version 15. 'Box and whisker' plots present the data in a consistent statistical format by dividing the entire data set into four quartiles [54]. In Minitab®, the top line of the box represents the third quartile of the data set (i.e., 75% of the data are less than or equal to that value). The middle line of the box is the median of the entire data set, and the bottom line of the box represents the first quartile of the data set (i.e., 25% of the data are less than or equal to that value). The vertical lines (whiskers) extend to the maximum data points

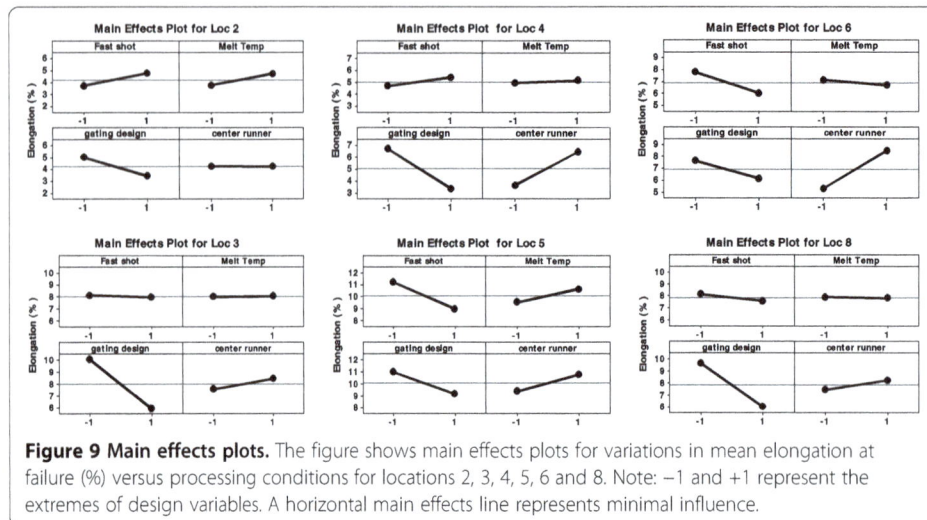

Figure 9 Main effects plots. The figure shows main effects plots for variations in mean elongation at failure (%) versus processing conditions for locations 2, 3, 4, 5, 6 and 8. Note: −1 and +1 represent the extremes of design variables. A horizontal main effects line represents minimal influence.

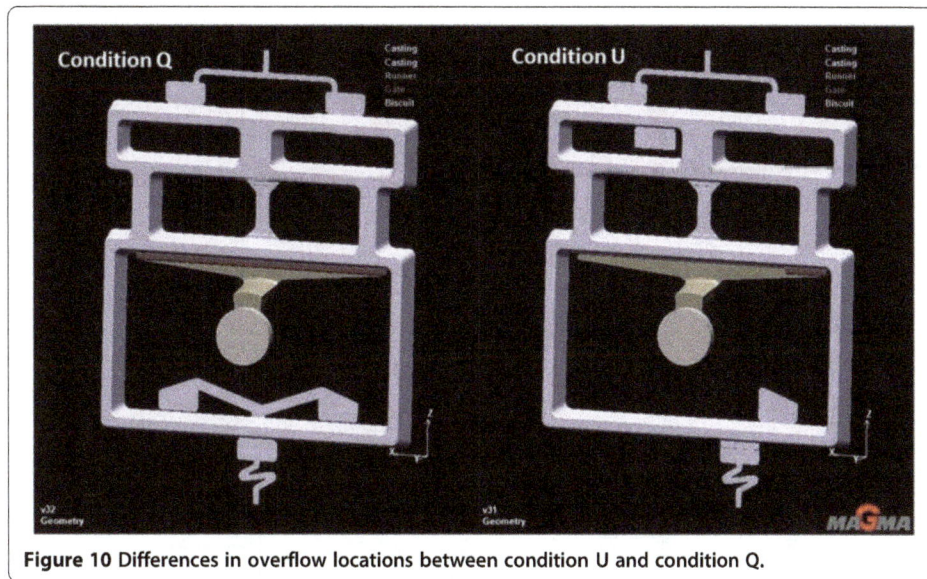

Figure 10 Differences in overflow locations between condition U and condition Q.

in the set that are within 1.5 times the third quartile number and the minimum data points in the set that are within 1.5 times the first quartile number. Stars represent data that lie outside that range [55]. The average elongation at failure for all the samples tested in condition U was 0.08 ± 0.02. An analysis of variation (ANOVA) of the means at each location gave a p value of <0.05, indicating that there was a statistically significant difference in the total elongation at failure in the different locations.

Figure 13 shows the true stress versus true strain curves for a representative sample; in this case, location 2 from condition U is shown. The shapes of stress/strain curves for all samples are virtually identical except for the value for elongation at failure. Location 2 shows elongation at failure variation from 7% to almost 13%. This level of variance in the measured ductility was typical of all of the samples tested for a given location and processing condition.

Box plots of the condition I tensile property results are shown in Figure 14a,b,c. Location 4 is not noted in the data for condition I because all of the samples failed outside the gage length. If the results from condition I are compared to the results of

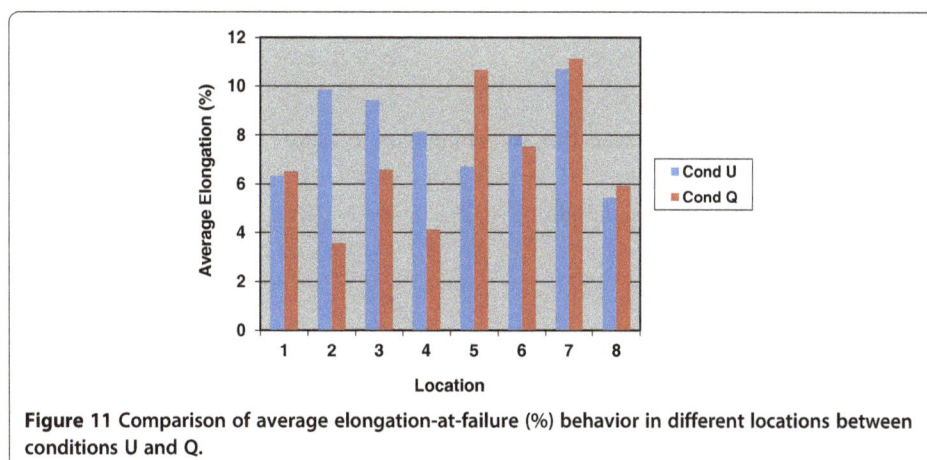

Figure 11 Comparison of average elongation-at-failure (%) behavior in different locations between conditions U and Q.

a)

b)

c)

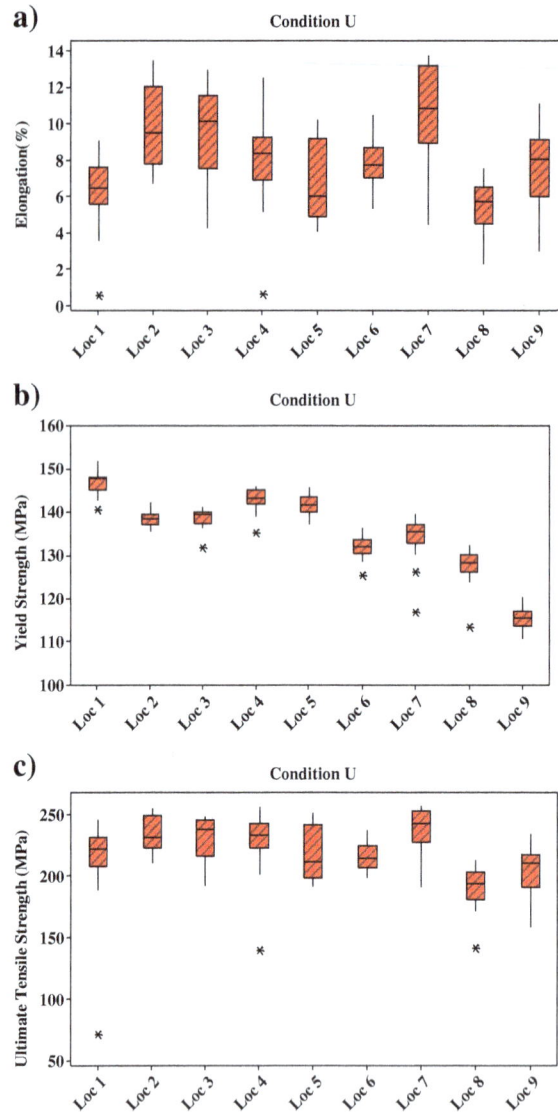

Figure 12 Box plots of tensile results. **(a)** Elongation at failure (%), **(b)** 0.2% offset yield strength, and **(c)** ultimate tensile strength versus location for tensile samples from condition U. Location numbers correspond to Figure 5. Middle horizontal lines represent median values, upper and lower edges of red boxes represent upper and lower quartile values, and vertical lines extend to minimum and maximum values in data set. Stars (*) signify outlier values.

condition U (Table 4), it can be seen that the elongation at failure is different for several locations, most notably location 2. Location 2 in condition U has an average elongation at failure of 10%, while in condition I, the same location fails at an average elongation at failure of 4%.

Component-level testing

Component-level testing was conducted on two different configurations: a pole test and an axial crush test of the lower part of the GFC casting. All experiments were conducted on a high-speed test frame at the Ford Safety Test Labs in Dearborn, MI, USA.

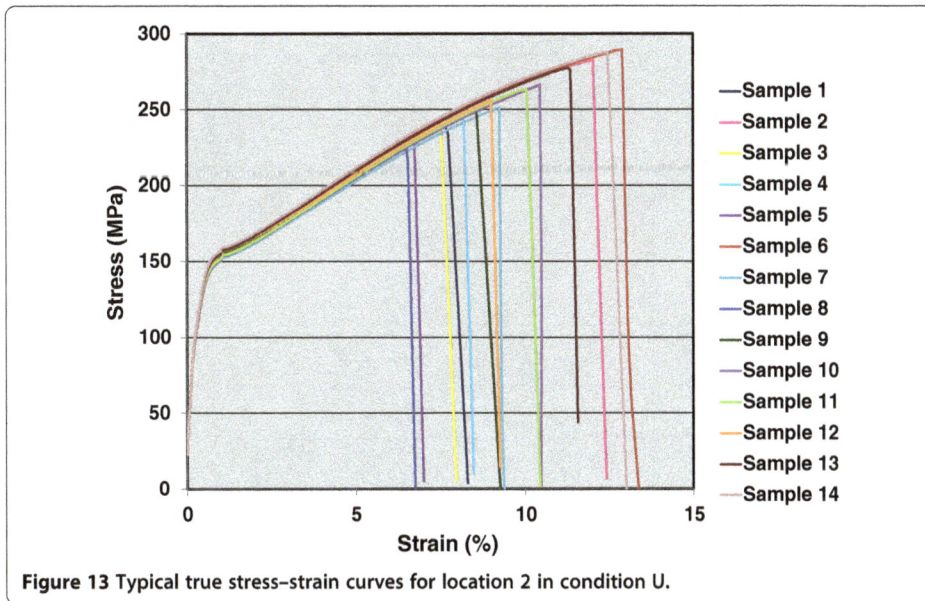

Figure 13 Typical true stress–strain curves for location 2 in condition U.

To run the tests, fixtures were designed to hold the parts stationary while a striker pole was activated to strike the sample at a particular location. The striker pole was loaded to provide a specified force throughout the test. The parts were tested to fracture, and the load and displacement data were recorded as a function of time.

A schematic diagram of the testing configuration for the pole test is shown in Figure 15. The side rails were specifically designed not to constrain the part during deformation. Knife edges were placed in the side rails to hold the part in place prior to the testing. Tape was used to secure the GFC panel to the testing fixture. The round striker pole was 10 cm in diameter. The pole was attached to a sled and then struck the GFC at 3 m/s speeds. The total load impacting the GFCs was 126 kg. A set of approximately 13 tests were performed to determine statistical variability in the testing for condition U. A set of 10 tests were also run for condition I.

Figures 16 and 17 show the force versus displacement curves for conditions U and I, respectively. In both cases, the data from multiple component tests show very consistent behavior. Although the initial peak load is similar for both conditions, the load drops immediately with only 15 to 20 mm of displacement in condition I. Condition U, however, does not experience appreciable load drop until approximately 40 mm of displacement. This leads to considerably less energy absorption for the condition I castings than the condition U. This behavior can be related to the tensile property results (Table 4). Locations 2, 5, and 9 were directly in the path of the striker pole in the pole test. When the values for local ductility are compared for condition U versus condition I, it is observed that condition U had a considerably higher average elongation at failure in location 2 and slightly higher in location 9, which presumably contributed to the better energy absorption under component loading.

Figure 18 shows a comparison of the castings post-test. The condition U casting remains intact, although there are large tears in the middle and lower wall sections and outer wall of the upper aperture is completely fractured through. The condition I casting is broken completely into two pieces.

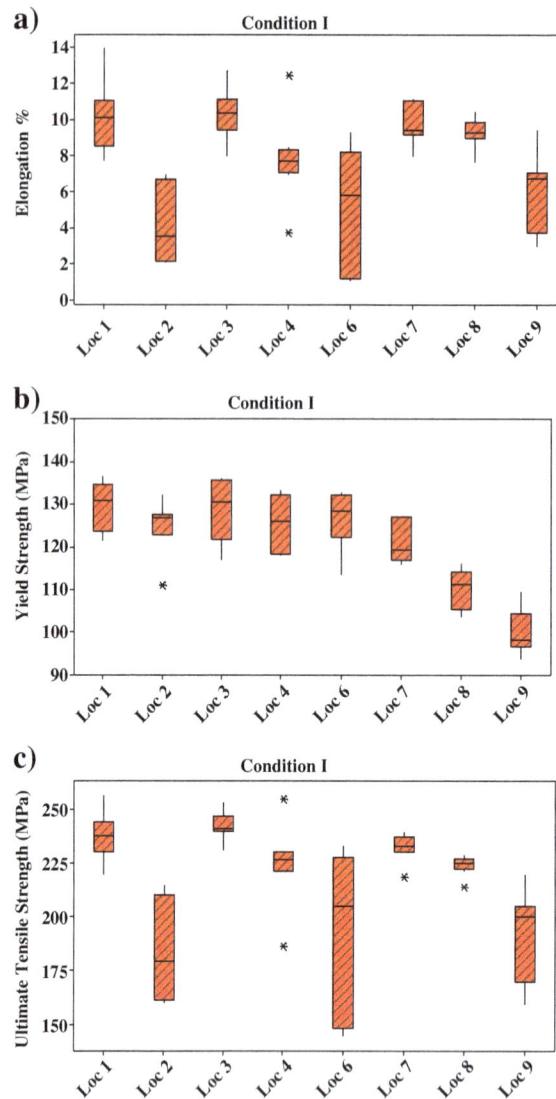

Figure 14 Box plots of tensile property results. **(a)** Elongation at failure (%), **(b)** 0.2% offset yield strength, and **(c)** ultimate tensile strength versus location for tensile samples for condition I. All samples from location 4 failed outside the gage section and thus were omitted. Location numbers correspond to Figure 5. Middle horizontal lines represent median values, upper and lower edges of red boxes represent upper and lower quartile values, and vertical lines extend to minimum and maximum values in data set. Stars (*) signify outlier values.

For the axial crush testing, only condition U was tested. The GFCs were machined into two sections; the upper window was separated from the lower set of windows, and only the lower window section was tested in axial crush loading. Figure 19 shows a schematic diagram of the testing setup. The casting was placed in a holder with a snug but not clamped fit. The holder contained a draft angle to allow for the casting to be

Table 4 Comparison of uniform elongation in excised samples in conditions U and I

	Average uniform elongation at failure (%)		
	Location 2	Location 5	Location 9
Condition U	9.8 ± 2.0	6.7 ± 2.2	7.5 ± 2.3
Condition I	4.3 ± 0.8	7.8 ± 0.8	6.0 ± 0.8

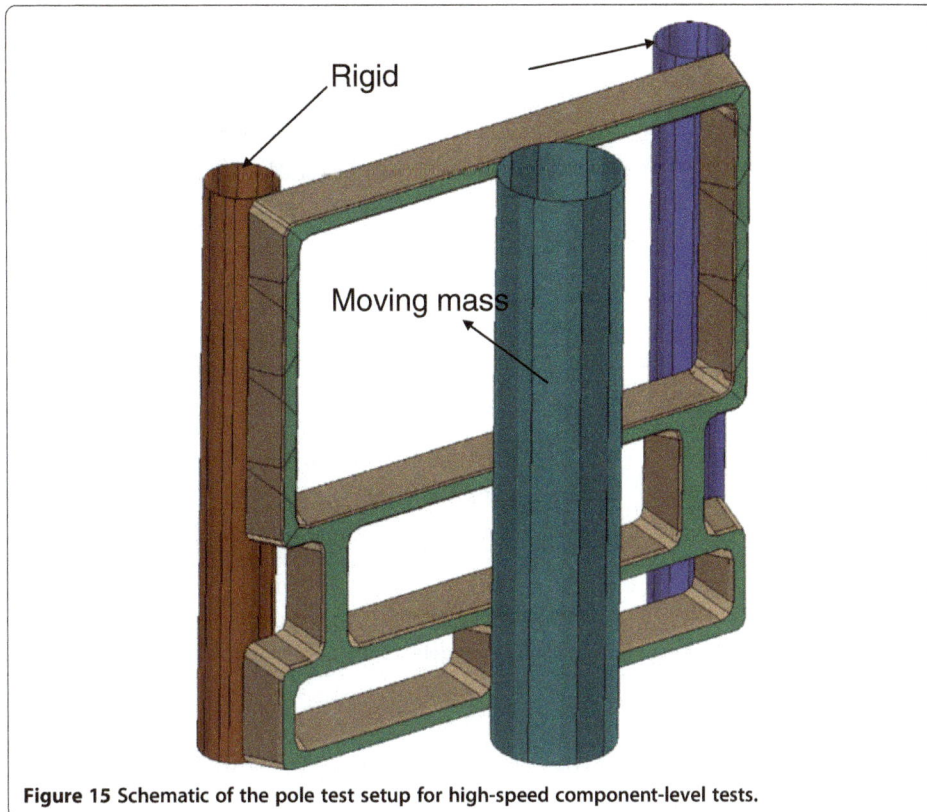

Figure 15 Schematic of the pole test setup for high-speed component-level tests.

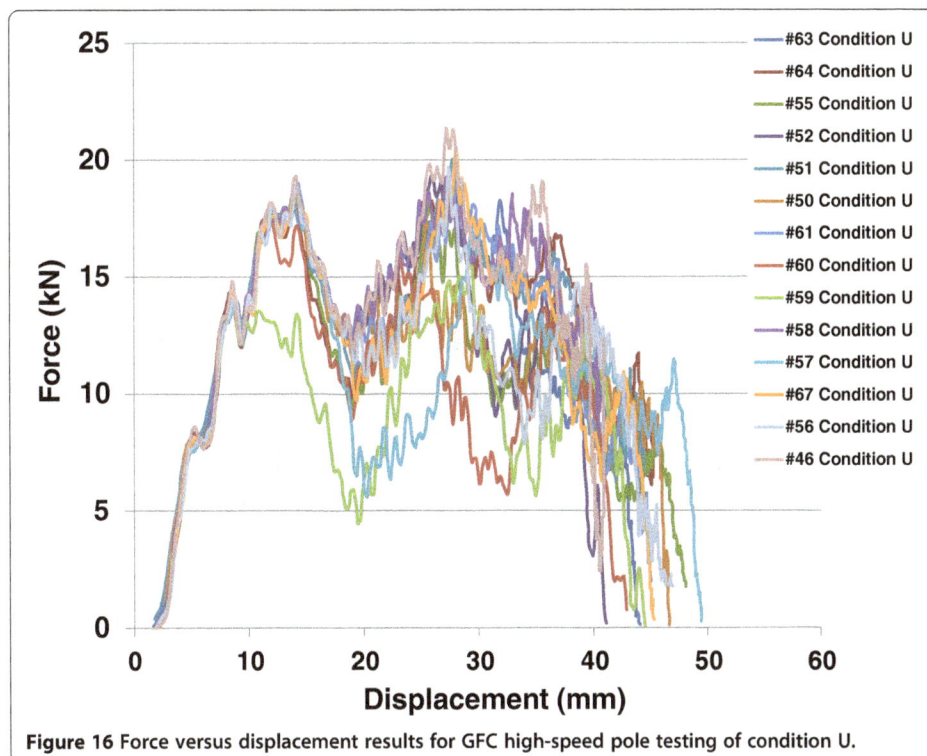

Figure 16 Force versus displacement results for GFC high-speed pole testing of condition U.

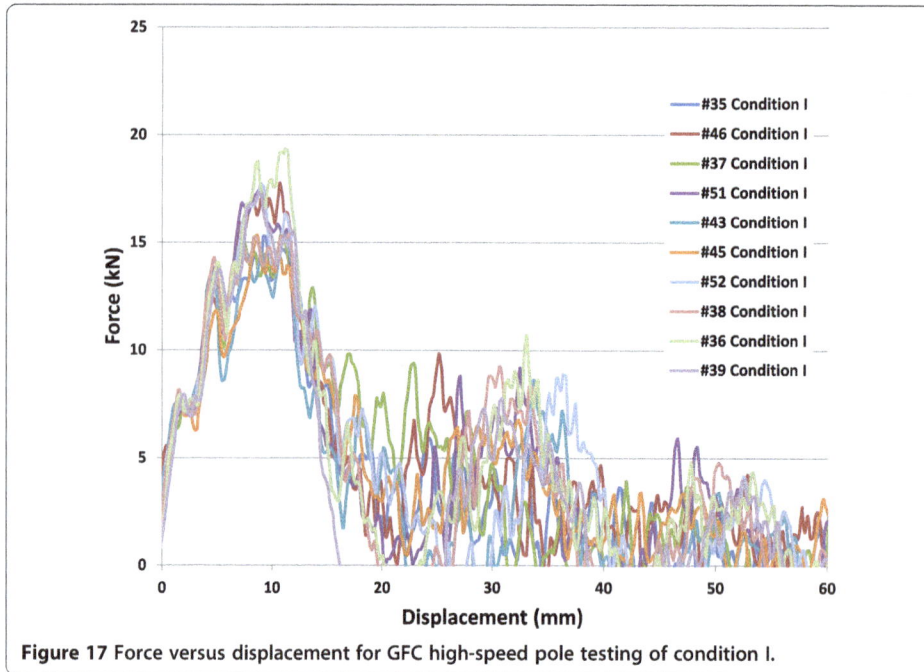

Figure 17 Force versus displacement for GFC high-speed pole testing of condition I.

held in the frame such that the top edge of the casting was parallel to the crosshead. Four tests were conducted on a Bendix™ machine with a speed of 3 m/s. A striker bar was again used to impact the sample, similar to the pole test although the carriage weight in this case was 94 kg. The striker bar impacted the entire surface of the sample, providing a uniform force.

The force versus displacement results are shown in Figure 20. The four tests are very consistent in their behavior, and generally, a peak load of 18 to 20 kN is observed. Only limited cracking was observed in the axial crush tests. Figure 21 shows a typical part after the test. The figure shows that the part deformed in an asymmetrical manner. The column on the left side of the photo indicates more deformation than on the right. This is consistent with the variation in ductility noted in condition U in this location.

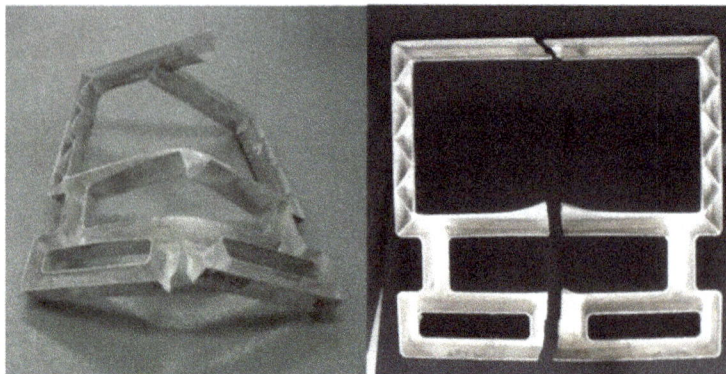

Figure 18 Condition U (left) and condition I (right) castings after testing at 3 m/s. Note higher levels of deformation in condition U.

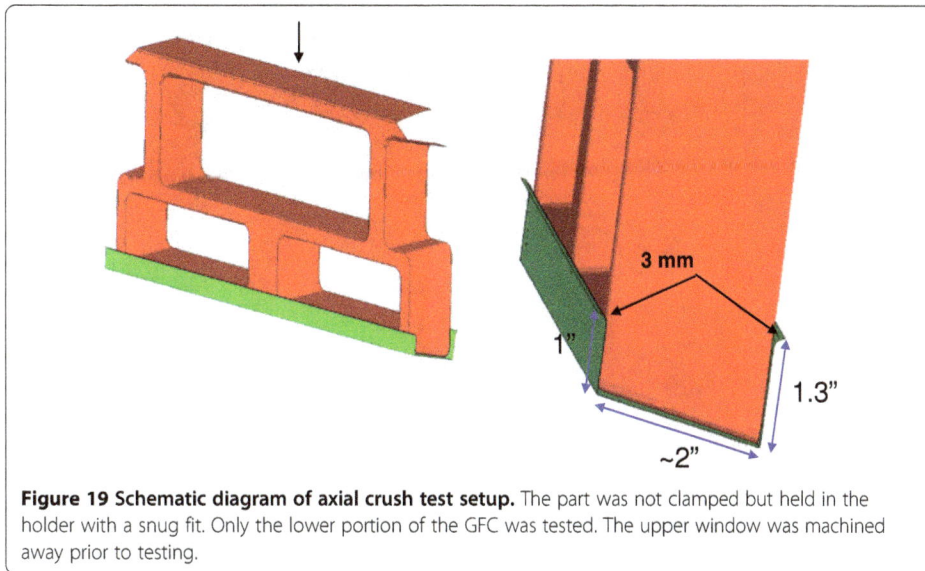

Figure 19 Schematic diagram of axial crush test setup. The part was not clamped but held in the holder with a snug fit. Only the lower portion of the GFC was tested. The upper window was machined away prior to testing.

Derivation of the quality mapping equations

For this study, the quality mapping equation uses several criteria functions identified by the MAGMASOFT® casting simulation code to map the local elongation-at-failure and yield strength predictions onto the cast component geometry. The equation takes the following form:

$$\text{Quality Index} = C_1 * CF_1{}^{C_2} + C_3 * CF_2{}^{C_4} + C_5 * CF_3{}^{C6}$$

where CF is a criteria function that is predicted through the MAGMASOFT® simulation and C_i is a set of constants. MAGMASOFT® has several criteria functions that numerically describe different behaviors in each meshed cell. The criteria functions are based either on flow calculations or on solidification calculations [56].

The criteria functions that were investigated were flow length (FL), solidification time (ST), air contact (AC), temperature at 100% fill (T100%FILL), and air entrapment (AE)

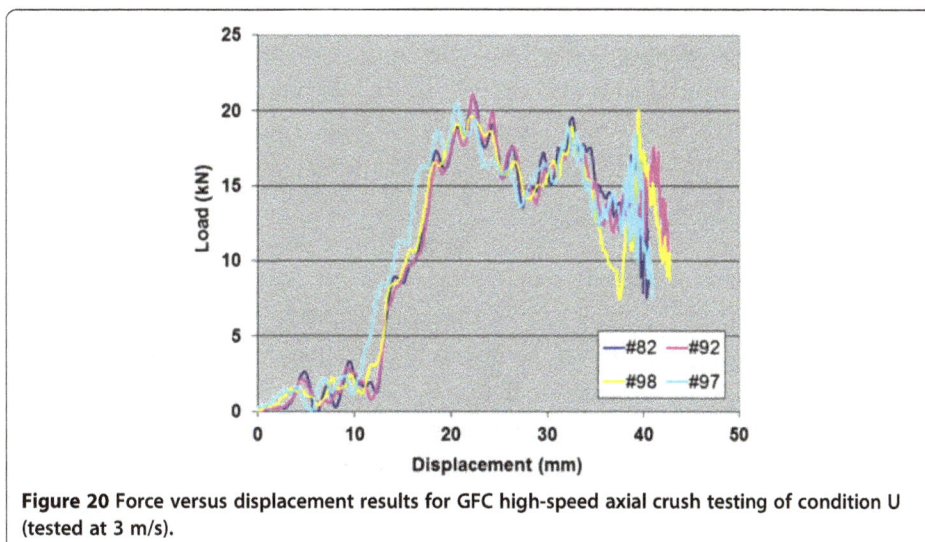

Figure 20 Force versus displacement results for GFC high-speed axial crush testing of condition U (tested at 3 m/s).

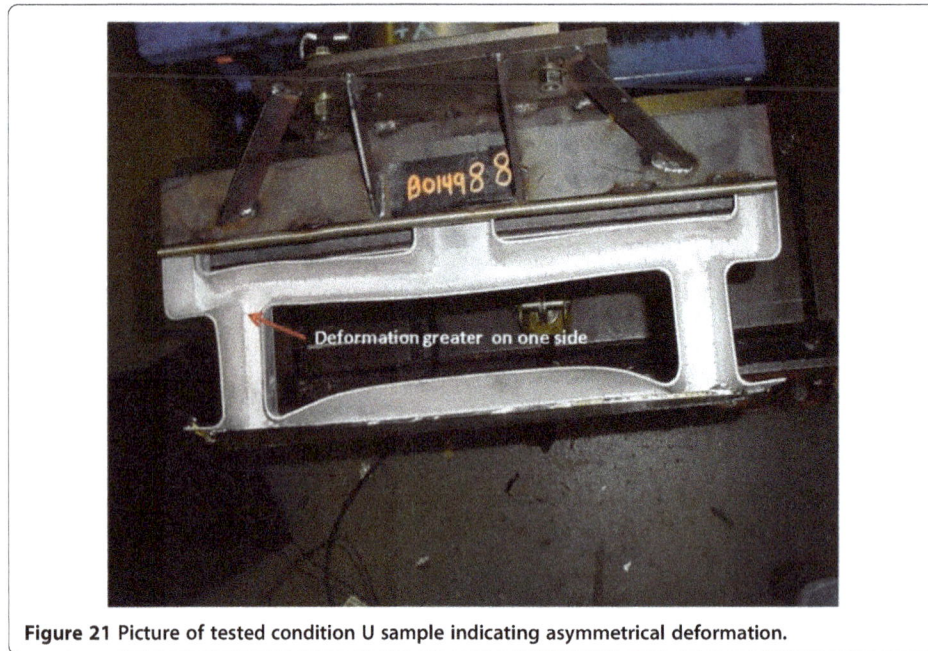

Figure 21 Picture of tested condition U sample indicating asymmetrical deformation.

based on the authors' previous experience and some earlier work from other researchers [4,6,13,14]. Flow length is the length a virtual marker travels in the die cavity before it solidifies. A typical flow length for a frame casting such as a door, instrument panel, or seat could be up to 2 m long depending on gating and overflow positions. Also, if there is significant turbulence, a particle could have a long and tortuous flow length even though the part itself is not large. Solidification time indicates the amount of time required for the temperature in a region to drop from the liquidus to the solidus. Generally, thicker areas will have longer solidification times than thinner areas. Air contact is a proxy for melt front. It gives the amount of time a metal particle is in contact with the air during the casting process. Areas where melt fronts come together tend to have higher times for air contact. Air entrapment estimates the amount of entrapped gas, and temperature at 100% fill indicates the temperature of the melt once filling is complete and the solidification calculations begin. A map of the criteria functions can be observed in the postprocessor, and there is the option to map user results, as well. This output is referred to a USER_Result in the postprocessor. A subroutine and a user input deck were developed to include the empirical equation into the mapping software. This was enabled by using the MAGMASOFT_API code available for version 4.4 of the software.

Separate simulations were conducted using the processing parameters of all of the casting conditions listed in Table 3. In order to have a realistic starting approximation of the mold cavity temperature during casting, the standard practice is to use a simulation procedure in which four 'warm-up' thermal-only simulations were performed first and then the temperature values from the fourth simulation were used as the starting temperature conditions for the flow simulation of the high-pressure die casting process. A total cycle time 45 s was used for the simulation. This allowed for the shot, opening of the die, removal of the part, and then closing of the die. Oil lines were also included in the simulation and were set to a temperature of 190°C in the cover half of the die and 260°C in the ejector half. These factors were chosen based on the actual casting trial conditions. The MAGMASOFT® HEATMED setting was used for thermophysical properties of the oil, and the interfacial heat transfer settings and the oil lines remained active the entire simulation. The oil lines also remained active during the actual HPDC process of this component.

MAGMASOFT® settings for thermophysical properties and interfacial heat transfer coefficients (IHTCs) of the tool steel for the die were also used (MAGMASOFT® X40CrMoV5 designation). The parameters for the tool steel were temperature dependent. The temperature-dependent IHTCs for each alloy with the tool steel were generated from an internal study. The thermophysical parameters chosen for the Mg alloys were derived from Thermocalc® calculations based on the actual compositions of the alloy used. Temperature-dependent values for heat capacity, thermal conductivity, latent heat, and solidus to liquidus temperature curve were calculated for the AM60 alloy.

The MAGMASOFT® solver uses a finite volume approach. To optimize run time, different parts of the mesh utilized different maximum cell sizes. The casting, gating, and overflows contained cells approximately 0.8 mm × 0.8 mm × 0.8 mm. The ingates to the casting were meshed slightly finer in the flow direction with a cell size of 0.8 mm × 0.5 mm × 0.8 mm to ensure at least three cells across the flow area. Other parts of the casting simulation such as the mold were given a much coarser cell size of 5 mm × 5 mm × 5 mm. Even so, the total number of cells for the simulations was approximately 59 million. Each simulation for the GFC castings required approximately 4 days to complete on a workstation with a single processor. Once each simulation was completed, the results were analyzed and several criteria functions were manually collected in addition to temperature and flow information at the center of each excised tensile sample location. Figure 22 shows an example for the air entrapment output. The color shadings indicate different values for each criteria function in different locations in the casting. The locations of the excised samples and values for the average elongation at failure are indicated as well. In most cases, there was little or no gradient in the criteria functions for a given area so manual selection of the criteria function value in a specific location was straightforward. However, in cases where there was a large change in value for a given criteria function in a particular location, the highest (or in the case of the temperature criteria function, the lowest) value in the center of the wall within a 25-mm area in length was chosen. It was assumed that a sample would fail in the weakest region so a criteria function that could contribute to the weakness would be assumed to be the most extreme value in the region.

The values of several of the criteria functions had quite different orders of magnitude. For example, typical flow length values are between 30 and 200 cm for this study, while most of the values for air contact were less than 0.02. This necessitated the use of normalization factors in the casting. Each criteria function was divided by a value specific to the criteria function. The maximum values for the GFC casting were used in most cases. ST was normalized to 4 s (the maximum solidification time in this casting), air contact was normalized to 0.01 s, and air entrapment was normalized to 0.30. In the case of flow length, the longest distance from the middle of the gate to the overflow on the upper window was used. In this case, it was 90 cm for the GFC. The temperature at 100% fill was normalized to the liquidus temperature of AM60 or 620°C. The

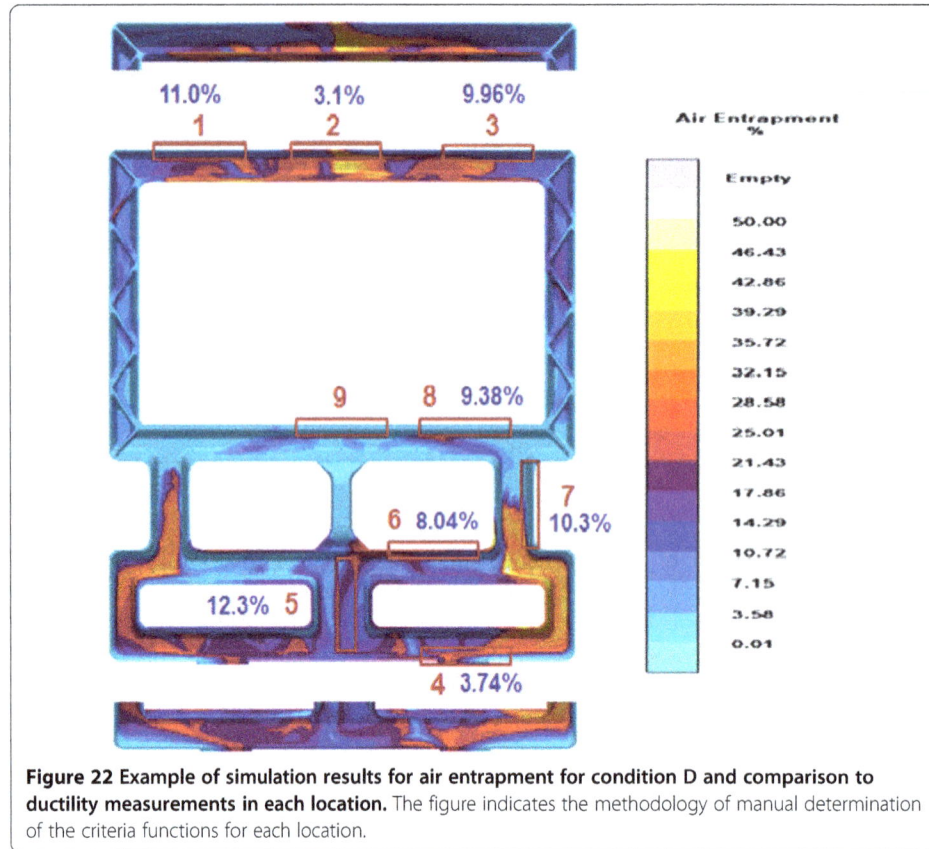

Figure 22 Example of simulation results for air entrapment for condition D and comparison to ductility measurements in each location. The figure indicates the methodology of manual determination of the criteria functions for each location.

assumption in this case is that the higher the value over the liquidus temperature, the less chance that areas would be 'mushy' and feeding or flow would be a problem.

Finally, an initial 'base' elongation at failure of 9% was assumed for the ductility prediction equation. This was an average value of strain at failure associated with an area of the AM60 casting that is considered a desirable or standard target ductility for this alloy. In the case of the yield stress equation, the base value for yield stress for AM60 was considered to be 140 MPa. The regression analysis was performed by comparing the criteria functions generated at each location with the average measured value of the elongation at failure and the yield strength at that location.

The data collected from the simulations (Table 5) was analyzed using the program SigmaPlot®. A linear regression analysis using a least squares fit was performed to obtain the values for each of the eight coefficients in the following equation:

$$Strain = 9 + C_1 * FLnorm^{C_2} + C_3 * AEnorm^{C_4} + C_5 * STnorm+^{C_6} + C_7 * T100\%FILLnorm^{C_7}$$

Where

$$FLnorm = FL/90$$
$$AEnorm = AE/30$$
$$STnorm = ST/2.5$$
$$T100\%FILLnorm = \frac{T100\%FILL - 620}{620}$$

Table 5 Locations and their average values for 0.2% offset yield strength and elongation-to-failure with their corresponding criteria functions

Location	Number of samples	Average 0.2% offset yield strength (MPa)	Average elongation (%)	Solidification time (s)	Flow length (cm)	Air entrapment (%)	Temperature at 100% fill
G - 1	18	143 ± 3	8.8 ± 1.8	1.5	90	4	714
G - 4	17	140 ± 2	2.4 ± 1.1	1.64	141	15	697
G - 5	9	146 ± 4	9.5 ± 3.6	1.8	77	12	714
G - 6	14	136 ± 6	4.5 ± 2.3	1.6	84	19	698
G - 7	13	146 ± 7	8.2 ± 2.0	1.64	58	0	706
G - 8	15	137 ± 3	6.4 ± 2.1	1.6	30	19	726
Q - 1	15	143 ± 1	6.5 ± 1.1	1.53	92	4	710
Q - 4	18	143 ± 2	4.1 ± 1.2	1.7	115	15	703
Q - 5	6	146 ± 1	10.8 ± 5.1	1.9	78	12	706
Q - 6	18	120 ± 12	7.5 ± 1.6	1.5	62	15	703
Q - 7	13	141 ± 2	11.1 ± 1.7	1.6	58	17	710
Q - 8	16	130 ± 4	5.9 ± 1.7	1.6	43	19	717
M - 1	14	145 ± 2	9.3 ± 1.6	1.43	94	6.	661
M - 4	11	142 ± 2	4.8 ± 3.5	1.38	106	13	653
M - 5	10	146 ± 2	10.4 ± 3.7	1.64	72	10	661
M - 6	17	141 ± 2	9.2 ± 1.8	1.47	72	13	653
M - 7	18	140 ± 2	8.5 ± 1.3	1.38	60	2	661
M - 8	13	137 ± 2	6.5 ± 1.7	1.47	29	27	669
A - 1	19	140 ± 2	11.5 ± 2.5	1.3	89	6	657
A - 2	16	125 ± 10	3.1 ± 1.4	1.6	112	18	650
A - 3	10	141 ± 2	10.0 ± 1.8	1.3	94	6	657
A - 4	11	139 ± 2	3.7 ± 2.5	1.5	100	19	647
A - 5	16	144 ± 3	12.3 ± 3.2	1.6	80	13	652
A - 6	15	136 ± 3	8.0 ± 1.6	1.4	63	19	655
A - 7	16	134 ± 2	10.3 ± 1.8	1.4	45	0	667
A - 8	15	118 ± 2	9.4 ± 1.3	1.6	31	0	672
J - 1	18	144 ± 2	10.6 ± 1.1	1.43	90	8.	673
J - 2	17	135 ± 2	5.5 ± 1.4	1.54	110	23	667
J - 3	16	146 ± 1	10.2 ± 1.8	1.43	95	12	673
J - 4	10	143 ± 3	9.1 ± 3.3	1.53	80	8	670
J - 5	16	136 ± 3	9.4 ± 2.2	1.8	70	2	670
J - 6	9	138 ± 2	7.9 ± 2.6	1.6	60	10	673
J - 7	17	127 ± 3	12.6 ± 1.6	1.54	45	0	682
J - 8	17	117 ± 2	10.4 ± 2.3	1.7	30	0	688
K - 1	15	142 ± 2	11.5 ± 1.2	1.48	92	8	690
K - 2	19	136 ± 2	4.3 ± 1.6	1.7	130	23	678
K - 3	16	145 ± 2	10.2 ± 2.0	1.48	98	12	690
K - 4	10	143 ± 2	7.8 ± 4.1	1.7	88	8	681
K - 5	12	137 ± 4	12.7 ± 2.6	1.857	55	2	690
K - 6	8	138 ± 2	9.4 ± 3.4	1.59	66	10	693
K - 7	20	130 ± 3	11.0 ± 1.3	1.64	49	0	702
K - 8	15	120 ± 3	10.2 ± 1.4	1.857	33	0	708
D - 1	12	144 ± 2	8.0 ± 2.0	1.5	91.4	4	712
D - 2	16	142 ± 2	7.2 ± 1.6	1.7	109	27	697
D - 3	19	144 ± 2	10.1 ± 1.4	1.5	91.4	8	708

Table 5 Locations and their average values for 0.2% offset yield strength and elongation-to-failure with their corresponding criteria functions (Continued)

D - 4	11	142 ± 3	6.3 ± 2.4	1.7	90	11	705
D - 5	18	134 ± 2	9.7 ± 1.5	1.8	85	15	708
D - 6	12	131 ± 11	5.3 ± 3.4	1.6	80	17	708
D - 7	16	117 ± 3	12.5 ± 1.3	1.6	45.7	0	722
D - 8	13	113 ± 1	8.7 ± 2.2	1.9	32.7	0	726

The coefficients for the ductility equation are listed in Table 6. The fit for the prediction equation was $R^2 = 0.60$. Figure 23 shows the measured average strain values and their standard deviations versus the predicted values for all of the data used in the linear regression analysis. The low value of 0.6 for the correlation coefficient is indicative of the many statistical factors that influence ductility that are not captured perfectly in the quality map approach. Despite this low R^2 value, the relationship did provide directional guidance on relative values of high and low ductility values within the castings examined and could still provide some guidance on design (e.g., identify areas of potential weakness). As Figure 23 also shows, this regression is conservative, which was an important consideration for component performance.

A quality map equation for the local yield strength was also determined using a similar procedure. In the case of the yield strength, the differences between the locations were not very significant. There was really only one location that was markedly different in yield strength from the other locations: location 8. This was presumably because this location was closest to the ingate and the solidification time was longer in that region. Therefore, only locations 1 and 8 were used for the linear regression. The average of the yield strength was used for the linear regression. The following equation was developed:

$$\text{Yield stress} = 140 + C_1 * \text{STnorm}^{C_2} + C_3 * \text{FLnorm}^{C_4} + C_5 * \text{T100\%FILLnorm}^{C_6}$$

Where

$$\text{STnorm} = \text{ST}/2.5$$
$$\text{FLnorm} = \text{FL}/90$$
$$\text{T100\%FILLnorm} = \frac{\text{T100\%FILL} - 620}{620}$$

Table 6 Coefficients for the empirical ductility quality map equation developed for MAGMASOFT® version 4.2

	Criteria function	Value
C1	FLnorm	−0.29
C2	Flnorm	4.51
C3	AEnorm	−7.03
C4	AEnorm	0.65
C5	STnorm	−8.06
C6	STnorm	1.95
C7	T100%Fillnorm	4.83
C8	T100%Fillnorm	0.097

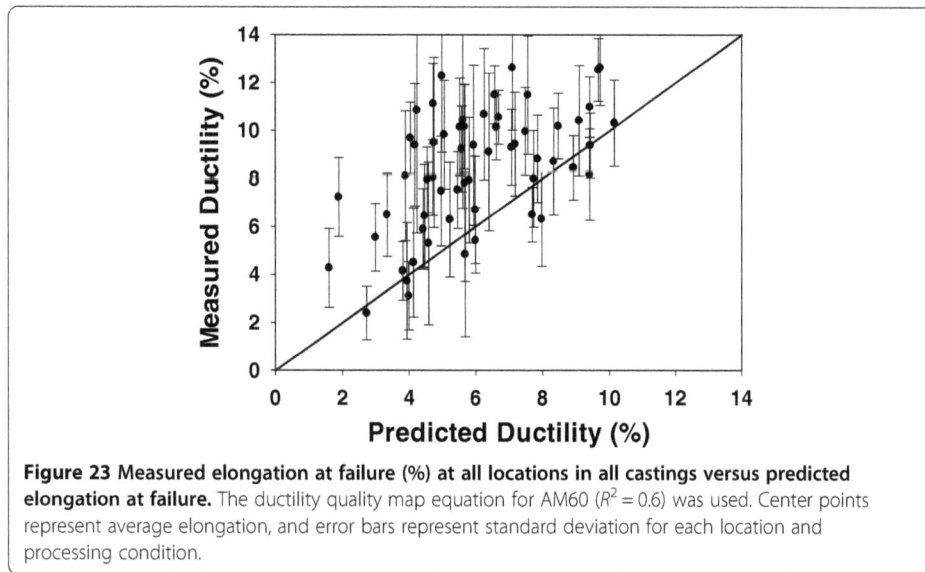

Figure 23 Measured elongation at failure (%) at all locations in all castings versus predicted elongation at failure. The ductility quality map equation for AM60 ($R^2 = 0.6$) was used. Center points represent average elongation, and error bars represent standard deviation for each location and processing condition.

The coefficients are listed in Table 7. Figure 24 shows the measured average yield strength and standard deviation for the various locations produced under different casting conditions versus the estimated yield strength using the quality map equation for yield strength. The fit for the yield strength equation produced $R^2 = 0.84$.

After the equations were developed, they were used as MAGMASOFT® input as user results to map the prediction throughout the casting. They were graphically displayed in the same manner as other results were presented in the program and provided a map of the local property of interest.

The quality mapping equations were developed using the results from tests conducted under the partial factorial DOE. Next, the quality mapping procedure and equations were used to examine local property differences between conditions U and I. First, the differences in the criteria functions as a function of location were compared for the two casting conditions. Additional files 2 and 3 show an animation comparison of the flow of melt in front and back views in the two castings, illustrating the differences in pattern between the two conditions. Figures 25, 26, and 27 indicate the local differences in air entrapment, flow length, and temperature at 100% fill, respectively.

Figure 28 shows the predicted elongation at failure as mapped onto condition U and condition I. The comparison indicates that the ductility prediction captures the

Table 7 Coefficients for the empirical yield strength quality map equation developed for MAGMASOFT® version 4.2

	Criteria function	Value
C1	STnorm	−271.33
C2	STnorm	8.84
C3	Flnorm	67.53
C4	Flnorm	0.19
C5	T100%Fillnorm	−47.13
C6	T100%Fillnorm	−0.11

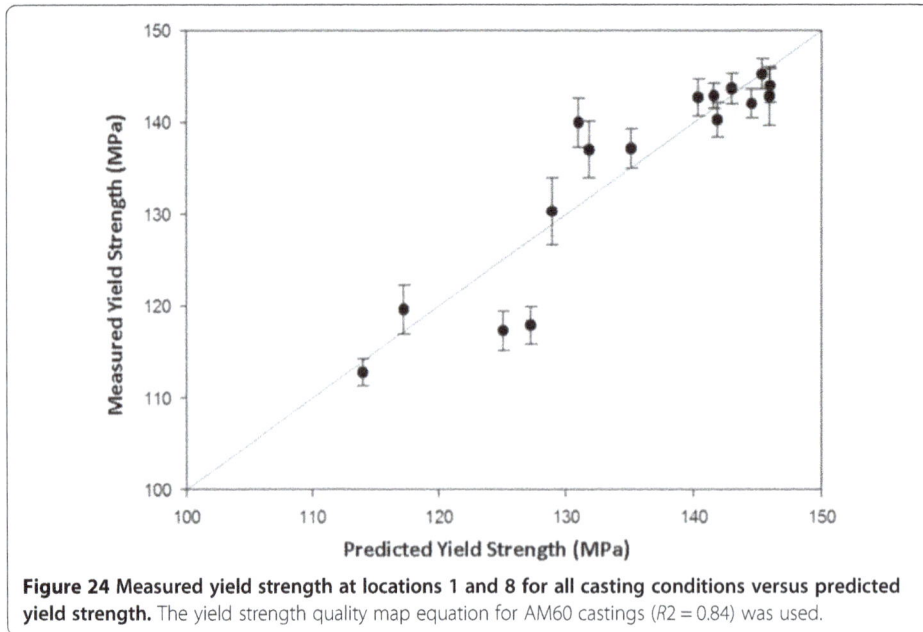

Figure 24 Measured yield strength at locations 1 and 8 for all casting conditions versus predicted yield strength. The yield strength quality map equation for AM60 castings ($R2 = 0.84$) was used.

differences between condition U and condition I at location 2 very well. Other locations show good comparison to the prediction. Figure 29 shows a comparison of the yield strength predictions for condition U and condition I. Both conditions indicate a trend of increasing yield strength with distance from gating, as also seen in the experimental results. However, the difference in yield strength between condition U and condition I is not captured in the prediction.

It must be noted that these equations and coefficients were specifically developed for use in MAGMASOFT® version 4.2. They have not yet been validated with other versions of MAGMASOFT® or other casting simulation codes.

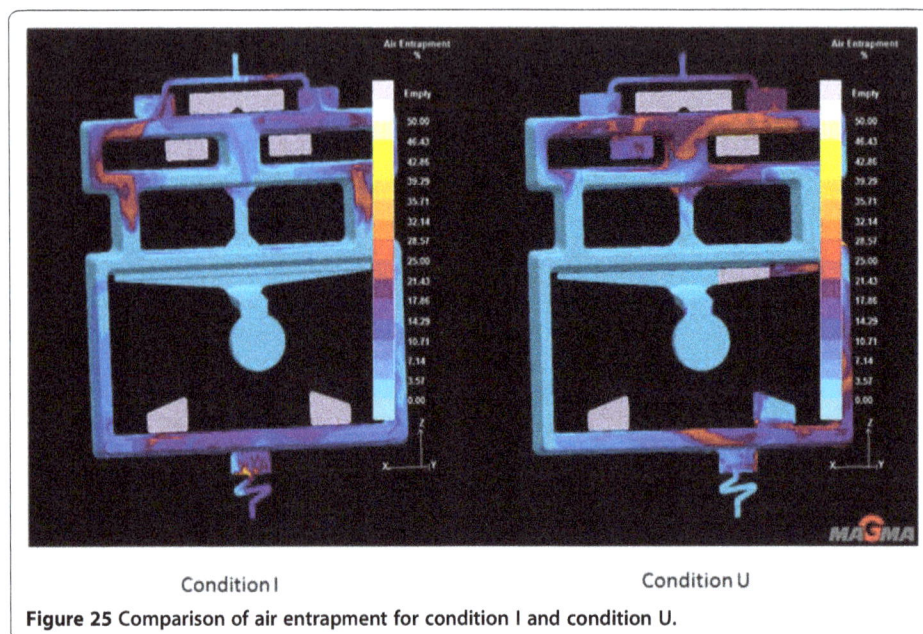

Condition I   Condition U

Figure 25 Comparison of air entrapment for condition I and condition U.

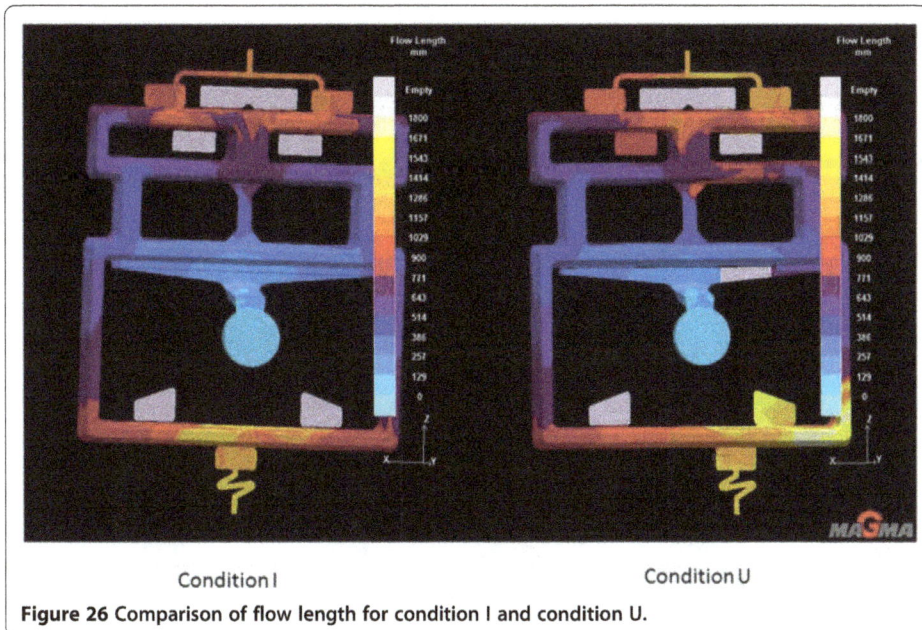

Figure 26 Comparison of flow length for condition I and condition U.

Application of the quality mapping for local properties

The GFC casting was specifically designed to be an experimental cast part substitute for a large thin-walled frame casting. Therefore, in order to test the quality mapping equation for ductility, a prototype liftgate inner casting produced by Meridian Lightweight Technologies (Strathroy, ON, Canada) was analyzed. The casting was produced in AM60 on a 4,200-ton HPDC production machine at a different facility from the GFC production. Figure 30 shows a schematic diagram of the prototype casting examined along with the locations of excised tensile samples. Flat sub-sized tensile bars were excised from 15 locations on the casting with a total of 10 castings machined. Tensile testing was conducted

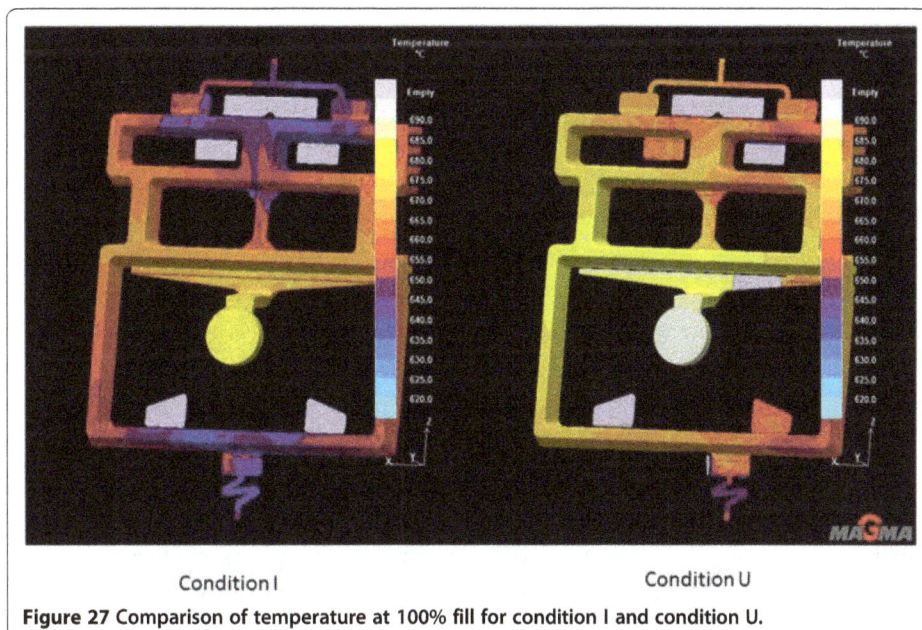

Figure 27 Comparison of temperature at 100% fill for condition I and condition U.

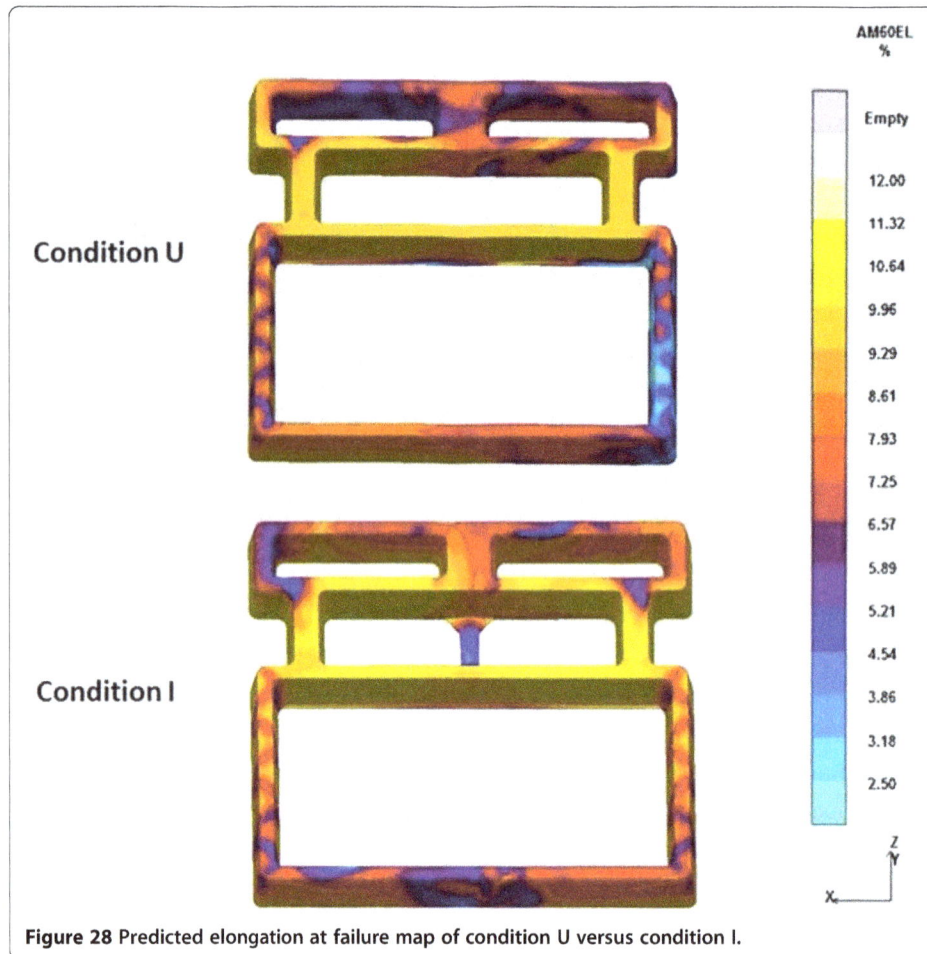

Figure 28 Predicted elongation at failure map of condition U versus condition I.

using a similar procedure and protocol to that previously described for the GFC testing. A complete MAGMASOFT® version 4.2 simulation was performed using the production processing parameters provided by the prototype casting supplier.

The quality mapping equation was then applied using the criteria function outputs from the simulation. The coefficients applied to the postprocessor were the same as those that were generated for the GFC. However, some changes were made to the normalization factors to accommodate the fact that the casting was much larger. Most notably, the flow length criteria function normalization parameter was changed from 90 to 200 cm to represent the longest flow length for that casting.

Figure 31 shows a plot of average measured elongation and standard deviation versus predicted elongation for the prototype casting. The predictions show good agreement for all locations with the exception of locations 6 and 7. In location 6, many of the samples failed at or close to yield. Examination of the fracture surface of many of the samples indicated defects and discolored material in the center of the sample possibly indicative of oxide films. One possible explanation for the failure of the model to predict the poor performance in this area may be due to a condition in this casting that was not present in the GFC. The GFC did explore different gating configurations; however, in both cases a single gate was used. The prototype casting tooling used six separate gates to fill the casting. These multiple gate melt fronts are not quantifiably

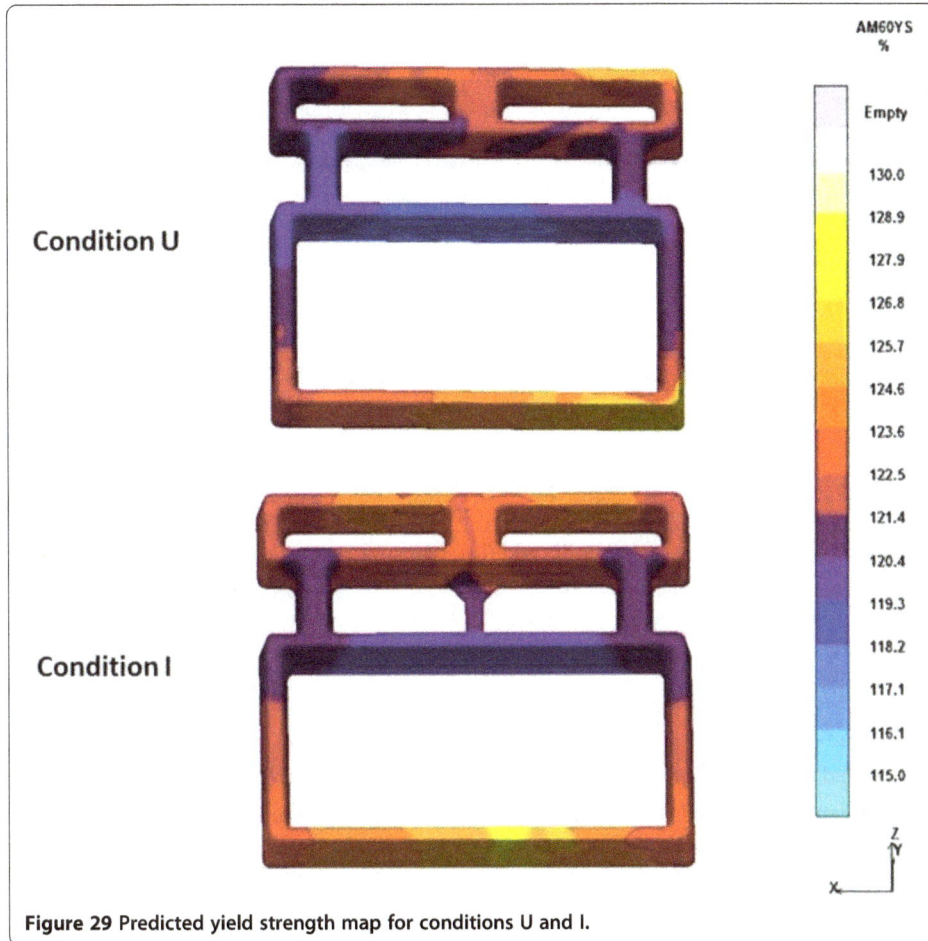

Figure 29 Predicted yield strength map for conditions U and I.

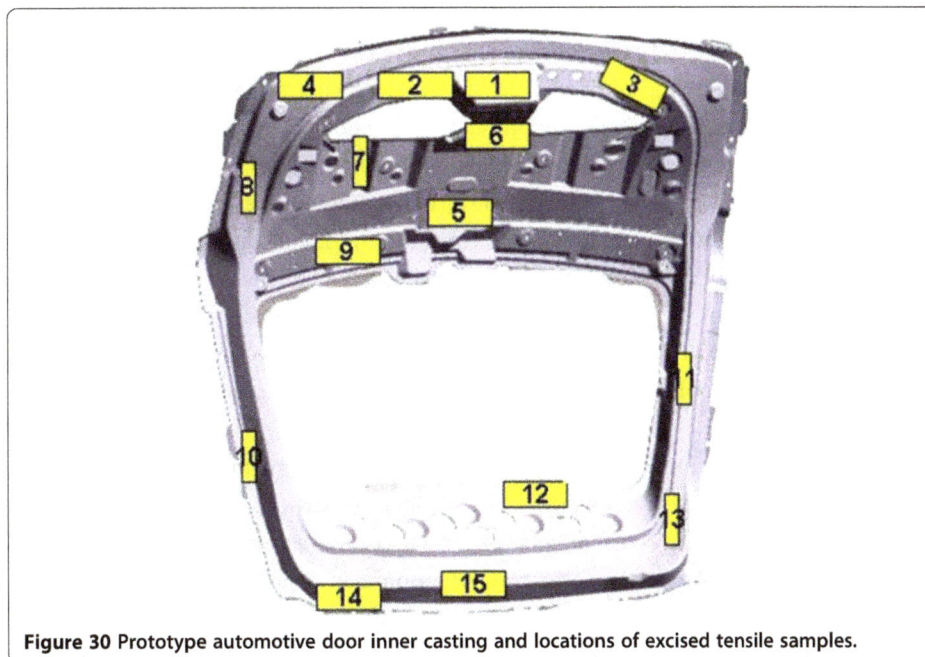

Figure 30 Prototype automotive door inner casting and locations of excised tensile samples.

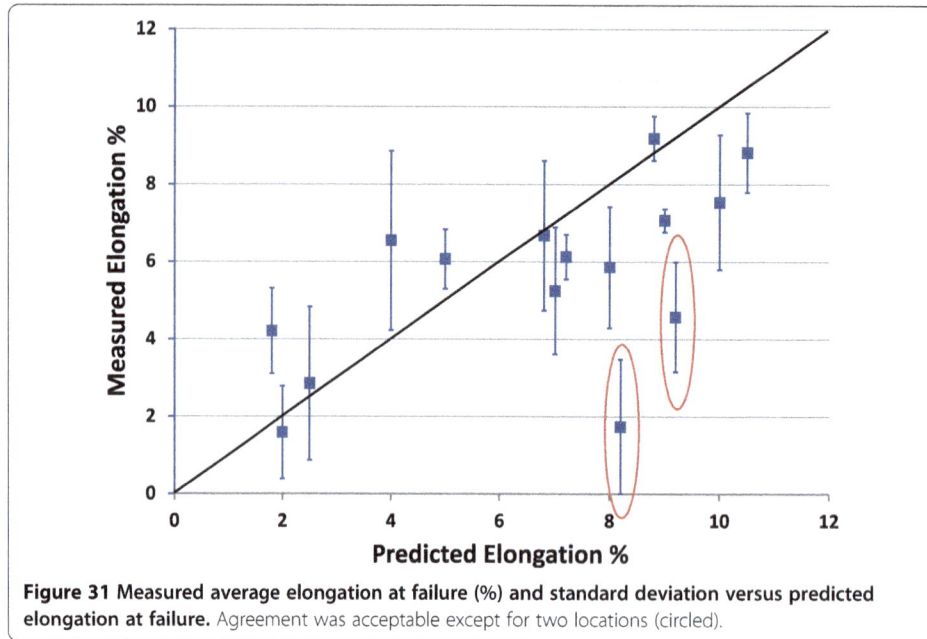

Figure 31 Measured average elongation at failure (%) and standard deviation versus predicted elongation at failure. Agreement was acceptable except for two locations (circled).

described in the casting simulation code. Location 6 is in a region where the different gate melt fronts meet and that may have influenced the behavior in this region. More work is needed to identify an appropriate criteria function to determine these effects, or an index can be assigned based on the program outputs to quantify these interactions. Examination of location 7 indicated no obvious reason for the discrepancy between the model and the prediction. However, the prototype casting exhibited variation of wall thickness across the part from 2.5 up to 5 mm, a condition not present in the areas examined in the GFC. This could have influenced flow conditions in a way that cannot be accounted for in the empirical quality mapping equation.

Utilizing local property predictions for performance predictions

In the ICME methodology, the ultimate use of manufacturing history sensitive local property prediction is in component performance prediction. Establishing the spatial variability in local properties for different casting conditions was important for understanding the overall behavior of the component during component performance prediction and confirmation testing. The predicted elongation-at-failure and yield strength values of a GFC geometry were used in a high-speed component performance simulation to establish the extent to which the component performance could be predicted or influenced by the local properties. In this study, the finite element solver code used was LS_DYNA®. These simulation results were compared to high-speed pole and axial crush testing described earlier.

The LS_DYNA® simulations were conducted using shell elements approximately 5 mm in length. The casting was divided into 17 different sections, and each section was meshed (shown in Figure 32). A base stress/strain curve was provided for the material using an average stress/strain curve from location 2 in condition U (curve from sample 10 in Figure 13). That curve was modified for each section in the GFC mesh based on the quality map-predicted local properties from the casting simulation.

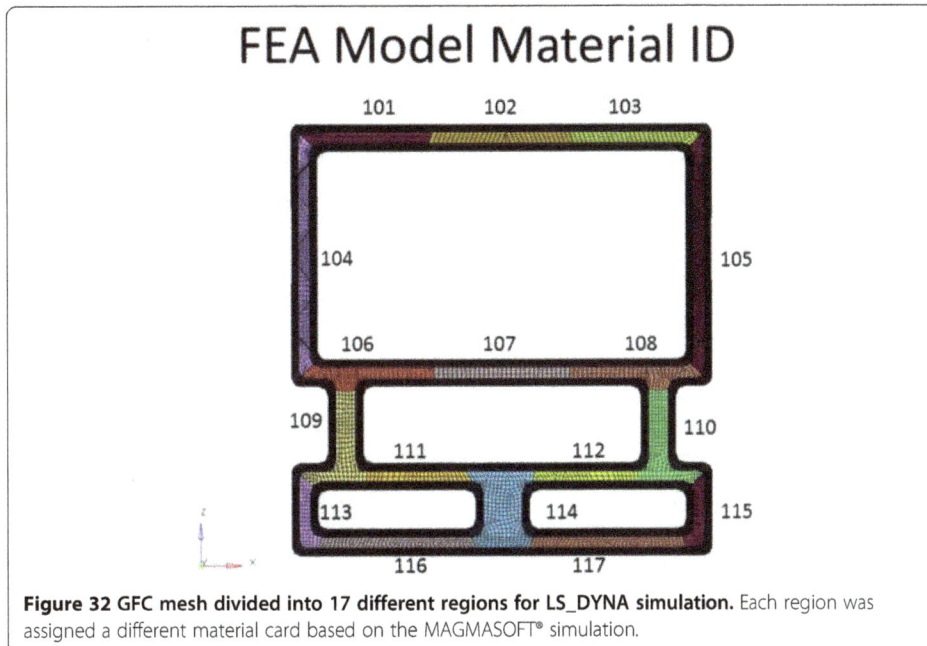

Figure 32 GFC mesh divided into 17 different regions for LS_DYNA simulation. Each region was assigned a different material card based on the MAGMASOFT® simulation.

Tables 8 and 9 show how the curve was modified based on the local property predictions to generate the input data for the material input deck for each meshed section. As indicated in Table 8, the yield strength was only changed in one location on the GFC. This reflected the region immediately adjacent to the gate that had a longer solidification time and a lower measured yield strength than other areas in the casting. The material input deck used in this simulation was MAT_24 (MAT_PIECEWISE_LINEAR_PLASTICITY).

Table 8 Elongation at failure values used for each GFC section in LS_DYNA® pole test and axial crush test simulations

Mat ID	Condition I	Condition U
101	0.08	0.03
102	0.03	0.06
103	0.08	0.11
104	0.08	0.04
105	0.08	0.07
106	0.09	0.10
107	0.05	0.07
108	0.09	0.09
109	0.09	0.06
110	0.09	0.10
111	0.09	0.04
112	0.09	0.10
113	0.08	0.03
114	0.05	0.05
115	0.08	0.10
116	0.06	0.05
117	0.06	0.07

Table 9 Stress factor used to modify the local yield strength prediction for each GFC section

Mat ID	Condition I	Condition U
101	1.0	1.0
102	1.0	1.0
103	1.0	1.0
104	1.0	1.0
105	1.0	1.0
106	1.0	1.0
107	0.875	0.875
108	1.0	1.0
109	1.0	1.0
110	1.0	1.0
111	1.0	1.0
112	1.0	1.0
113	1.0	1.0
114	1.0	1.0
115	1.0	1.0
116	1.0	1.0
117	1.0	1.0

Failure was defined by the total strain at failure in the stress–strain curve defined for all elements in the region. When the failure strain was reached in that element, it was deleted.

Force versus displacement curves were generated from the simulations for both condition I and condition U in the pole test case. Those results are shown in Figure 33. There is good agreement in both cases between the CAE predictions and the actual test results. Animations of the tests are shown in Additional files 4 and 5. The animations indicate the earlier total failure of the condition I GFC when compared to the condition U GFC.

A force versus displacement curve was also generated from the LS_DYNA® simulation for the axial crush testing performed on the condition U sample (see Figure 34). An animation of the behavior during testing indicates buckling on one side of the test (Additional file 6). This indicates that the local properties in that area caused the region to yield and fracture earlier in the deformation than the other side and are an indication of the value of the ICME approach for product development.

Discussion

A comparison of the different casting parameters and their associated tensile properties at various locations indicated that, in the case of the thin-walled frame casting investigated in this study, the flow characteristics dominated the local tensile properties for both experiments and predictions. Changing the casting parameters that impacted the flow conditions either globally or locally within the casting had the most significant effect on the ductility, whereas processing parameters such as melt temperature, which would affect solidification, appeared to play a lesser role. Gating configuration, a

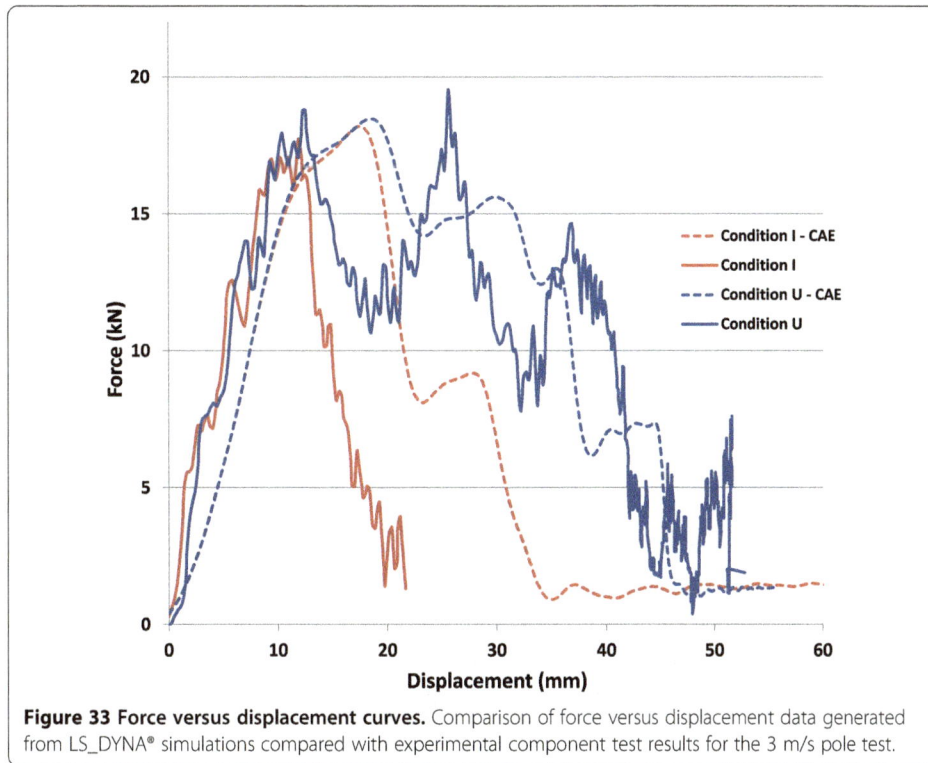

Figure 33 Force versus displacement curves. Comparison of force versus displacement data generated from LS_DYNA® simulations compared with experimental component test results for the 3 m/s pole test.

geometrical parameter, consistently shows a larger influence on the ductility for most of the locations than most of the other casting parameters examined, as can be observed in Figure 9. Figure 11 illustrates the influence of overflow placement on the flow characteristics in a casting. Conditions U and Q are virtually identical except for the placement of the overflows. Those overflows are significant because they capture the melt fronts and remove them from the casting itself. This is evident not only in the

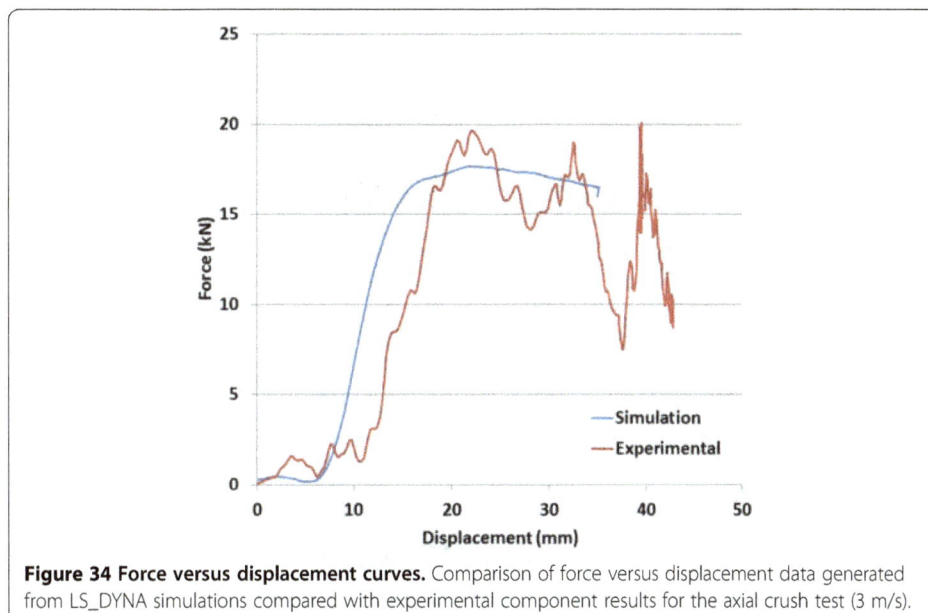

Figure 34 Force versus displacement curves. Comparison of force versus displacement data generated from LS_DYNA simulations compared with experimental component results for the axial crush test (3 m/s).

experimental results, but it is also captured by the quality mapping ductility model (Figure 35). Comparing these two conditions illustrates the importance of melt fronts and melt flow turbulence in the prediction of general ductility in areas of the casting.

As shown in Figure 9, in locations where melt fronts meet (e.g., location 2 and location 5), processing parameters such as melt temperature and shot speed appeared to have some influence over the ductility. A lower initial melt temperature tended to result in a lower ductility for these locations because the lower temperature could result in lower melt feeding into those areas that were last to fill, resulting in more shrinkage-related porosity and a higher likelihood that metal closer to the liquidus temperature would have lower fluidity and less ability for that melt to be pushed out of the casting into the overflows. A high shot speed could either contribute positively to the ductility in a location by pushing poor metal out of the casting quicker (location 2) or negatively by contributing to local turbulence and defect creation (location 5). A faster shot speed, however, was only an improvement in the ductility in the symmetrical gating configuration. In the asymmetrical gating configuration, it actually reduced the ductility in location 2. This could be due to potentially increased turbulence in the region for the asymmetrical gating configuration.

Once the inputs to the casting simulation were understood, the outputs of the casting simulation were also important to evaluate. The criteria functions of flow length, melt

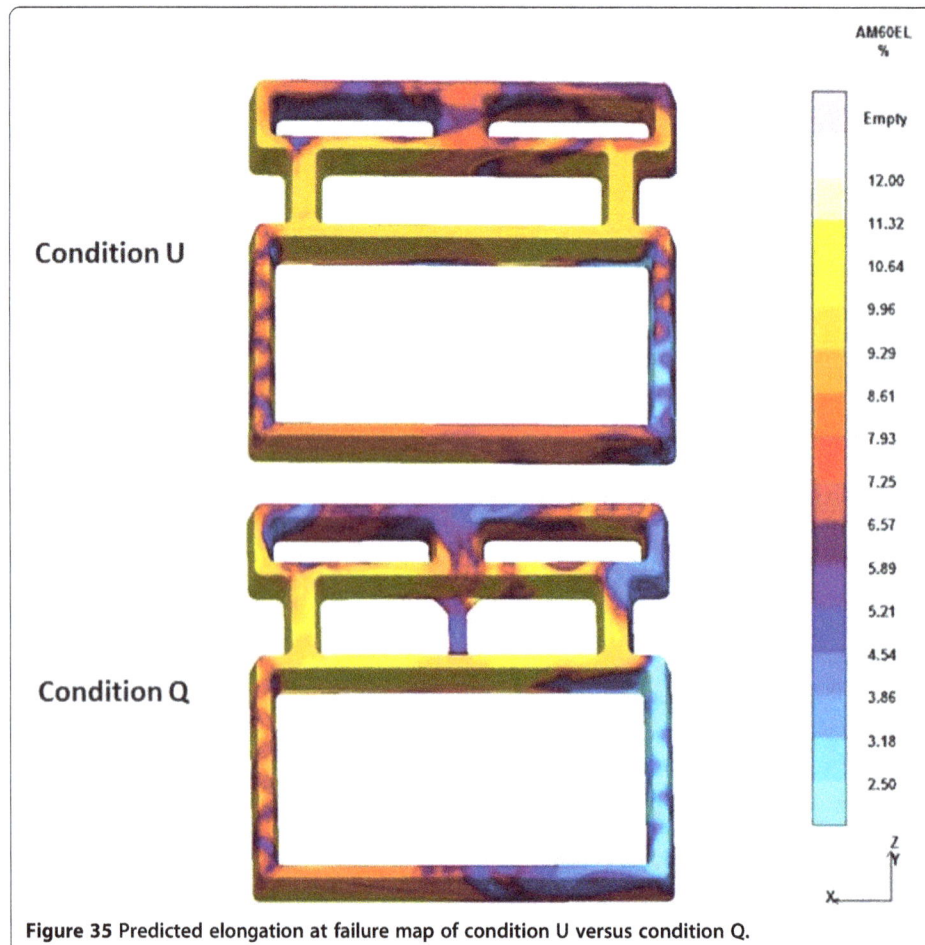

Figure 35 Predicted elongation at failure map of condition U versus condition Q.

temperature at 100% fill, air entrapment, and solidification time were used in the quality mapping procedure in part because they were independently calculated parameters in the MAGMASOFT code that can represent the flow and solidification behavior observed in cast parts. The criteria function, solidification time, is a good indicator of solidification gradients which can have an effect on yield strength. Indeed, the most heavily weighted parameter in the yield strength quality map equation was solidification time, with only minor contributions coming from flow length or temperature at 100% fill. Unfortunately, the effects of melt flow turbulence and melt fronts coming together, key areas that have been shown to have a significant influence over the local ductility in this study, are less well-defined using the criteria functions available. Normalized flow length, air entrapment, and temperature at 100% fill were proxies for the melt flow conditions that could contribute to differences in ductility from the simulation. These criteria functions were used together with similar weighting to describe local ductility. While the use of these particular criteria functions did appear to at least capture trends in the local ductility in the GFC casting, there was some disagreement when the equation was used to predict local ductility in the large prototype casting. This could potentially indicate the need to identify an additional criteria function or functions within the code to capture some additional effects (such as melt flow behavior from multiple gates).

One important issue throughout this work was the need to recognize that uncertainty exists throughout the casting, simulation, mechanical testing, and modeling trials performed. Panchal et al. [57] have highlighted this as a key issue in the practical application of ICME approaches. There will be some variability in the processing parameters from casting to casting in a single trial (e.g., slight changes in shot profile or minor changes in melt temperature). The casting simulation process relies on the accurate input of all casting parameters and thermophysical properties. While the authors attempted to be as accurate as possible and measured all process parameters carefully during each trial, some input parameters, of necessity, were averages or estimates. The starting melt temperature, for example, was averaged across the trial based on the temperature measured in the shot sleeve. A fast shot speed was also estimated from overlays of many individual shot profile curves for each condition. The heat transfer coefficients between a solidifying melt and the die wall can only be estimated based on embedded thermocouples in a die, and while the ones used in this case were temperature dependent, no accommodation was made for whether the casting was freezing onto the mold wall or freezing away. Fluidity is also an estimated value based on limited measurements of similar alloys.

The experimental results of the excised samples indicated a great deal of statistical variability particularly in the ductility measurements. This made it difficult to determine a single value of the mechanical property to use in the regression analysis for the modeling (in this case, the authors simply used the average). It was not clear if there was inherent variability in the sample or if the variation seen was, in part, caused by the excising of the test bar from the cast part. Alain et al. [58] proposed that the variability observed in tensile tests was due to the removal of the as-cast surface during machining of the sample and much of the resulting scatter in the results was due to the exposure of internal porosity or other defects due to the machining. Weiler et al. [13,14] and Yang et al. [59-61] have reported the pronounced decreases in mechanical properties in Mg alloys between areas closest to the surface ('skin') and areas in the middle of the cross section ('core'). Therefore, the act of excising and testing a tensile sample from a component could indicate more mechanical property variation, particularly in elongation at failure, than actually exists in the cast component. Indeed, though there was significant statistical variability in the excised sample testing for a given location, the force versus displacement behavior of the overall component in the loading situations examined in this study was repeatable within a very tight range. The average peak load for the component-level high-speed pole testing for condition U was 17.8 kN with a standard deviation of 1.3 kN, and the average peak load for the condition I component-level testing was 16.6 kN with a standard deviation of 1.5 kN. The load versus displacement curves also indicated consistent displacement values for final failure (40 mm for condition U and 20 mm for condition I).

The quality mapping process also produced some uncertainty. Selection of the criteria functions relied on a manual analysis of each region of interest in each of the eight simulations. While, most of the time, the region of interest contained a single value of a particular criteria function, there were some instances where a judgment was made because of large gradients in the outputs. Further, a simple linear regression analysis was used to develop the model. The authors recognize that other techniques exist to produce a correlation that could potentially more accurately predict the mechanical properties of interest.

It is also important to point out that this case study involved transferring one prediction from one simulation code into another simulation code to perform another prediction. One of the issues with doing this is the change in scale of the mesh itself. In the casting simulation code, the cell size of the mesh must be small to accurately predict the flow of the melt, and the mesh size in this study was typical of what would be used in an actual component casting simulation. However, a component performance simulation would utilize a much larger mesh size. This poses an issue when trying to transfer a local ductility prediction to material input deck of a code such as LS_DYNA. The authors tried to avoid this issue by manually dividing the casting into regions of interest and then determining a low value for each region based on the quality mapping output from the casting simulation. This introduces some uncertainty in the selection of mechanical properties for a given location but allows for simplification of the performance simulation to address areas of interest in a computationally efficient manner.

Finally, the current investigation deliberately avoided the microstructure prediction to go to a direct processing to property prediction. Measured property values were used to compare to the simulation results, and the quality map approach infers that the microstructural variables that affect properties are captured in the combined and weighted criteria functions. This assumption does not appear to be unreasonable in the case where the casting has a fairly uniform wall thickness (even with a complex shape). However, many automotive castings have large variations in wall thickness. This can result in enough variation in microstructural features (such as skin thickness and microporosity locations or segregation of elemental composition) that those features need to be independently predicted and their impact fundamentally understood. Indeed, the failure of the quality mapping equation to predict the ductility in all of the tested locations in the large prototype frame casting may have also been due to limited ability to predict local microstructure in those regions. Further work is required to improve upon the current state of the art.

Conclusions

- A thin-walled frame high component was produced by high-pressure die casting to investigate the effect of processing conditions on the local and component-level mechanical behavior of AM60.
- A casting DOE was conducted to determine the effects of casting parameters on local ductility. Castings were produced and detailed processing parameters such as die temperature and shot speed were collected.
- Tensile properties, specifically elongation at failure and yield strength, were characterized from a subset of the casting DOE, and the main effects were determined for each location. The effect of geometrical parameters was the most important for all locations, but melt temperature and shot speed were important for certain locations.
- MAGMASOFT® version 4.2 simulations were conducted for all of the conditions from the DOE and two additional conditions (condition U and condition I), and several criteria functions from the local areas corresponding to the excised tensile samples were determined and used in development of an empirical quality map for prediction of local properties.
- The quality mapping equation was utilized in the local property predictions of three separate castings: two frame castings (conditions U and I) and a prototype automotive door inner casting.
- The quality map-predicted local properties were used in an LS_Dyna® model to simulate two different component tests. The LS_Dyna® prediction showed good correlation with experimental component test results.
- Quality mapping represents an important ICME tool for predicting local properties in HPDC Mg components and can provide a good estimate of local property trends for frame-like castings. This methodology can be utilized to incorporate the manufacturing history into casting design upfront and potentially reduce the number of physical prototype iterations required to verify component performance.

Additional files

> **Additional file 1:** A schematic diagram of a typical cold-chamber high-pressure die cast machine.
> **Additional file 2:** Flow comparison of conditions I and U from the front side.
> **Additional file 3:** Flow comparison of conditions I and U from the rear side.
> **Additional file 4:** Animation of condition I pole test.
> **Additional file 5:** Animation of condition U pole test.
> **Additional file 6:** Animation of condition U axial crush test.

Competing interests
The authors declare that they have no competing interests.

Authors' contributions
JF, ML, and JA developed the overall quality mapping approach for Mg components and design of program. JWZ and LG designed GFC casting, developed the casting design of experiment, and conducted all casting trials. JF and JZ conducted all casting simulations, and JF conducted and analyzed all tensile testing and assisted with the design and analysis of component-level trials. JF, ML, and JZ conducted regression analysis and developed quality mapping equations, and JZ conducted simulation sensitivity analyses. JF and ML supervised LS_DYNA® simulations. All authors read and approved the final manuscript.

Acknowledgements

JF would like to acknowledge Sheila Ryzyi of Ford Motor Company for assistance with tensile testing, Ari Caliskan of Ford Motor Company for assistance with the design of the component tests, and the Ford Safety Lab for the high-speed testing. JWZ and LG would like to acknowledge the support of Mag-Tec Casting in Jackson, MI for assistance with the casting trials. We acknowledge helpful discussions with and support of X. Sun, Pacific Northwest National Laboratory (Batelle Memorial Institute). We would also like to acknowledge Meridian Lightweight Technologies, Inc., for assistance with the prototype casting simulation and mechanical testing. JZ and JA wish to acknowledge the financial support of Battelle Memorial Institute and US Department of Energy under Contract No. DE-AC05-76RL01830. Their work was funded by the Department of Energy Vehicle Technologies Office under the Automotive Lightweighting Materials Program managed by William Joost.

Author details

[1]Materials Research Department, Ford Motor Company, Research and Innovation Center, MD3182, P.O Box 2053, Dearborn, MI 48121, USA. [2]Department of Materials Science and Engineering, University of Michigan, 2300 Hayward St., Ann Arbor, MI 48109, USA.

References

1. Friedrich HE, Mordike BL (2006) Magnesium Technology: Metallurgy, Design Data, Application. Springer, Berlin
2. Schumann S (2005) Paths and strategies for increased magnesium applications in vehicles. Mater Sci Forum 488–489:1–8
3. Schumann S, Friedrich H (2003) Current and future use of magnesium in the automobile industry. Mater Sci Forum 419–422:51–56
4. Grebetz JC (1993) A Comparison of the Impact Characteristics of Several Magnesium Die Casting Alloys. SAE Tech. Pap, Ser, 930417
5. Padfield CJ, Padfield TV (2002) Plane stress fracture toughness testing of die cast magnesium alloys. SAE Tech. Pap., Ser. 2002-01-0077
6. Chen X, Wagner DA, Houston DQ, Cooper RP (2004) Elongation variability of AM60 die cast specimens. 2004 ASME International Mechanical Engineering Congress.
7. Coultes BJ, Wood JT, Wang G, Berkmortel R (2003) Mechanical properties and microstructure of magnesium high pressure die castings. In: Kaplan HI (ed) Magnes. Technol. 2003. The Minerals, Metals, and Materials Society, pp 45–50.
8. Weiler JP, Wood JT, Klassen RJ, Maire E, Berkmortel R, Wang G (2005) Relationship between internal porosity and fracture strength of die-cast magnesium AM60B alloy. Mater Sci Eng A A395:315–322
9. Wood JT, Klassen RJ, Gharghouri MA, Maire E, Wang G, Berkmortel R (2005) Mechanical properties of AM60B die castings: a review of the AUTO21 program on magnesium die-casting. SAE Tech. Pap., Ser. SAE-2005-0725
10. Carlson BE (1995) The effect of strain rate and temperature on the deformation of die cast AM60B. SAE Tech. Pap., Ser. SAE-950425
11. Li M, Zindel J, Godlewski L, Allison J (2006) Prediction of porosity defects and mechanical properties of high pressure die cast A380 aluminum alloy components. TMS Lett 3:31–32
12. Horstemeyer MF, Wang P (2003) Cradle-to-grave simulation-based design incorporating multiscale microstructure-property modeling: reinvigorating design with. J Comput Mater Des 10:13–34
13. Weiler JP, Wood JT, Basu I (2012) Process-structure–property relationships for magnesium alloys. Mater Sci Forum 706–709:1273–1278, doi:10.4028/www.scientific.net/MSF. 706–709.1273
14. Weiler JP, Wood JT, Klassen RJ, Berkmortel R, Wang G (2006) Variability of skin thickness in an AM60B magnesium alloy die-casting. Mater Sci Eng A 419:297–305, doi:10.1016/j.msea.2006.01.034
15. Laukli HI, Lohne O, Arnberg L (2005) High Pressure Die Casting of Aluminum and Magnesium Alloys - some comparisons of Microstructure Formation. Shape Cast. John Campbell Symp, TMS Annual Meeting
16. Dahle AK, Lee YC, Nave MD, Scha PL, StJohn DH (2001) Development of the as-cast microstructure in magnesium-aluminum alloys. J Light Met 1:61–72
17. Rodrigo D, Murray M, Mao H Brevick J, Mobley C, Chandrasekar V, Esdaile R (1999) Effects of section size and microstructural features on the mechanical properties of die cast AZ91D and AM60B magnesium alloy test bars. SAE Tech. Pap., Ser. 1999-01-0927
18. Rodrigo D, Murray M, Mao H, Brevick J, Mobley C, Esdaile R (1999) Characteristic microstructural features of die cast magnesium alloys. World Die Cast. 20th Int. Die Cast. Congr. Expo. NADCA, pp 219–225.
19. Rodrigo D, Ahuja V (2000) Effect of casting parameters on the formation "pore/segregation" bands in magnesium die castings. In: Aghion E, Eliezer D (eds) Magnes. 2000 Proc. 2nd Isr. Conf. Magnes. Sci. Technol. Magnesium Research Institute Ltd, Dead Sea, Israel, pp 97–104
20. Mao H, Brevick J, Mobley C Chandrasekar V, Rodrigo D, Murray M (1999) Microstructural characteristics of die cast AZ91D and AM60 magnesium alloys. SAE Tech. Pap., Ser. 1999-01-0928
21. Chadha G, Allison JE, Jones JW (2007) The role of microstructure on ductility of die-cast AM50 and AM60 magnesium alloys. Metall Mater Trans A 38A:286–296
22. Dahle AK, StJohn DH (1999) Rheological behavior of the mushy zone and its effect on the formation of casting defects during solidification. Acta Mater 47:31–41
23. Dahle AK, StJohn DH (1999) The origin of banded defects in high pressure die cast magnesium alloys. World Die Cast T99–062:205–211
24. Bowles AL, Griffiths JR, Davidson CJ (2001) Ductility and the skin effect in high pressure die cast Mg-Al alloys. In: Hryn J (ed) Magnes. Technol. 2001. The Metals Mineral and Materials Society, pp 161–168
25. Dahle AK, Sannes S, St. John DH, Westengen H (2001) Formation of defect bands in high pressure die cast magnesium alloys. Journal of Light Metals 1:99–103
26. Sannes S, Gjestland H, Westengen H, Laukli HI, Lohne O (2003) Die casting of magnesium alloys - the importance of controlling die filling and solidification. SAE Tech Pap., Ser SAE–2003–01–0183

27. Wang G, Froese B, Bakke P (2003) Process and property relationships in AM60B die castings. Magensium Technol. 2003 The Metals Mineral and Materials Society.

28. Cao H, Wessen M (2004) Effect of microstructure on mechanical properties of As-Cast Mg-Al alloys. Metall Mater Trans A 35A:309–319

29. Cao H, Wessen M (2005) Characteristics of microstructure and banded defects in die cast AM50 magnesium components. Int J Cast Met Res 18:377–384

30. Cao H, Wessen M (2003) Modeling of microstructure - mechanical property relations in cast Mg-Al alloys. In: D. M. Stefanescu, Warren J, Jolly M, Krane M (eds) Model. Cast. Welding, Adv. Solidif. Process. X. The Minerals, Metals, and Materials Society, pp 165–172.

31. Forsmark JH, Boileau J, Houston D, Cooper R (2012) A microstructural and mechanical property study of an AM50 HPDC magnesium alloy. Int J Met 6:15–26

32. Lee CD, Shin K (2007) Effect of microporosity on the tensile properties of AZ91 magnesium alloy. Acta Mater 55:4293–4303, doi:10.1016/j.actamat.2007.03.026

33. Lee CD (2007) Tensile properties of high-pressure die-cast AM60 and AZ91 magnesium alloys on microporosity variation. J Mater Sci 42:10032–10039, doi:10.1007/s10853-007-2003-1

34. Choi KS, Li D, Sun X, Li M, Allison J (2013) Effects of pore distributions on ductility of thin-walled high pressure die-cast magnesium. SAE Tech Pap, Ser. doi:10.4271/2013-01-0644

35. Unigovski Ya, Tuman E, Eliezer A, Riber L, Koren Z (2000) Correlation of mechanical properties of die cast magnesium alloys with processing conditions. In: Aghion E, Eliezer D (eds) Magnes. 2000 Proc. 2nd Isr. Int. Conf. Magnes. Sci. Technol. Magnesium Research Institute, Dead Sea, Israel, pp 105–111

36. Gjestland H, Sannes S, Svalestuen J, Westengen H (2005) Optimizing the magnesium die casting process to achieve reliability in automotive applications. SAE Tech. Pap., Ser. SAE 2005–0333

37. Gertsberg G, Nagar N, Lautzhe M Bronfin B, Moscovitch N, Schumann S (2005) Effect of HPDC parameters on the performance of creep resistant alloys MRI153M and MRI230D. SAE Tech. Publ., SAE 2005–0

38. Greve L (2004) Development of a PAM-CRASH material model for die casting alloys. Magnes. Proc. 6th Int. Conf

39. Sannes S, Gjestland H, Westengen H, (2005) The use of quality mapping to predict performance of thin-walled magnesium die casting. SAE Tech Pap. Ser. 2005-01-0332.

40. Dorum C, Hopperstad OS, Langseth M, Lademo OG (2005) Numerical modeling of the structural behavior of thin-walled cast magnesium components using a through-process approach. SAE Tech. Pap., Ser. 2005-01-07

41. Dorum C, Hopperstad OS, Landemo O-G, Langseth M (2005) Numerical modelling of the structural behavior of thin-walled cast magnesium components. Int J Solids Struct 42:2129–2144

42. Dorum C, Hopperstad OS, Lademo O-G, Langseth M (2006) An experimental study on the energy absorption capacity of thin-walled castings. Int J Impact Eng 32:702–724

43. Dorum C, Hopperstad OS, Lademo O-G, Langseth M (2007) Energy absorption capacity for thin-walled AM60 castings using the shear-bolt principle. Comput Struct 85:89–101

44. Dorum C, Hopperstad OS, Berstad T, Dispinar D (2009) Numerical modelling of magnesium die-castings using stochastic fracture parameters. Eng Fract Mech 76:2232–2248

45. Weiss U, Bach A (2011) Magnesium HPDC Crash CAE. La Met Ital 103(11–12):1–9

46. Li N, Osborne R, Cox B, Penrod D (2005) Magnesium Engine Cradle - The USCAR Structural Cast Magnesium Development Project. SAE Trans., 2005-01-03

47. Balzer JS, Dellock PK, Maj MH, Cole GS, Reed D, Davis T, Lawson T, Simonds G (2003) Structural magnesium front end support. SAE Tech. Pap., Ser. 2003-01-0186

48. Blanchard PJ, Bretz GT, Subramanian S, deVries JE, Syvret A, MacDonald A, Jolley P (2005) The application of magnesium die casting to vehicle closures. SAE Tech. Pap., Ser. 2005-01-0338

49. Gibbs S (2010) Magnesium structural part parts with myth. Met Cast Des Purch 12:29–33

50. Herman EA (1996) Gating and Die Casting Dies, E-514th ed. NADCA Publications, Rosemont, IL, USA

51. North American Die Casting Association (1998) Magnesium Die Castings Handbook, #201 ed. NADCA Publications, Rosemont, IL, USA

52. Box GEP, Hunter WG, Hunter JS (1978) Statistics for Experimenters. John Wiley & Sons, New York

53. ASTM E8/E8M - 13a (2015) Standard test methods for tension testing of metallic materials 1., pp 1–28, doi: 10.1520/E0008

54. Khan S (2012) Box-and-whisker plots. In: Khan Acad., https://www.khanacademy.org/math/probability/descriptive-statistics/box-and-whisker-plots/v/reading-box-and-whisker-plots

55. Minitab 15 statistical software statguide (2012), Minitab Inc., State College, PA USA

56. MAGMASOFT version 4.4 user manual, MAGMA Foundry Technologies Inc., Aachen, Germany.

57. Panchal JH, Kalidindi SR, McDowell DL (2013) Key computational modeling issues in Integrated Computational Materials Engineering. Comput Des 45:4–25, doi:10.1016/j.cad.2012.06.006

58. Alain R, Lawson T, Katool P, Wang G, Jekl J, Berkmortel R, Miller L, Svalestuen J, Westengen H (2004) Robustness of large thin wall magnesium die castings for crash applications. Reprinted from: Magnesium for automotive components. SAE Tech Pap., Ser 2004–01-0131

59. Yang KV, Cáceres CH, Easton MA (2013) A microplasticity-based definition of the skin in HPDC Mg–Al alloys. Mater Sci Eng A 580:355–361, doi:10.1016/j.msea.2013.05.018

60. Yang KV, Easton MA, Cáceres CH (2013) The development of the skin in HPDC Mg–Al alloys. Mater Sci Eng A 580:191–195, doi:10.1016/j.msea.2013.05.017

61. Yang KV, Cáceres CH, Easton MA (2014) Strengthening micromechanisms in cold-chamber high-pressure die-cast Mg-Al alloys. Metall Mater Trans A 45:4117–4128, doi:10.1007/s11661-014-2326-x

6

Numerical model of fiber wetting with finite resin volume

Michael Yeager and Suresh G Advani[*]

* Correspondence: advani@udel.edu
Department of Mechanical
Engineering and Center for
Composite Materials, University of
Delaware, Newark, DE 19716, USA

Abstract

The partial wetting of cylindrical surfaces is encountered in many industrial applications such as composites manufacturing, MEMS, hair care products, and textile engineering. Understanding the impact of key parameters such as resin and fiber surface interaction properties and the geometric arrangement of the fibers on wetting would lead to tailoring a desired interface between the resin and the fiber surface. A three-dimensional model of resin wetting a single fiber is presented. This model is then extended to study a finite volume of resin wetting fibers in square and triangular packing arrangements. The impact of changing wetting properties and fiber volume fraction is examined for each packing arrangement.

Keywords: Wetting; Contact angle; Computational multiphase flow; Composites

Background

The partial wetting of cylindrical surfaces by a finite volume of resin is an important phenomenon in many industrial applications such as composites manufacturing, MEMS, and textile engineering. A constitutive equation governing the partial wetting of a finite volume of liquid on a flat plate has been formulated and reported [1]. The equilibrium shape of resin on single fibers has also been studied in depth [2-5]. Carroll [2] was the first to develop an analytical solution for the equilibrium shape of a resin drop on a single fiber. A drop at rest on a fiber will either conform to a "barrel" geometry, where the drop wraps around the fiber, or a "clamshell" geometry in which the fiber rests on the fiber's surface without wrapping around it. A phase diagram predicting which of these configurations a particular drop will adopt has been constructed [5].

An analytical solution for the equilibrium shape of a liquid drop on a fiber surface was first derived by Carroll [2]. Wu and Dzenis later developed an analytical solution to this problem using an energy approach [4]. Both of these solutions assume an axisymmetric shape. The equations necessary to determine the maximum height of the drop on the fiber and the length of contact between the fiber and resin are given in [2,4] and reproduced below:

$$y^2 = y_0{}^2\left(1 - k^2 sin^2\phi\right) \tag{1}$$

$$x = \pm[\lambda r F(k,\phi) + y_0 E(k,\phi)] \tag{2}$$

Here, x is the location on the axis of the fiber measured from the center of the drop, and y is the height of the drop, measured from the axis of the fiber. y_0 is the maximum height of the drop measured from the axis of the fiber, and r is the fiber radius. $E(k,\phi)$ and $F(k,\phi)$ are Legendre's elliptical functions of the second and first kind, respectively. Here λ and k are defined as follows:

$$\lambda = \frac{y_0\cos\theta - r}{y_0 - r\cos\theta} \tag{3}$$

$$k^2 = 1 - \lambda^2\left(\frac{r}{y_0}\right)^2 \tag{4}$$

Here, θ is the static contact angle between the fiber and resin.

The final wetted length, L, which is also defined in Figure 1, can be calculated using the known volume V once y_0 is solved for with the above equations and:

$$L = 2[\lambda r F(k,\phi_0) + y_0 E(k,\phi_0)] \tag{5}$$

$$V = 2\pi\int_r^{y_0} \frac{y^2(y^2 + \lambda r y_0)}{\sqrt{(y_0{}^2 - y^2)(y^2 - \lambda^2 r^2)}} dy - \pi r^2 L \tag{6}$$

ϕ_0 is found through setting $y = 0$ in Equation 1.

There have also been other investigations with resin spreading within multiple fibers, for example final resin configuration between two parallel fibers has been studied with relation to static contact angle, filament spacing, resin volume, and fiber diameter [6]. The axial wetting of a single fiber from a reservoir of resin has been experimentally examined and constitutive equations have been developed to describe this phenomenon [7]. The dynamics of a finite volume of resin spreading on a single fiber has yet to be explored and is the subject of this paper. Trends seen with the dynamics of resin wetting a single fiber hold true for systems with multiple fibers.

A numerical model is presented using the level set method to study the movement and spreading of a finite volume of resin on any planar or curvilinear surface. The method and accuracy is verified by comparing the model results with experiments conducted of a drop spreading on a flat plate. The method is then used to describe the wetting dynamics of a finite drop of resin on a single fiber. The model is further extended to investigate the flow of finite volume resin within multifiber unit cells representing square and hexagonal fiber packing arrangements, which are commonly used

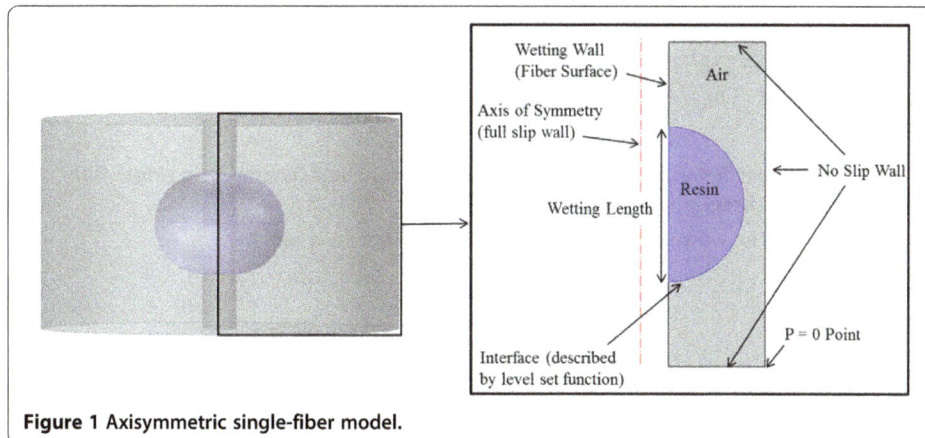

Figure 1 Axisymmetric single-fiber model.

in composites. The use of a finite volume of resin is necessary because there are situations where it is desirable to strategically create microvoids to increase the composite's energy-absorption capabilities. Each tow would be a porous structure comprised of a series of microstructures represented by the unit cells. The model studies the impact of key composite-processing properties on the fiber-matrix interfacial area. Through examination of the interfacial area instead of the contact length, one is better able to understand the impact of manipulating processing parameters on the resulting composite properties. This investigation should prove useful in tailoring the interface properties between fibers and resin as a function of the resin and fiber surface properties and the fiber arrangement.

Methods

Model setup

The numerical models were developed using the COMSOL Multiphysics and the Microfluidics Module to investigate the dynamics of wetting over a single fiber and within unit cells of multiple fibers with a finite volume of resin and are presented below. A model was also constructed of a drop spreading on a flat plate with the goal of experimentally validating the solution method.

Axisymmetric single-fiber model

Figure 1 shows the axisymmetric single-fiber model. In this model, a spherical drop of resin is initially enveloping the fiber. This simplifies the resin movement to be along the fiber surface in the axial direction. An axis of symmetry is utilized to increase computational efficiency. The axis of symmetry is the center of the fiber with a full-slip condition (symmetry condition), which sets the derivative of the tangential velocity equal to zero along the axis. The no-slip boundary condition is applied along the walls shown in Figure 1 where the velocity is set to zero. The pressure is set equal to zero, the reference pressure, at a single point to ensure that the pressure solution is unique [8]. This is needed because the Navier–Stokes equations only solve for the gradient of pressure. The wetting wall is the surface of the fiber along which the resin moves and employs a slip length and the final static contact angle to drive the wetting and spread the resin, both of which will be discussed in a later section. The fiber diameter was 9 μm. These baseline values, shown in Table 1, were selected based on resin and fiber systems used in composites processing.

Three-dimensional single-fiber model

The three-dimensional representation of a drop of resin spreading on a single fiber is depicted in Figure 2. The resin drop diameter to fiber radius ratio in this model is

Table 1 Baseline properties used for parametric studies for the axisymmetric single-fiber model as well as the three- and four-fiber unit cells

Baseline properties used for axisymmetric single-fiber model and fiber unit cells	
Resin density	1.17 g/cm^3
Resin viscosity	9.5 Pa·s
Static contact angle	30°
Resin surface tension	0.07 J/m^2
Slip length (β)	0.1 μm

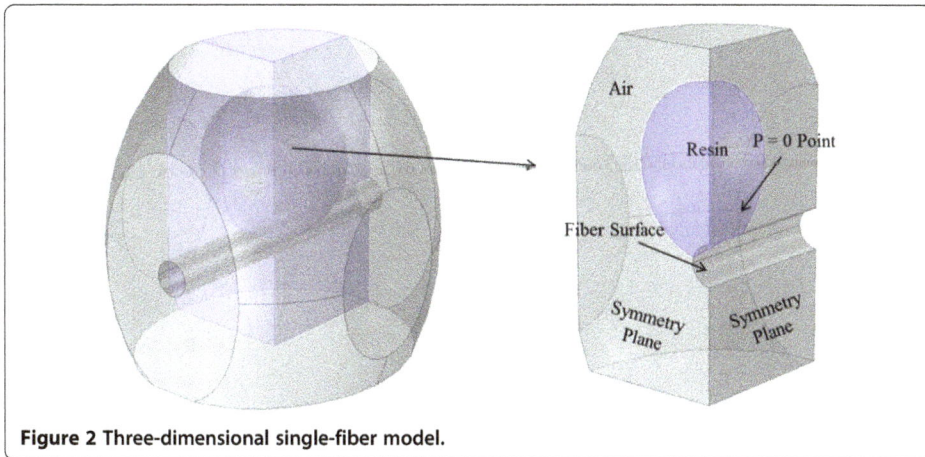

Figure 2 Three-dimensional single-fiber model.

intentionally large. This is because we desire the equilibrium position of the resin to be in the "barrel" shape. A fiber radius of 3 μm, resin volume of 2,280 μm^3, and contact angle of 15° were selected to ensure that the final shape is a "barrel" guided by the studies performed by Eral et al. [5]. The other important properties are the same as the baseline values described in Table 1.

Unit cell with square and triangular packing arrangements

The square and triangular fiber packing-arrangement unit cells are shown in Figure 3. Flow in each unit cell is simplified through the use of symmetry planes to increase computational efficiency. The four-fiber unit cell has three planes of symmetry due to the assumption that gravity is negligible, which will be discussed later. The fiber surfaces are wetting walls, and walls other than fiber surfaces and symmetry planes are no-slip walls. Both models also include a point where the pressure is set equal to zero. The pressure is set equal to zero at a reference location, to ensure that the pressure solution is unique. The fiber radius was 4 μm. The baseline values for the resin and interaction parameters can be found in Table 1.

Resin spreading on a flat plate

A numerical model of the drop of glycerin, shown in Figure 4, was developed to compare the numerical solution and experimental results. The glycerin properties were found using traditional characterization techniques, described and reported in the "Experimental setup

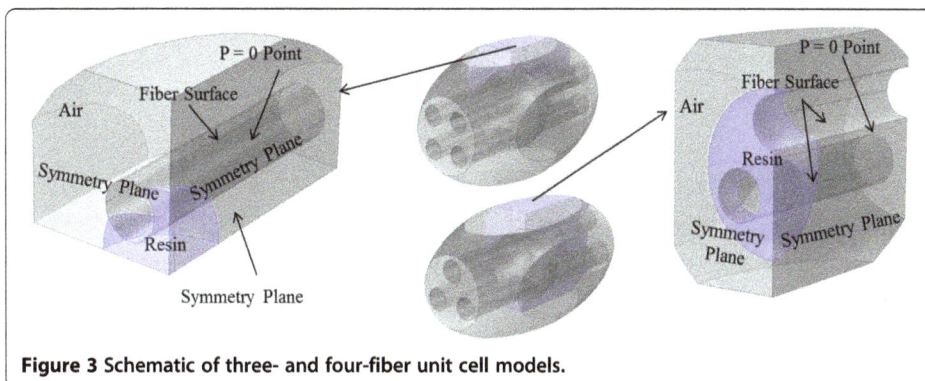

Figure 3 Schematic of three- and four-fiber unit cell models.

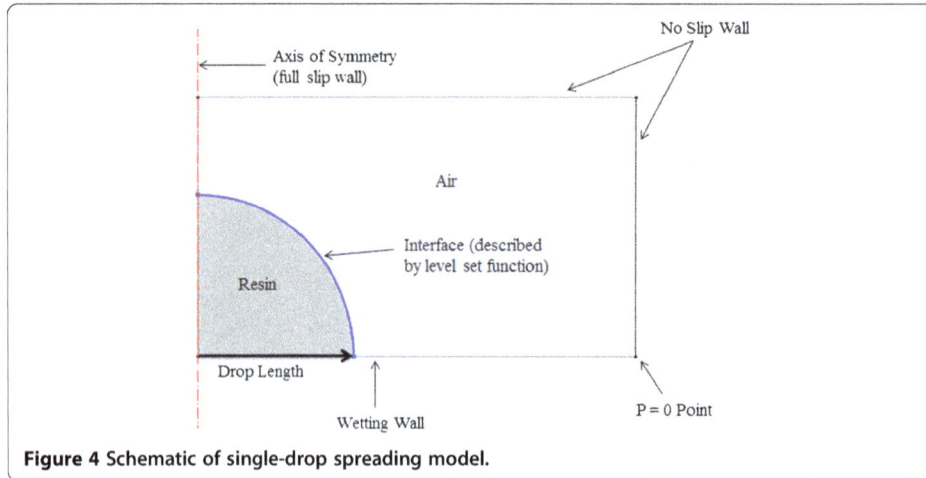

Figure 4 Schematic of single-drop spreading model.

and procedure" section. The properties for the air phase were taken from the COMSOL material library. The two-dimensional model takes advantage of the axisymmetric property of the process being modeled.

Assumptions

The Reynolds number in this problem is on the order of 10^{-8}; thus, it can be assumed that the inertial forces are negligible relative to the viscous forces and the Navier–Stokes equations can be reduced to the Stokes flow equations. The ratio of gravitational to capillary forces, represented by the bond number, is on the order of 10^{-6}, making it acceptable to neglect gravity. For the axisymmetric model, due to the geometry being symmetric about the fiber axis and our assumption of no gravity, the flow is considered axisymmetric about the axis of the fiber. It is assumed that the resin used does not cure during the wetting process, allowing us to maintain a constant viscosity value during the flow. It is also assumed that the fibers are rigid and do not move as the resin flows.

Governing equations

The governing equations (Equations 7 and 8) in the model are the Stokes and continuity equations. The interface between the two fluids is tracked using the level set method [9]. The level set method creates an interface with a finite thickness, described by the level set variable (ϕ), which continuously changes from 0 to 1 across the interface using a smeared out Heaviside function [10]. These equations (Equation 9), modified to account for the stated assumptions, are given by [11]:

$$\rho \cdot \boldsymbol{u}_t = -\nabla \cdot \boldsymbol{p} + \mu \nabla^2 \boldsymbol{u} + \boldsymbol{F}_{st} \tag{7}$$

$$\nabla \cdot \boldsymbol{u} = 0 \tag{8}$$

$$\phi_t + \boldsymbol{u} \cdot \nabla \phi = \sigma \nabla \cdot \left(\varepsilon \nabla \phi - \phi (1-\phi) \frac{\nabla \phi}{|\nabla \phi|} \right) \tag{9}$$

Where $\boldsymbol{u}$ is the velocity vector, the subscript denotes the partial derivative with respect to that variable, μ is the viscosity, $\boldsymbol{p}$ is the pressure, $\boldsymbol{F}_{st}$ is the force due to surface tension, σ is the reinitialization parameter for the interface, ε is the interface thickness, and ϕ is the level set variable. To minimize computational cost, the interface thickness is set to one half of the largest element length [11,12]. The interfacial tension term is implemented using the continuous surface force formulation, given by:

$$F_{st} = \nabla \cdot \left(\gamma \left(I - nn^T \right) \right) \delta \tag{10}$$

$$\delta = 6 |\nabla \phi| |\phi(1-\phi)| \tag{11}$$

Here, γ is the surface tension of the resin-air interface, and δ is a Dirac delta function.

The density and viscosity within the interface between the resin and air are found using rule of mixtures [11]:

$$\rho = \rho_{Resin} + \left(\rho_{Air} - \rho_{Resin} \right) \phi \tag{12}$$

$$\mu = \mu_{Resin} + \left(\mu_{Air} - \mu_{Resin} \right) \phi \tag{13}$$

Fiber and resin parameters

The properties of the resin and the fiber-resin interactions play an important role in the wetting of the fibers by the resin. The viscosity of the resin has a large impact on the rate of wetting, but not a significant effect on the final shape of the drop. The bond number is the ratio of surface forces to body forces, providing a good indication if the resin flow is driven by surface forces or gravity. This study focuses on flows with low-bond numbers. The contact angle between the fiber and resin, largely impacted by the surface tension of the resin, represents the principle force driving wetting at the microscale. The fiber diameter and resin droplet size will be important geometrical parameters when investigating drops spreading on the fiber surfaces. When the model is extended to include multiple fibers, the fiber spacing and packing arrangement will influence the wetting dynamics.

Static contact angle between fiber and resin

Wetting describes the spreading of a liquid on a solid substrate [13]. The wettability of a substrate by a liquid wetting rate and region is quantified by the static contact angle, a force balance at the line of contact between the fiber surface (solid (s)), resin (liquid (l)), and air (vapor (v)) and is given by Young's equation [13]:

$$\cos\theta = \frac{\gamma_{sv} - \gamma_{sl}}{\gamma_{lv}} \tag{14}$$

In Young's equation, γ_{ij} represents the surface energy at the i-j interface. As shown in Equation 14, the final static contact angle takes into account both the resin surface tension and the difference in interfacial energies of the solid-vapor and solid–liquid interfaces. The solid-vapor and solid–liquid surface energies can be manipulated by modifying the fiber sizing, which is a coating that is applied to the fiber surface. The final static contact angle of the resin on the fiber surface has been shown to have a direct relationship with the interfacial shear strength of the resulting composite [14].

Resin viscosity

The viscosity of the resin does not affect the final position of the resin on the fibers since it is assumed to be constant. As the Stokes solution is linear, the time it takes to wet the fiber surface will be directly proportional to the viscosity of the resin.

Slip length

There are stress and velocity singularities at the three-phase contact line when solving the Stokes equations with a no-slip condition at the solid surface [15]. A way to handle this boundary condition is to move the "no-slip" condition to a plane located a distance β (slip length) below the solid surface and assume simple shear flow in the region between the wall and the no-slip plane [16]. The frictional force at the wall is scaled with the slip length [11]. Not unlike viscosity, changing the slip length will influence the wetting rate, but not the final distribution and configuration of the resin on the fiber surface. The slip length is a parameter that models the interactions at the fiber-resin interface.

Fiber volume fraction and packing

The fiber volume fraction is an important property of composites when considering their strength and stiffness. The fiber volume fraction will be controlled in this study through manipulation of the distance between fibers, measured from axis to axis. Two common packing arrangements for fibers are square and hexagonal packing. The hexagonally packed fibers are modeled with a unit cell in which lines connecting the center of each fiber would form an equilateral triangle. The relationship between the fiber volume fraction, (v_f), fiber radius (r), and distance between fiber axes, (d), will be:

$$v_f = \pi \left(\frac{r}{d}\right)^2 \tag{15}$$

for fibers in a square packing arrangement. For fibers in a triangular packing arrangement, the relationship will be:

$$v_f = \frac{2\sqrt{3}}{3} \pi \left(\frac{r}{d}\right)^2 \tag{16}$$

Experimental setup and procedure

The experimental setup, shown in Figure 5, includes a substrate on which a drop of liquid can be deposited by using a thin wire and a camera to capture time-stamped images of the process.

This experiment was performed by depositing a glycerin drop on a flat glass substrate. Glycerin was used as the test liquid because it has similar properties to the epoxy ultimately being used in the drop-spreading experiment. The surface tension of the glycerin was measured with a dynamic contact analyzer to be 0.07 N/m. A

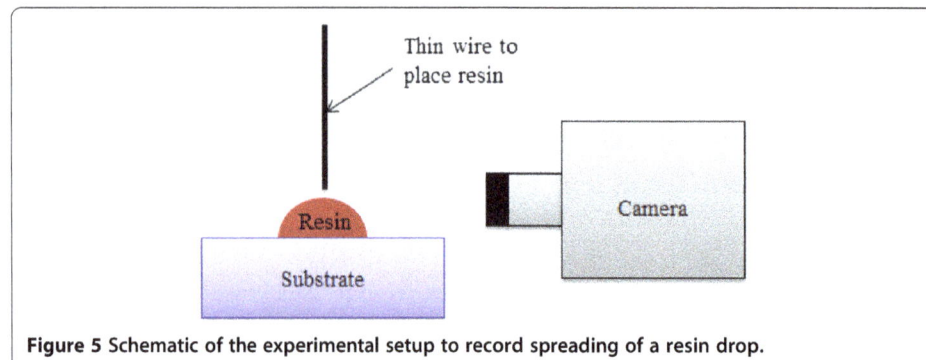

Figure 5 Schematic of the experimental setup to record spreading of a resin drop.

Brookfield DV-E viscometer measured the viscosity of the glycerin to be 0.674 Pa·s. The density of the glycerin, measured using a precision scale and flask, was 1.236 g/cm^3. Sample images of the drop spreading are shown in Figure 6. The static contact angle between the glass and glycerin, measured using image analysis software on the drop in equilibrium, is 28.5°.

Results and discussion

First, experimental and analytical validation of the model used will be provided in the next section before parametric studies are conducted to investigate the dynamics of resin spreading.

Validation of the numerical model

The physics involved in the preceding models is multiphase fluid flow with a high surface to gravitational force ratio. A model of a drop spreading on a flat plate was developed to experimentally verify that the governing equations could predict an acceptable numerical solution to a multiphase wetting dynamics dominated by surface forces.

Comparison between experimental and numerical results

Two experimental trials were conducted of a drop spreading on a flat plate. The first experiment was used to determine the value of β which defines the resin fiber surface characteristics. A β value of zero would correspond to the case where the liquid will not wet the substrate, and an infinite value would describe the scenario where the liquid would reach its final configuration instantaneously. The β value in real systems will fall between the preceding extreme cases and can be determined experimentally by comparing the numerical and experimental solutions using a range of β values. As β is increased or decreased, the wetting rate in the numerical solution will become higher or lower. The β value that describes the liquid-substrate system is found by adjusting the β value until the dimensionless length, defined as the drop length at time t divided by the final length of the drop, matches the experimental results. This experiment used

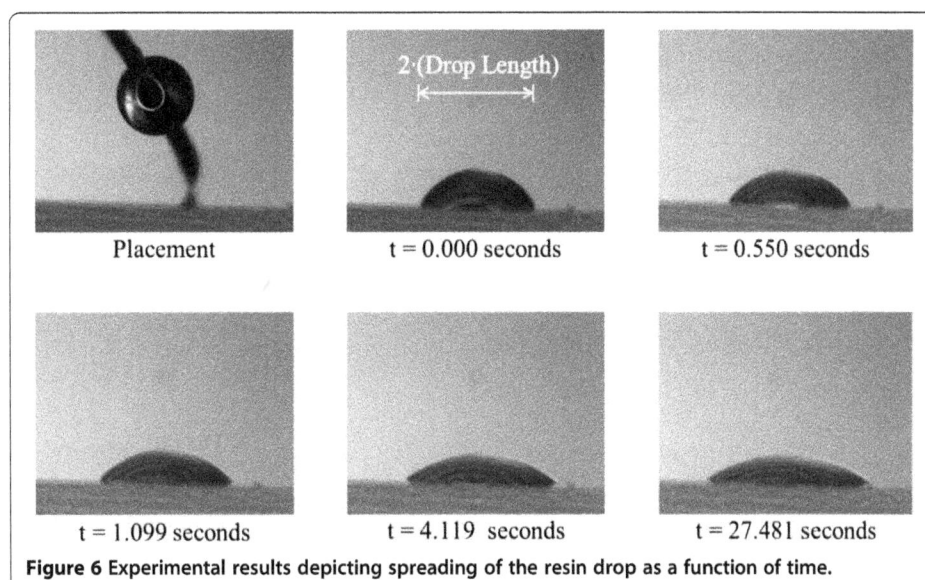

Placement   t = 0.000 seconds   t = 0.550 seconds

t = 1.099 seconds   t = 4.119 seconds   t = 27.481 seconds

Figure 6 Experimental results depicting spreading of the resin drop as a function of time.

the resin volume of 0.057 mm^3. The β value for slip length from this case was determined to be 0.25 µm. The experimental and COMSOL results for this β value are shown in Figure 7 along with an inset that describes distance β (slip length) below the solid surface and assumes simple shear flow in the region between the wall and the no slip plane.

Having determined the β value, the next experiment was conducted with a drop volume of 0.140 mm^3. The dimensionless length of the COMSOL simulation is compared to the experimental dimensionless length in Figure 7 using the characterized β value. Comparing the results verify the numerical model used to describe the dynamics of resin spreading on a surface for a large surface force to body force ratio.

Mesh-refinement study

A mesh-refinement study was performed to ensure that the numerical results converged as the number of elements in the mesh was increased. Wetting length, as shown in Figure 1, will be used as the characteristic output parameter studied for the axisymmetric model of resin wetting a single fiber. The area of the fiber-resin interface will be used as the characteristic output parameter for the three-dimensional model of resin spreading on a single fiber. Comparing the four solutions for each, depicted in Figure 8, confirms that the numerical output converges and the lowest mesh density used provides an acceptable result.

Comparison of final drop shape with an analytical solution

The equilibrium solution for the axisymmetric model of resin spreading on a fiber was compared to the resin configuration predicted by Carroll [2]. In the numerical solution, resin volume, fiber diameter, and final contact angle are all known. These values were substituted into Equations 3, 4, and 5 and then substituted into Equation 6 to create an equation with one unknown, allowing one to solve for y_0. Once y_0 is known, it can be substituted back into Equations 1 to 4 to develop a parametric equation for x and y. φ was varied for values corresponding to $y > r$. The resin-air interface shape at the yz-plane is solved by using this method and is compared to the numerical solution in

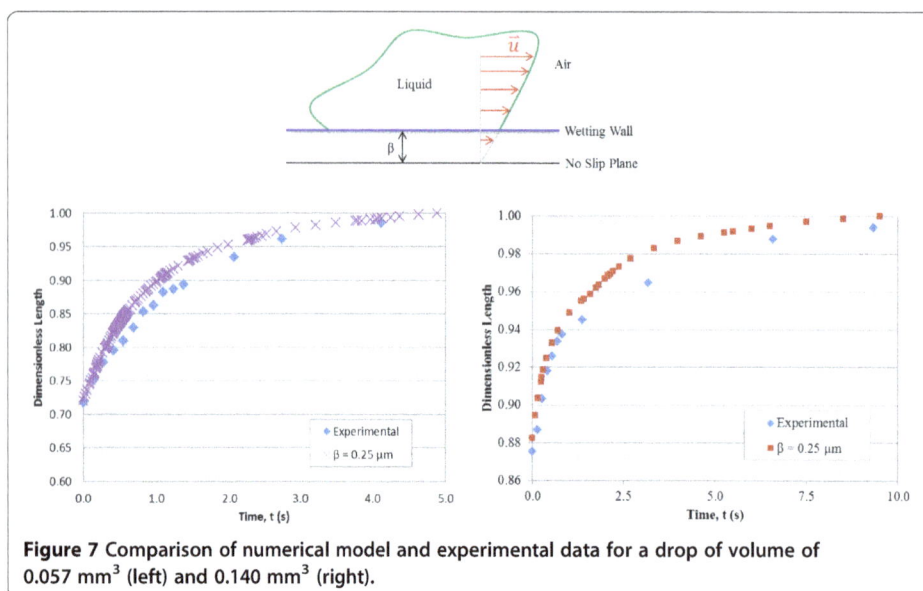

Figure 7 Comparison of numerical model and experimental data for a drop of volume of 0.057 mm^3 (left) and 0.140 mm^3 (right).

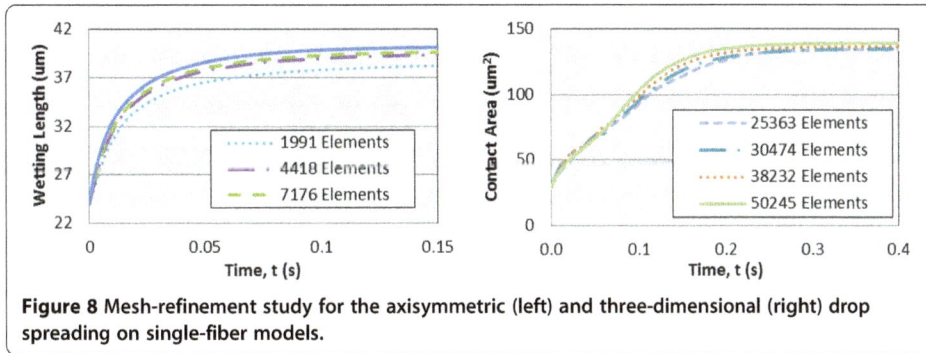

Figure 8 Mesh-refinement study for the axisymmetric (left) and three-dimensional (right) drop spreading on single-fiber models.

Figure 9. The closeness of the two solutions provides further validation of the numerical model.

Parametric study of axisymmetric model

The wetting physics in the axisymmetric model was influenced by the static contact angle, slip length, fiber and resin geometry, and viscosity.

Static contact angle between fiber and resin

The evolution of non-dimensional wetting length over time is shown in Figure 10 for a range of static contact angles. The baseline values are used for all other properties. Here, the non-dimensional wetting length is normalized by the initial wetting length. Fiber-resin interface surface properties with high-contact angles reach their equilibrium position at a lower time because the resin does not travel very far. As the wetting properties are increased, evidenced by a lower contact angle, the amount of the fiber surface covered by the resin increases. The trends found in the contact angle study can be translated to changes in the surface energy of the solid-resin, solid-air, or resin-air through Equation 14.

Resin volume

The volume of the resin impacted both the final wetted length and the rate of wetting as can be seen from Figure 11. The fiber and resin properties are equal to the baseline values. The initial wetting rate was similar for the different resin volumes, but it can be

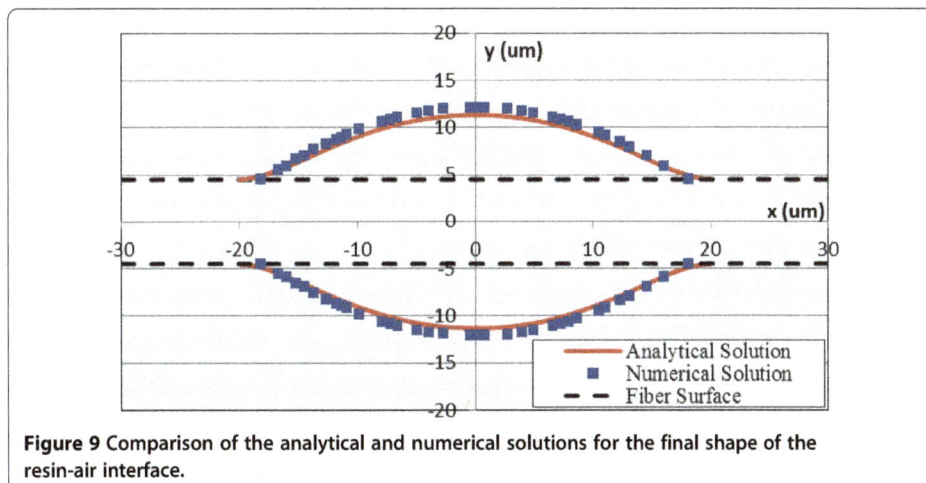

Figure 9 Comparison of the analytical and numerical solutions for the final shape of the resin-air interface.

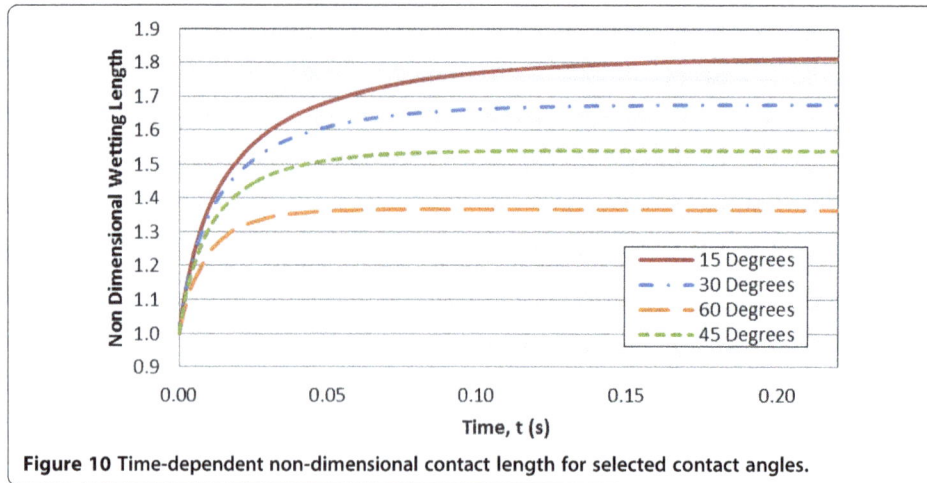

Figure 10 Time-dependent non-dimensional contact length for selected contact angles.

seen that smaller resin drops reached equilibrium faster. This is a result of the resin having to travel a shorter distance and the surface force to resin volume increasing.

Fiber radius

The non-dimensional wetting length, as a function of time for various fiber radii is depicted in Figure 12. The baseline values are used for the fiber and resin properties. All of the initial wetting lengths were slightly different due to the radius of the resin drop changing slightly to keep the resin volume constant for a varying fiber radius. At low times, the resin moves at a similar rate for all trials. With increasing time, the resin reaches equilibrium on the smaller fibers first because it has to travel less and the capillary forces are stronger. The final wetting length increases as the fiber radius was increased due to the resin trying to minimize its surface area.

Slip length

With the exception of the slip length, all properties were equal to their baseline values for this study. The slip length, which characterizes the fiber-resin interface property, did not impact the final wetting length of the resin on the fiber. It did impact the wetting rate as shown in Figure 13. As one would expect, the system reached equilibrium at a faster rate when the slip length was increased due to the increase in slip velocity at

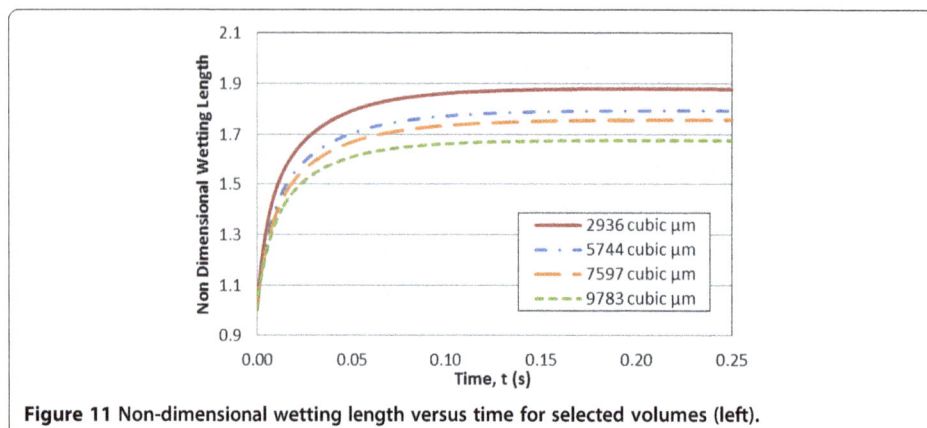

Figure 11 Non-dimensional wetting length versus time for selected volumes (left).

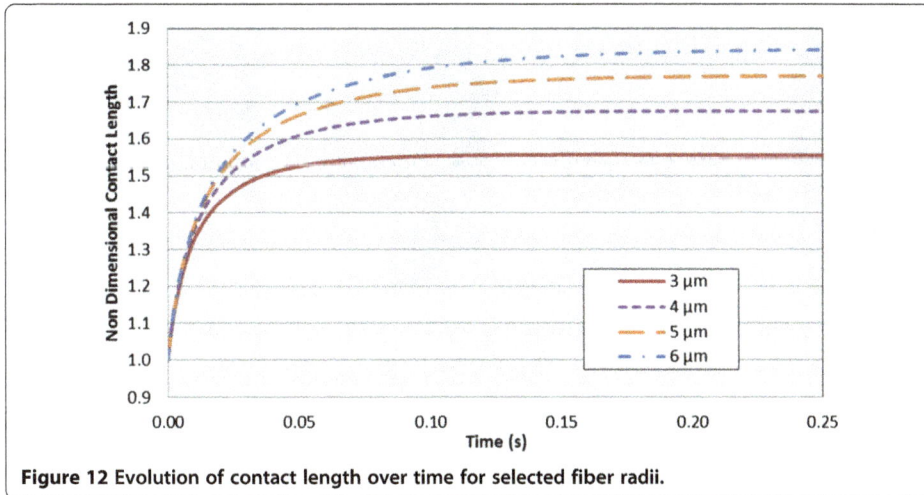

Figure 12 Evolution of contact length over time for selected fiber radii.

the fiber surface. The slip length found in the experimental validation of the physics would fall close to the middle of the values studied.

Three-dimensional single-fiber model

In the three-dimensional model, resin wets the top of the fiber faster than it wets around the circumference of the fiber, depicted in Figure 14. The curvature of the surface slows the rate of wetting on the outside of a concave surface because more resin surface area is created per unit length traveled. The wetted length on the top of the fiber decreases slightly after the resin begins to spread along the bottom of the fiber, the time of which is indicated by the plateau of the circumferential spreading curve.

Square and hexagonal packing fiber unit cells

The spreading of a finite volume of resin within three- (hexagonal packing) and four-fiber unit cells (square packing) with a fiber volume fraction of 30%, static contact angle of 30°, and fiber radius of 4 μm is shown in Figure 15. The interface is described by the level set function, described by the scale bar where a value of one represents purely

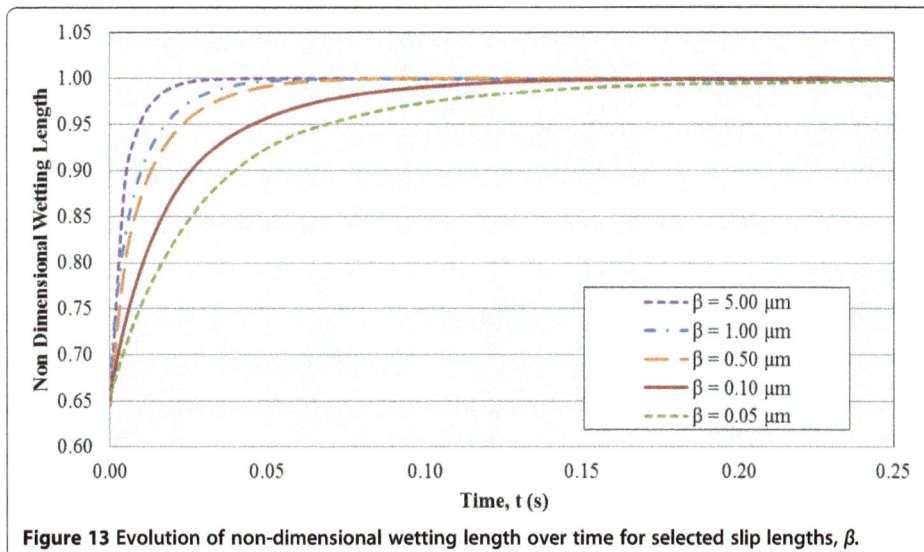

Figure 13 Evolution of non-dimensional wetting length over time for selected slip lengths, β.

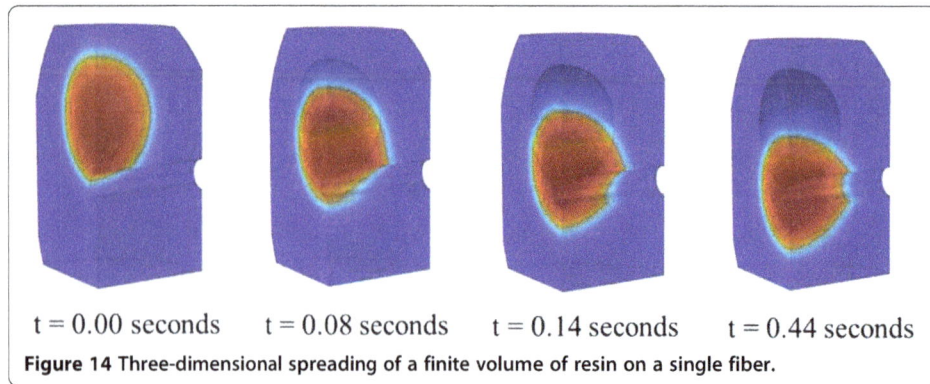

| t = 0.00 seconds | t = 0.08 seconds | t = 0.14 seconds | t = 0.44 seconds |

Figure 14 Three-dimensional spreading of a finite volume of resin on a single fiber.

resin. The resin spreads axially and circumferentially along the fiber. The four-fiber unit cell spreads and reaches equilibrium at a much faster rate (0.38 s) than the three-fiber unit cell (1.47 s). This is a result of there being an increased fiber-resin contact area in the four-fiber unit cell.

Fiber volume fraction

For this study, the base parameters were used and the spacing between fibers was varied. The effect of changing fiber volume on the fiber-resin contact area increased as the fiber volume fraction was decreased, as shown in Figure 16. As the volume fraction is changed for the three- and four-fiber unit cells, the spacing between fibers changes at different rates, described in Equations 15 and 16. The change in fiber spacing for the two types of unit cells causes the capillary pressure to change, which can be modeled using the Young-Laplace pressure equation [13]. At larger fiber volume fractions, the capillary pressure increases at a much faster rate, leading to an increased wetted area. At lower fiber volume fractions, the rate of change of capillary pressure is not as high, resulting in a similar increase in wetted area for both types of unit cells. The triangular packing arrangement had a larger fiber-resin contact area per fiber; thus, it would be the preferred packing arrangement if one were to create a network of resin microdrops within a fiber tow with the goal of maximizing fiber-resin contact area.

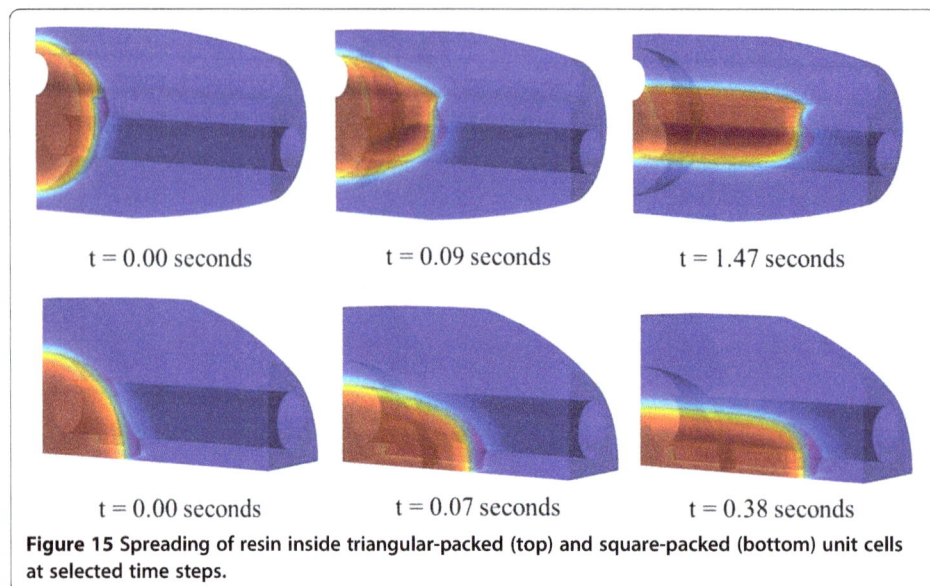

| t = 0.00 seconds | t = 0.09 seconds | t = 1.47 seconds |

| t = 0.00 seconds | t = 0.07 seconds | t = 0.38 seconds |

Figure 15 Spreading of resin inside triangular-packed (top) and square-packed (bottom) unit cells at selected time steps.

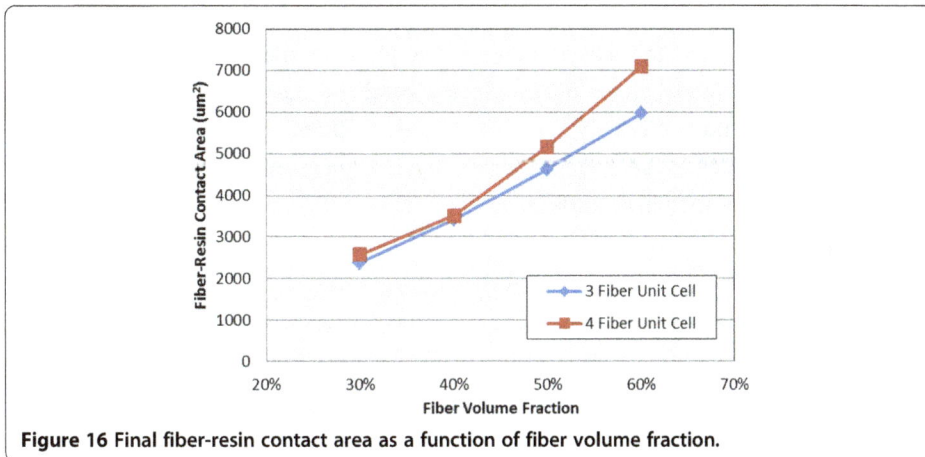

Figure 16 Final fiber-resin contact area as a function of fiber volume fraction.

Static contact angle

The static contact angle had a large effect on the final fiber-resin contact area. The contact area was linear with the cosine of the static contact angle, shown in Figure 17. A linear increase with $\cos(\theta)$ makes sense because the final results shown are with a fiber volume fraction of 30%. It is clear that for a given fiber volume fraction, the square packing arrangement is preferred for increasing fiber-resin contact area. For this particular combination of resin volume, fiber volume fraction, and fiber size, the ratio of fiber-resin contact area for the triangular and square packing arrangements ranged between 1.07 and 1.11 for the given static contact angles. The static contact angle did not have as significant of an impact on the resin spreading as the packing arrangement did. Resin reached its equilibrium position inside square-packed fibers in about 0.1 to 0.2 s as compared to the 1.1 to 1.5 s with these initial conditions. This indicates that when the same volume of resin wets fibers in a square packing arrangement, the resulting composite will have a higher fiber-resin contact area and faster processing time when compared to a triangular packing arrangement.

Limitations of the model

A limitation on this model is imposed by the assumption of a microscopic length scale. This is because when the diameter of the fiber or the volume of the resin is increased

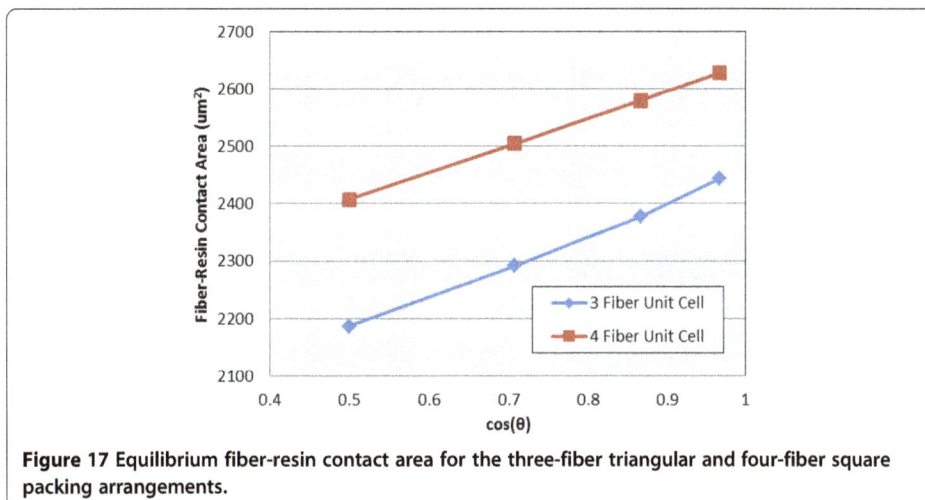

Figure 17 Equilibrium fiber-resin contact area for the three-fiber triangular and four-fiber square packing arrangements.

by a large amount, the inertial and gravitational forces are no longer considered negligible. This would invalidate the axisymmetric assumption in the axisymmetric fiber model. In the four-fiber model, one would no longer be able to use the symmetry plane orthogonal to the direction of gravity. The trends seen in these models may not hold for models with extremely large contact angles because they only examine the case where the liquid will wet the fiber's surface.

Conclusions

Numerical models describing the partial wetting of a finite volume of resin on a single fiber and in triangular- and square-packed unit cells was presented and validated. The static contact angle affected both the rate of axial spreading as well as the final fiber-resin contact area. The volume of resin impacted the final fiber-resin contact area and the wetting rate because larger volumes of resin travel farther. Both the wetting length and final fiber-resin contact area increased with increasing fiber diameter. This claim is only for the case when the resin is in a barrel shape around the fiber as the clamshell shape was not investigated. The slip length had a defined effect on the rate of wetting, but did not impact the final fiber-resin contact area. This indicates that the slip length will not impact the composite properties. Fiber volume fraction had a significant impact on fiber-resin contact area, being more influential at higher fiber volume fractions. The final fiber-resin contact area was larger for square-packed unit cells than triangular-packed unit cells. In unit cells with triangular or square packing arrangements, the static contact angle had a large impact on the final fiber-resin contact area. The effect of static contact angle on wetting rate was small compared to the impact of packing arrangement on wetting rate. These models can be used to predict the impact of manipulating fiber and resin surface properties, interaction, and geometry on the wetting of fibers by a finite volume of resin. By predicting the influence of processing parameters on fiber wetting, one can correlate the resulting microstructure in the unit cell with process and material parameters. The properties of the fibers and matrix can then be used to determine the mechanical properties of a unit cell with the predicted microstructure. The mechanical properties of the unit cells can be used to determine the composite properties.

Competing interests
The authors declare that they have no competing interests.

Authors' contributions
MY did the numerical simulations, analysis, and prepared the article. SGA discussed the numerical results; provided guidance, feedback, and expertise; and revised the article for technical content. All authors contributed significantly to the writing and reviewing. Both authors read and approved the final manuscript.

Acknowledgements
Research was sponsored by the Army Research Laboratory and was accomplished under Cooperative Agreement Number W911NF-12-2-0022. The views and conclusions contained in this document are those of the authors and should not be interpreted as representing the official policies, either expressed or implied, of the Army Research Laboratory or the U.S. Government. The U.S. Government is authorized to reproduce and distribute reprints for Government purposes notwithstanding any copyright notation herein.

References
1.   Lavi B, Marmur A (2004) The exponential power law: partial wetting kinetics and dynamic contact angles. Colloids Surf A Physicochem Eng Asp 250(1–3):409–414

2. Carroll BJ (1976) The accurate measurement of contact angle, phase contact areas, drop volume, and Laplace excess pressure in drop-on-fiber systems. J Colloid Interface Sci 57(3):488–495

3. Mchale G, Newton M (2002) Global geometry and the equilibrium shapes of liquid drops on fibers. Colloids Surf A Physicochem Eng Asp 206(1–3):79–86

4. Wu X-F, Dzenis YA (2006) Droplet on a fiber: geometrical shape and contact angle. Acta Mech 185(3–4):215–225

5. Eral HB, De Ruiter J, De Ruiter R, Jung Min O, Semprebon C, Brinkmann M, Mugele F (2011) Drops on functional fibers: from barrels to clamshells and back. Soft Matter 7(11):5138–5143

6. Bedarkar A, Xiang-Fa W, Vaynberg A (2010) Wetting of liquid droplets on two parallel filaments. Appl Surf Sci 256(23):7260–7264

7. Vega M-J, Seveno D, Lemaur G, Adão M-H, De Coninck J (2005) Dynamics of the rise around a fiber: experimental evidence of the existence of several time scales. Langmuir 21(21):9584–9590

8. Slawig, Thomas (2006) PDE-constrained control using COMSOL Multiphysics – control of the Navier–Stokes equations. Tech. no. 2005/26

9. Osher S, Sethian JA (1988) Fronts propagating with curvature-dependent speed: algorithms based on Hamilton-Jacobi formulations. J Comput Phys 79(1):12–49

10. Olsson E, Kreiss G (2005) A conservative level set method for two phase flow. J Comput Phys 210(1):225–246

11. COMSOL Inc., COMSOL Microfluidics User's Guide. 2013

12. Lafaurie B, Nardone C, Scardovelli R, Zaleski S, Zanetti G (1994) Modelling merging and fragmentation in multiphase flows with SURFER. J Comput Phys 113(1):134–147

13. De Gennes P-G, Brochard-Wyart F, Quéré D (2004) Capillarity and Wetting Phenomena: Drops, Bubbles, Pearls, Waves. Springer, New York

14. Bernet N, Bourban P-E, Maanson J-AE (2000) On the characterization of wetting and adhesion in glass fiber-PA12 composites. J Thermoplastic Compos Mater 13(6):434–450

15. Hocking LM (1976) A moving fluid interface on a rough surface. J Fluid Mech 76(04):801–817

16. Andrienko D, Dünweg B, Vinogradova OI (2003) Boundary slip as a result of a prewetting transition. J Chem Phys 119(24):13106–13112

Investment casting of nozzle guide vanes from nickel-based superalloys: part I – thermal calibration and porosity prediction

Agustín Jose Torroba[1], Ole Koeser[2], Loic Calba[2], Laura Maestro[3], Efrain Carreño-Morelli[1], Mehdi Rahimian[4], Srdjan Milenkovic[4], Ilchat Sabirov[4] and Javier LLorca[4,5]*

* Correspondence:
javier.llorca@imdea.org
[4]IMDEA Materials Institute, C/Eric Kandel 2, Getafe 28906, Madrid, Spain
[5]Department of Materials Science, Polytechnic University of Madrid, Madrid, Spain
Full list of author information is available at the end of the article

Abstract

Investment casting is the only commercially used technique for fabrication of nozzle guide vanes (NGVs), which are one of the most important structural parts of gas turbines. Manufacturing of NGVs has always been a challenging task due to their complex shape. This work focuses on development of a simulation tool for investment casting of a new generation NGV from MAR-M247 Ni-based superalloy. A thermal model is developed to predict thermal history during investment casting. Experimental casting trials of the NGV are carried out and the thermal history of metal, mold, and insulation wrap is recorded. Inverse modeling of the casting trials is used to define accurately some thermophysical parameters and boundary conditions of the thermal model. Based on the validated thermal model, another model is developed to predict porosity in the as-cast NGVs. The porosity predictions are in good agreement with the experimental results in the as-cast NGVs. The advantages and shortcomings of the developed modeling tool are discussed.

Keywords: Ni-based superalloys; Investment casting; Nozzle guide vanes; Thermal model; Thermal history; Porosity

Background

Nozzle guide vanes (NGVs) are important structural parts of gas turbines [1]. NGVs are typically made from Ni-based superalloys because they have to withstand very high temperatures and aggressive environments [2]. Investment casting in vacuum, also often referred to as lost-wax process, is the only commercially used manufacturing route of these parts that have very complex shapes [3]. Large improvements in turbine efficiency can be achieved with improved designs of the NGVs that normally lead to more complex shapes and thinner geometries. However, these innovations are hindered by the complexity of the manufacturing process, which leads to an increasing number of defects (mainly porosity) during investment casting of parts with complex shapes and very thin elements. As a result, the development of investment casting routes for the new generation of NGVs is carried out via a 'trial and error' approach or, in other words, via experimental casting trials. But this strategy is very expensive and time consuming and thus dramatically limits the rate of innovation.

Presently, a paradigm shift is underway in which the experimental casting trials are partially replaced by the numerical simulation of the investment casting process to overcome the limitations of standard 'trial and error' approach [3]. Reconfiguration of the mold that was made on the foundry floor can now be made on a computer and simulated. In addition, the thermal history of a casting can be examined by means of simulations, and the effect of the casting parameters on the microstructure and quality of the as-cast parts can be evaluated. For example, Anglada et al. [4] and Rafique and Iqbal [5] successfully performed the simulation of heat transfer during investment casting of prototypes from Ni-based superalloys. A short description of the modeling tools developed to date and their application to casting of Ni-based superalloys is provided below.

Models developed for porosity prediction

Porosity is known to be the most common defect found during investment casting and dramatically limits the strength and fatigue life of aerospace components [6]. Thus, investment casting foundries strive to minimize, if not eliminate, this insidious and persistent defect. The available modeling strategies for the prediction of porosity can be classified into three main groups, which are briefly described below.

Analytical models

Computer models describing the formation of microporosity on the scale of the casting are based on volume-averaging methods for the calculation of the local temperature and pressure fields in the inter-dendritic liquid. These quantities are then used to estimate the level of gas segregation and to determine if the conditions for the nucleation of a pore are met. Most of these approaches originate from the pioneering work by Piwonka and Flemings [7], who developed analytical models that range from exact mathematical solutions to asymptotic approximations using 1D Darcy's law for pore nucleation. A constant solidification velocity together with a constant thermal gradient were assumed in these models. In order to obtain a more accurate prediction of the pore size, the gas pressure within the pores was included in the model, leading to a reasonable agreement with experimental results.

Criteria function models

Criteria functions were developed in the 1950's for dimensioning the size of risers and prevent inter-dendritic centerline shrinkage and porosity in steel plates [8]. Among the different criteria functions proposed, the Niyama criterion is the most widely used in metal casting to predict feeding-related shrinkage porosity caused by shallow temperature gradients [9]. The Niyama function N_y is a local thermal parameter defined as $N_y = G/\dot{T}$, where G is the local temperature gradient and $\dot{T}$ the local cooling rate. It is assumed that shrinkage porosity will form in regions in which the Niyama parameter is below a given threshold, which should be experimentally determined for each alloy. A dimensionless form of the Niyama function was presented in [10] that accounts for not only the thermal parameters but also the properties and the solidification characteristics of the alloy and it is able to predict the shrinkage pore volume fraction from the solid fraction-temperature curve and the total solidification shrinkage of the alloy.

Numerical models

The first model for porosity prediction was developed by Kubo and Pehlke in two-dimensions (2D) [11] and was based on the relationship between the fraction of porosity and local pressure. Lee and Hunt [12] simulated the growth of pores due to hydrogen diffusion in Al-Cu alloys using a 2D continuum diffusion model, combined with a stochastic model of pore nucleation. Although the model did not include the effect of pressure drop due to shrinkage, it showed a good correlation with *in situ* observations of pore growth. Later, Lee et al. [13] developed a multi-scale model of solidification in Al-Si-Cu alloys, including microsegregation and microporosity. ProCAST was used to solve the energy, momentum, and continuity equations to determine the temperature and pressure evolution with time. This information was coupled to a mesoscale model for microstructural development. Carlson et al. proposed a volume-averaged model for finite rate diffusion of hydrogen in Al alloys [14]. They coupled the calculation of the micro/macroscale gas species transport in the melt with a model for the feeding, flow, and pressure field. This was the first work considering hydrogen diffusion in the growth of pores for three-dimensional (3D) calculations. Pequet et al. developed a 3D microporosity model based on the solution of Darcy's equation and microsegregation of gas [15]. The model coupled microporosity with macroporosity and pipe-shrinkage predictions in a coherent way, with appropriate boundary conditions. Later, this approach was improved by developing a porosity model for multi-gas systems in multi-component alloys, including a realistic model for pore pinching [16,17].

Porosity prediction in casting of Ni-based superalloys

Most of the research on porosity prediction has been focused in Al alloys and steels [18,19], and the work on Ni-based superalloys is more limited. Simulation of solidification to predict porosity in investment castings from Ni-based superalloys started a long time back; though, very simple geometries were considered in the earlier works. Overfelt et al. [20] developed a computer solidification model for the castings of plates with thicknesses of 2.54, 12.7, and 25.4 mm made from the In-718 Ni-based superalloy. The model was used to validate and disprove various phenomenological criteria for predicting porosity. The computer model was shown to be effective in predicting unfed centerline shrinkage in the 25.4-mm-thick plates, but it did not provide precise results for the thinner plates. Monastyrskiy [21] proposed a modeling tool based on liquid metal deformation due to solidification to model shrinkage porosity formation in a GS 32 Ni-based superalloy with low gas content. The model predicted the volume fraction and size of the shrinkage porosity. Nucleation of pores depended on the local stress level in the melt and the pore growth was driven by stress relaxation after pore nucleation. Numerical studies of directional solidification under an imposed temperature gradient and cooling rate were in good agreement with experimental data on porosity formation in Ni-based superalloys [21].

Modeling of investment casting of complex-shape parts has shown to be a more challenging task. Kang et al. [22] applied a model based on the dimensionless Niyama criterion to predict the formation of microporosity in a Ni-based superalloy casting containing complex shapes with thin walls. The theoretical predictions of microporosity showed reasonable agreement with the experimental results, though they underestimated the porosity content in the complex thin-wall regions. However, the model was not suitable for the shrinkage porosity prediction in the thick parts of the casting, since those sections often formed isolated liquid pools.

In this work, an advanced modeling approach is applied to the development of a new generation of NGVs with complex shape for aero engines. The objective of this work is twofold. Firstly, a thermal model capable of predicting the thermal history during investment casting of the new generation NGVs is developed. The principles of the thermal model were earlier described by Calba and Lefebvre [23]. Once the developed thermal model is validated against experimental results, the overall casting process can be analyzed in detail. The second aim of the present work is to simulate the development of defects in the as-cast NGVs (such as shrinkage porosity) and the final grain structure. The present manuscript consists of two parts and this (first) part focuses on the development and validation of the thermal model and the porosity prediction tool.

Material and experimental procedures

Investment casting of the NGVs was carried out using MAR-M247 Ni-based superalloy. The chemical composition of the material is presented in Table 1. The MAR-M247 superalloy is characterized by high temperature strength and excellent corrosion and oxidation resistance at elevated temperatures [24].

The ceramic molds for the experimental casting trials were prepared using the standard manufacturing route. The wax pattern for the NGV was prepared via injection molding and then assembled with a wax feeding system. The obtained wax cluster was immersed into a ceramic slurry and allowed to dry, and this step was repeated until the desired thickness of ceramic mold was reached. The assembly was dewaxed in an autoclave for 15 min at elevated temperature and high pressure. To burn the wax remains, the ceramic cluster was fired at 900°C for 1 h. Finally, the interior of the ceramic cluster was thoroughly rinsed.

The ceramic cluster was wrapped by an insulation layer (made from kaolin wool), having a thickness of 13 mm and was preheated to 1,200°C. The geometry and mesh for the model, including different cross sections, are presented in Figure 1b. Before entering the casting furnace, the thermocouples for recording the thermal history during investment casting were quickly set on the assembly. The equipment for *in situ* temperature measurements consisted of K- and S-type thermocouples and a standalone data logger able to withstand high vacuum (10^{-3} mbar), magnetic fields (coming from the induction furnace), and thermal radiation due to the high temperature of the melt. Thermocouples were placed at defined points in the insulation wrap, ceramic mold, and metal. Temperature in the wrap was measured with a thermocouple placed in the center of the wrap layer (marked by a blue spot in Figure 1c). Shell temperature was measured with a thermocouple placed on the leading edge of one external airfoil. Three thermocouples were used to measure the temperature in the alloy but only the results of one of them are shown because the other thermocouples failed during investment casting. The location of each thermocouple is illustrated in Figure 1c, and it was identified by

Table 1 The chemical composition of the MAR-M247 Ni-based superalloy

Ni	C	Cr	Co	Mo	W	Ta	Al	Ti	Hf
Base	0.15	8.4	10	0.7	10	3.1	5.5	1.05	1.4

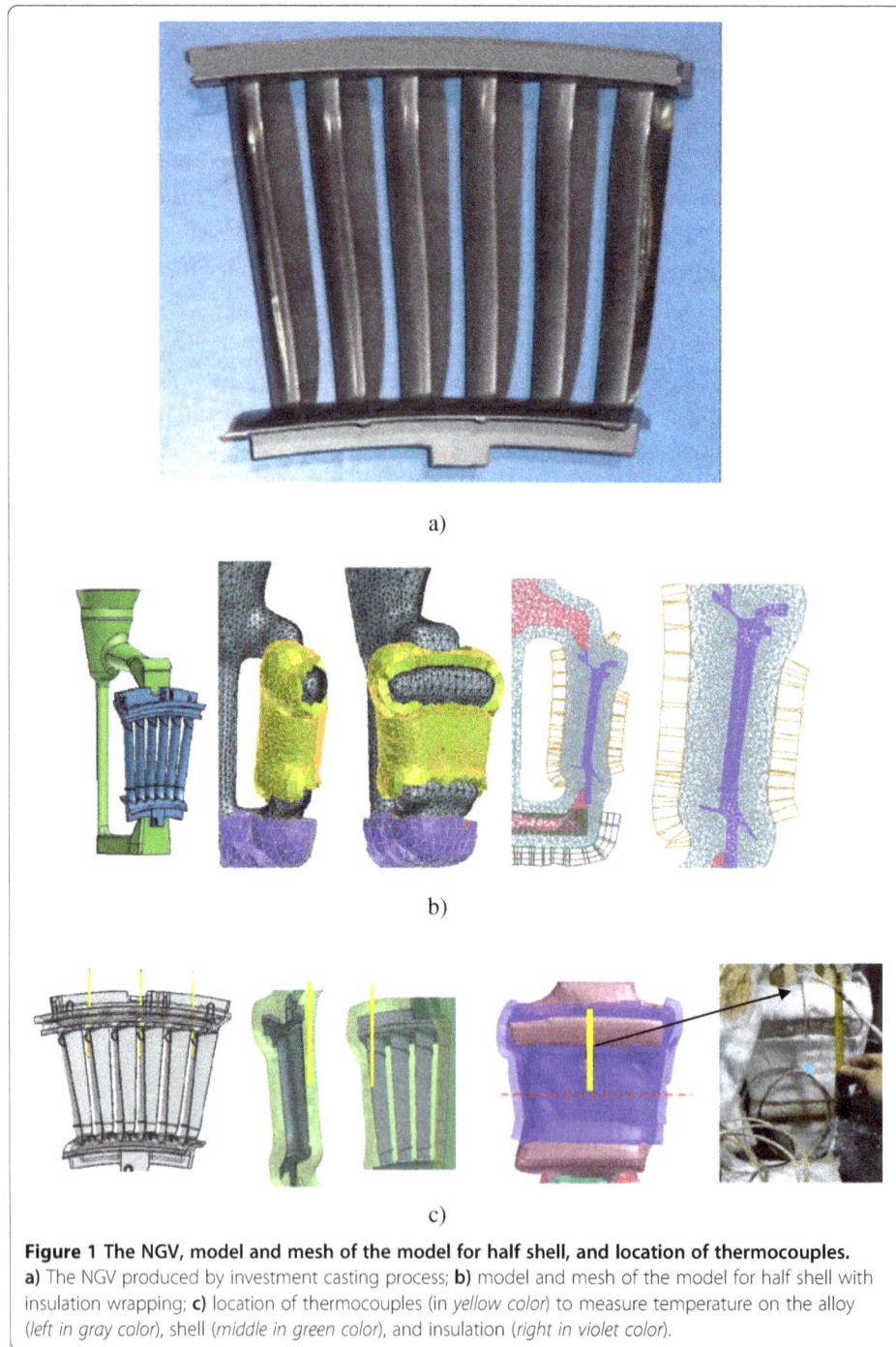

Figure 1 The NGV, model and mesh of the model for half shell, and location of thermocouples.
a) The NGV produced by investment casting process; **b)** model and mesh of the model for half shell with insulation wrapping; **c)** location of thermocouples (in *yellow color*) to measure temperature on the alloy (*left in gray color*), shell (*middle in green color*), and insulation (*right in violet color*).

nodes in the thermal model (see 'Development of the thermal model' section). The preheated assembly was placed in the vacuum casting furnace where the ceramic mold was filled by the molten metal poured at 1,549°C with a melt pouring velocity of 1,700 mm/s. The assembly was then removed from the furnace and allowed to cool. The thermal history at defined nodes of the metal, the ceramic mold, and the wrap was recorded.

The as-cast NGV (Figure 1a) was cut into smaller specimens for the analysis of porosity. The selected areas for porosity evaluation are shown in the 'Porosity characterization in the as-cast new generation NGV and experimental validation of the model' section. The specimens for porosity characterization were ground and polished to the mirror-like surface using standard metallographic techniques. The optical microscope OLYMPUS BX51 was used for porosity characterization. At least three images were taken from each area of interest. Quantitative analysis of porosity (pore size and porosity volume fraction) was performed using ANALYSIS software. The pore size was given by the equivalent circle diameter due to the complex shape of pores.

Modeling

A modeling approach to investment casting of the new generation NGVs should allow the definition of the cast component, gate, mold, and insulation wrapping configuration and geometries. Starting from the component geometry, the casting process can be gradually developed and optimized, and critical design decisions can be made. Such a model has to be able to cover issues such as heat transfer (including radiation, convection, and conduction), mass transfer (mainly fluid dynamics), and phase transformations, considering at any moment the conservation of mass, momentum, and energy. And it should be able to assess the influence of the geometric and physical parameters on the porosity and structure of the as-cast NGVs. For investment casting of the new generation NGVs, most geometrical parameters such as gating, mold thickness, and wrapping scheme are already defined by the manufacturer, but most of the physical parameters remain unknown. The development of the modeling tool and definition of the thermophysical parameters are described below.

Thermal model

Development of the thermal model

The basis for reliable modeling of investment casting is a very accurate prediction of the thermal history at each point of the cast. Development of the thermal model requires the optimal selection of the thermophysical parameters along with the proper establishment of boundary conditions as noted in the ASM Handbook [25]. It should also be noted that each manufacturing process has unique boundary conditions that have to be identified, understood, and characterized for the specific application being simulated. The boundary conditions can also be equipment specific, meaning that a furnace may not give rise to the same boundary conditions as another furnace under the same nominal processing conditions.

Mold filling during investment casting was modeled using the three-dimensional finite element solver ProCAST (a trademark of ESI group) by solving the conservation of mass, momentum, and heat flow equations [26]. Conservation of mass is enforced through the continuity equation

$$\frac{\delta \rho}{\delta t} + \frac{\delta(\rho u_i)}{\delta x_i} = 0 \tag{1}$$

where u_i is the corresponding component of the velocity and ρ stands for the density. The momentum equation as used in ProCAST is given by

$$\rho \frac{\delta u_i}{\delta t} + \rho u_j \frac{\delta u_i}{\delta x_j} + \frac{\delta}{\delta x_j}\left(p\zeta_{ij}(\mu + \mu_T)\frac{\delta u_i}{\delta u_j}\right) = \rho g_i - \frac{\mu}{Kp}u_i \tag{2}$$

where ζ_{ij} is the Kronecker delta, p the pressure, g_i the gravitational acceleration, μ the viscosity, μ_T the eddy viscosity, and Kp the permeability. These equations are solved under the assumption that the spatial derivatives of viscosity are small and that the fluid is nearly incompressible.

During investment casting, heat flows by conduction through the metal, ceramic mold, and insulation wrap and is removed from the surface by natural convection and radiation. The heat flow is transient, i.e. the temperatures in the casting, mold, and insulation wrap change with time. The governing partial differential equation of heat flow by conduction is expressed as

$$\rho \frac{\delta H}{\delta T} \frac{\delta T}{\delta t} - \nabla[K\nabla T] - S(r) = 0 \qquad (3)$$

where $\nabla = \frac{\partial}{\partial x} + \frac{\partial}{\partial y} + \frac{\partial}{\partial z}$, T stands for the temperature, t for the time, K for the thermal conductivity, $S(r)$ is a spatially varying heat source, and H the enthalpy of solidification, which encompasses both the specific heat term (c_p) and the evolution of latent heat (L) during solidification according to

$$H(T) = \int_0^T c_p dr + L(1 - f_s(T)) \qquad (4)$$

where f_s is the fraction of solid.

Initial and boundary conditions for the resolution of previous equations are applied on temperature, velocity, pressure, fixed turbulent kinetic energy, fixed turbulent dissipation rate, and specific, convective, and radiation heat flux. An iterative algorithm is used to simulate solidification by solving Equation 2, finding a coherent solution between enthalpy and temperature results. Further details about this strategy can be found in [27,28].

To solve the complex view factor radiation capability, ProCAST uses a net flux model, in which an overall energy balance for each participating surface is considered rather than tracking the reflected radiant energy from surface to surface. At a particular surface i, the radiant energy being received is denoted $q_{in,i}$. The outgoing flux is $q_{out,i}$. The net radiative heat flux is the difference of these two.

$$q_{net,i} = q_{out,i} - q_{in,i} \qquad (5)$$

Utilizing the diffuse gray-body approximation, the outgoing radiant energy can be expressed as:

$$q_{out,i} = \sigma \; \varepsilon_i T_i^4 \; + \; (1 - \varepsilon_i) \; q_{in,i} \qquad (6)$$

The first term of this equation represents the radiant energy which comes from direct emission. The second term is the portion of the incoming radiant energy which is being reflected by surface i.

The incoming radiant energy is a combination of the outgoing radiant energy from all participating surfaces being intercepted by surface i. Specifically, the view factor F_{i-j} is the fraction of the radiant energy leaving surface j which impinges on surface i. Thus,

$$q_{in,i} = \Sigma_{j=1}^N \; F_{i-j} \; q_{out,i} \qquad (7)$$

where N is the total number of surfaces participating in the radiation model and the view factors are calculated from the following integral.

$$F_{i \cdot j} = \frac{1}{A_i} \int_{Aj\,Ai} \frac{\cos\theta\, j\, \cos\theta\, i}{\pi r^2}\, dA_i dA_j \tag{8}$$

where A_i is the area of surface i, θ_i the polar angle between the normal to surface i and the line between i and j, and r the magnitude of the vector between surfaces i and j.

Traditionally, Equation 8 is evaluated by numerical integration, either in the form shown or converted into an equivalent line integral. In ProCAST, the view factors are computed using a proprietary technique.

Solving Eq. 6 for $q_{\text{in},i}$ yields:

$$q_{in,i} = \left[\frac{1}{1-\varepsilon_i}\right]\left(q_{out,i} - \sigma\,\varepsilon_i\,T_i^4\right) \tag{9}$$

Combining Equation 9 with Equation 7 gives a relationship involving the outgoing radiant fluxes only. These outgoing fluxes are known as radio sites. The final form is:

$$\sum_{j=1}^{N}\left(\zeta_{ij} - (1-\varepsilon_i)\,F_{i\cdot j}\right)q_{out,j} = \sigma\,\varepsilon_i\,T_i^4 \tag{10}$$

where the Kronecker delta ζ_{ij} has been included to incorporate the diagonal term. Since there are equations of the form (10), a simultaneous solution is required for a large non-symmetric system. Because of the reciprocity relation, $A_jF_{i\,-j} = A_jF_{j\,-i}$ can be transformed into a symmetric form which is more economical to solve. Multiplying (10) by

$$\frac{A_i}{1-\varepsilon_i} \tag{11}$$

yields

$$\sum_{j=1}^{N}\left(\frac{A_i}{1-\varepsilon_i}\,\zeta_{ij} - A_i\,F_{i\text{-}j}\right)q_{out,j} = \frac{A_i}{1-\varepsilon_i}\,\sigma\,\varepsilon_i\,T_i^4 \tag{12}$$

which is solved for the vector of radiosities, $q_{out,i}$. The net radiant flux is obtained by combining Equation 5 and Equation 9 that gives

$$q_{net,i} = \left[\frac{\varepsilon_i}{1-\varepsilon_i}\right]\left[\sigma\,T_i^4 - q_{out,i}\right] \tag{13}$$

This heat flux then appears as a boundary condition for the heat conduction analysis in ProCAST.

Several software packages were used to generate the thermal model. The NGV design was created with SolidWorks software (powered by Dassault Systems SolidWorks Corporation), while the feeding system was created with Unigraphics software (powered by Unigraphics Solutions Incorporation). Both packages are linked to the casting simulation software ProCAST. Surface and volume meshes were generated by Visual-Mesh (ProCAST software package), considering a maximum distance between nodes of 2 mm inside the NGV, and 6 mm for the gating system and pouring cup. The investment casting ceramic mold was composed of layers which were created by ProCAST 3D-Mesh. The ceramic mold has an average thickness of 13 mm. The thickness of the insulation wrap was also 13 mm and was created and meshed, following the same procedure. Only one half of the mold was considered due to symmetry (Figure 1b) in order to speed up the simulations.

Data from different sources were carefully analyzed to assign the thermophysical properties to all the components of the casting system. Those sources include the Pro-CAST database which was described in detail by Pequet et al. [15], experimental data from industrial companies (Precicast Bilbao and Precicast Novazzano), as well as technical references from previous similar exercises. The thermophysical properties of MAR-M247 Ni-based superalloy (including temperature-dependent thermal conductivity, density, specific heat, and viscosity) were extracted from the ProCAST database (Table 2). Figure 2 illustrates dependence of these properties with temperature. The values of the liquidus and solidus temperatures (1,366°C and 1,266°C, respectively) were also taken from the ProCAST database. It should be noted that a comparison with the Lever Rule model and Scheil model (both described in the ASM Handbook [29]) was made to confirm these data.

Regarding the ceramic mold and wrap insulation, the density and specific heat as functions of temperature were taken from the ProCAST database (Table 2). Values for the thermal conductivity as a function of temperature (Figures 3a and 4a) were obtained by an inverse simulation procedure by comparing the simulation results for simple casting geometries with experimental data generated earlier by Precicast Bilbao. The description of the inverse simulation procedure can be found in O'Mahoney and

Table 2 Thermophysical properties and boundary conditions used in the thermal model

Material	Property (units)	Value
MAR-M247	Thermal conductivity ($W \cdot m^{-1} \cdot K^{-1}$)	15–35[a]
	Density ($kg \cdot m^{-3}$)	7,300–8,600[a]
	Enthalpy (kJ/kg)	100–800[a]
	Viscosity ($kg \cdot m^{-1} \cdot s^{-1}$)	2–$3.25 \cdot 10^{-3}$[a]
	Liquidus temperature (°C)	1,366
	Solidus temperature (°C)	1,266
Mold	Thermal conductivity ($W \cdot m^{-1} \cdot K^{-1}$)	0.4–1.7[a]
	Density ($kg \cdot m^{-3}$)	1,860–1,915[a]
	Specific heat ($kJ \cdot kg^{-1} \cdot K^{-1}$)	0.7–1.3[a]
	Emissivity	0.7
Insulation wrap	Thermal conductivity ($W \cdot m^{-1} \cdot K^{-1}$)	0.1–0.5[a]
	Specific heat ($kJ \cdot kg^{-1} \cdot K^{-1}$)	0.9–1.3[a]
	Emissivity	0.7
Metal mold	HTC ($W \cdot m^{-2} \cdot K^{-1}$)	200–2,500[a]
Mold wrap	HTC ($W \cdot m^{-2} \cdot K^{-1}$)	100
Mold enclosure	HTC ($W \cdot m^{-2} \cdot K^{-1}$)	3
Wrap enclosure	HTC ($W \cdot m^{-2} \cdot K^{-1}$)	10.6
Enclosure	Emissivity	0.9
Others	*Units*	*Value*
Melt pouring velocity	(mm/s)	1,700
Melt temperature	(°C)	1,549
Preheating temperature	(°C)	1,200

[a]The value depends on the temperature.

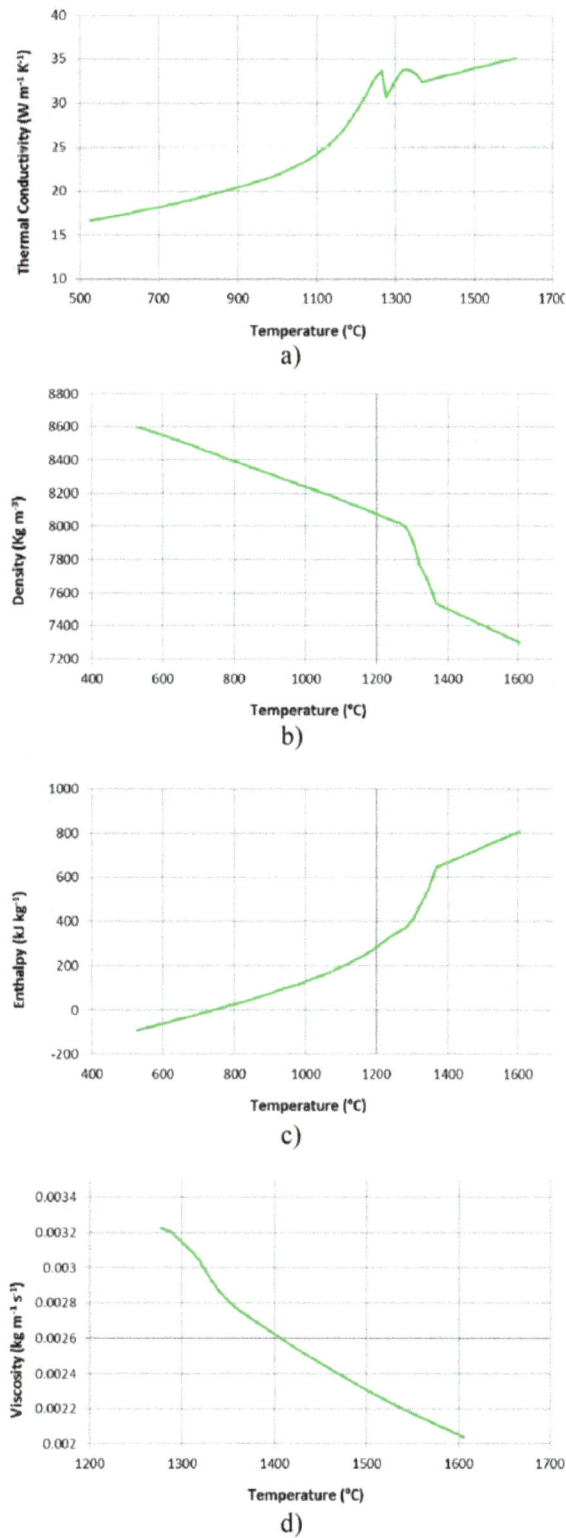

Figure 2 Properties of MAR-M247 Ni-based superalloy vs. temperature. a) Thermal conductivity,
b) density, **c)** enthalpy, and **d)** viscosity.

Browne [30]. Figures 3 and 4 illustrate the variation of these properties with temperature for the ceramic mold and wrap, respectively.

It is known that pouring of the melt at high temperature leads to radiation heat loss. As this heat loss is not always correctly taken into account during the modeling process, the value of mold conductivity at high temperatures should be increased to account for this phenomenon. Experimental studies on this topic were earlier carried out by Precicast Bilbao and the experimental data from earlier measurements using the laser flash method (according to the ASTM E1461-07 standard) were considered. Analysis of all available data led to a final interval of mold conductivity in the range from 0.4 to 1.75 $W \cdot m^{-1} \cdot K^{-1}$ (Table 2), which is in a very good accordance with the data provided by Konrad et al. [31] for low temperatures, and coincide with the experimental data measured by the laser flash method at high temperature (Figure 3a). The ProCAST database, data from the manufacturer of the kaolin wool, and Precicast Bilbao were considered to define the thermal conductivity of the insulation wrap. The final values of the thermal conductivity in the insulation wrap were in the range of 0.1 to 0.5 $W \cdot m^{-1} \cdot K^{-1}$ (Figure 4a and Table 2).

A suitable temperature-dependent functional form (shown in Figure 3d) was used to determine the values of heat transfer coefficient (HTC) at the metal-mold interface. It is known that the HTC at the metal-mold interface is influenced by many factors such as casting geometry, pouring and preheating temperature, mold thickness, etc. Inverse and direct simulations were carried to obtain the final form of this function, which is plotted in Figure 3d as a function of temperature. This function is slightly different from the one proposed by the ProCAST database as was demonstrated by Santos et al. [32] and Dong et al. [33]. Nevertheless, the final HTC at the metal-mold interface was in a very good agreement with the data reported in the literature for molten Ni-based superalloys in contact with ceramic molds. For example, Sahai and Overfelt [34] reported a HTC in the range 50–5,000 $W\ m^{-2}\ K^{-1}$ for IN-718 Ni-based superalloy. The HTC at the mold-wrap, mold-enclosure, and wrap-enclosure interfaces have less influence on the final result of the thermal model as shown by Yuang et al. [35]. Thus, it was assumed that they were constant with temperature and time, and the data from the ProCAST database were used (Table 2). Values of emissivity for mold, wrap, and enclosure were also taken from the ProCAST database, and the environmental conditions were fitted with those registered during experimental casting trials (Table 2). The pouring of the melt into the mold was introduced in the model by the definition of a planar surface on the top of the pouring cap, where a velocity to the liquid was applied to simulate the pouring process. The preheating temperature of the mold and temperature of melt poured into the mold were also specified (Table 2). The same filling steps performed during the experimental procedure were simulated by the software, using 2 s of filling time to introduce the molten alloy into the mold. Thus, the full solidification process was completed 830 s after the pouring. All the experimental data were taken into account during the simulations to synchronize the experimental data with the simulation results. The simulation process was operated by the ProCAST Parallel Solver with four processors (2.40 GHz) and took nearly 11 h to simulate the whole thermal history of the NGV investment casting process.

Experimental validation of the thermal model

Experimental casting trials were carried out for validation of the thermal model as described above. Figure 5 illustrates the experimental temperature-time plots for metal,

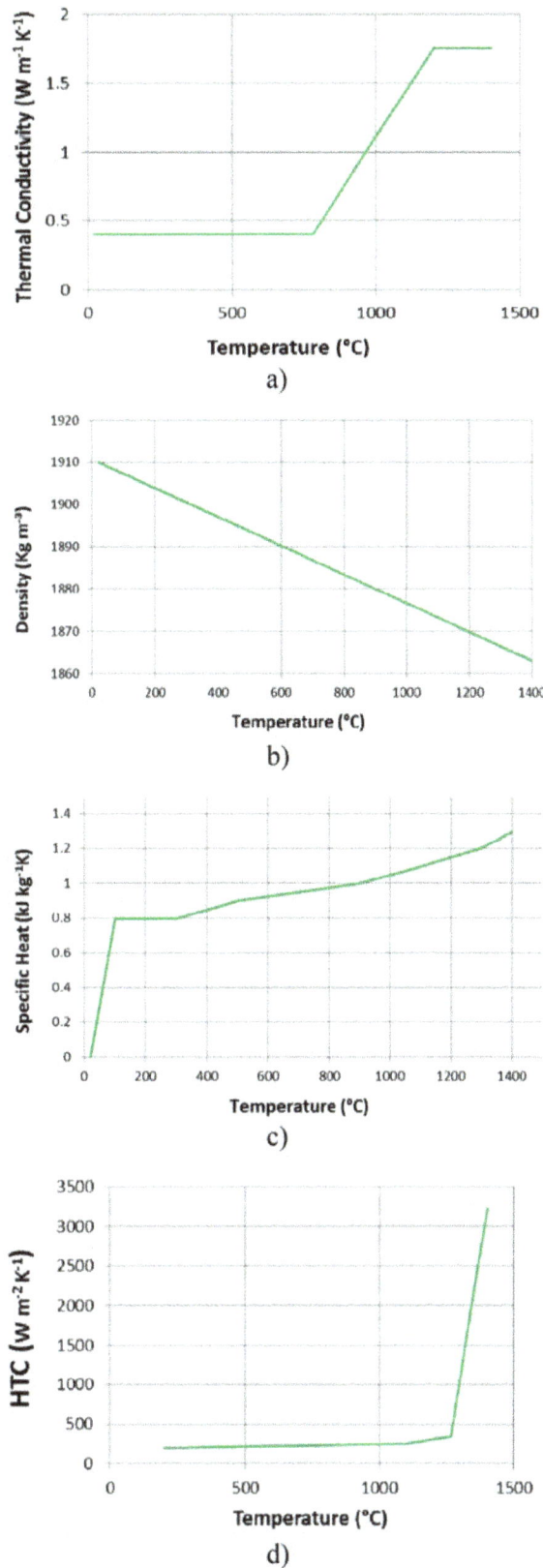

Figure 3 Properties of ceramic mold vs. temperature. a) Thermal conductivity, **b)** density, **c)** specific heat, and **d)** heat transfer coefficient (HTC) at the metal-mold interface.

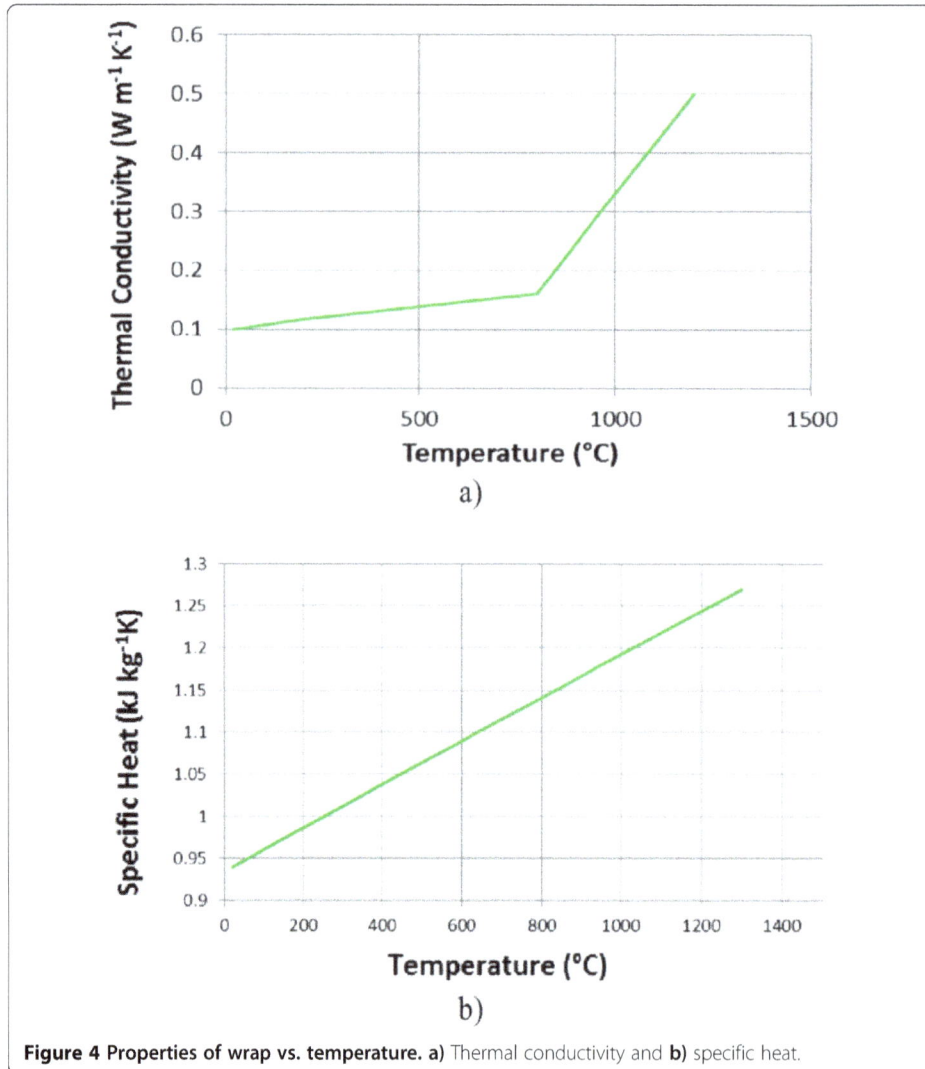

Figure 4 Properties of wrap vs. temperature. a) Thermal conductivity and **b)** specific heat.

ceramic mold, and insulation wrap during investment casting. Temperature recording was started once the thermocouples were located in the defined spots. The thermocouple placed in the metal is close to reach the preheating temperature 1,200°C, while the thermocouple placed in the mold records a temperature slightly over 1,100°C. In the readings from the thermocouple fixed to the insulation wrap, temperature rises up to 900°C. Significant difference of temperatures between metal and wrap was registered at the beginning since it took time to place correctly each thermocouple into its location. This loss of time leads to partial cooling of the mold that, in turn, increases the temperature gradient between metal and mold.

After the mold entered the vacuum casting furnace, vacuum was pumped and melt was poured into the ceramic mold. The thermocouples placed in the metal and wrap clearly registered this event by showing a rapid temperature rise, whereas the thermocouple located in the ceramic mold showed a temperature decrease (Figure 5). The liquidus-solidus transition in the metal can be easily identified in the experimental temperature-time plot because of the reduced slope. The cooling rate increases once the melt is solidified.

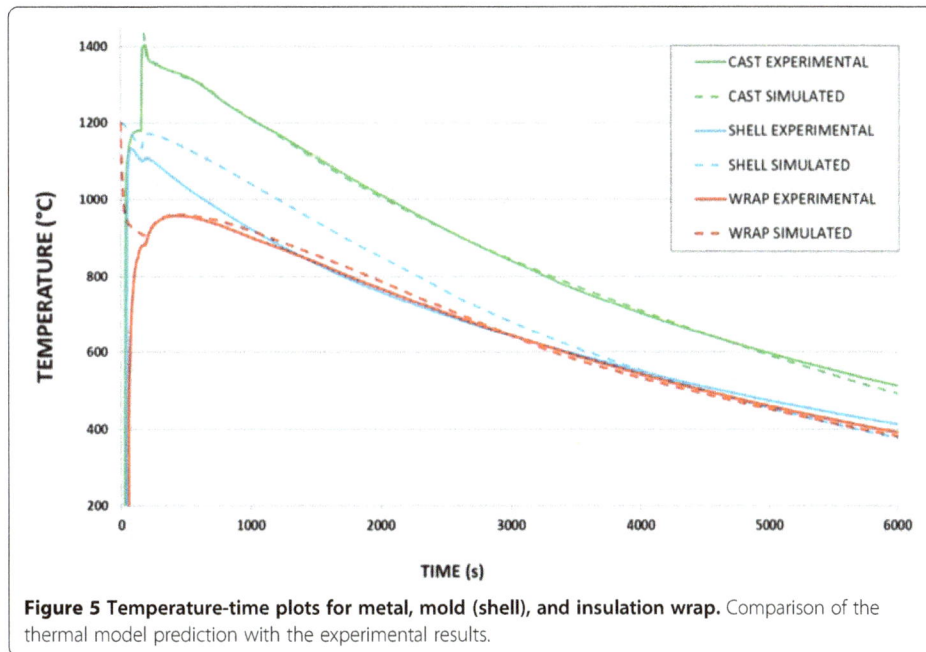

Figure 5 Temperature-time plots for metal, mold (shell), and insulation wrap. Comparison of the thermal model prediction with the experimental results.

The temperature-time plots generated by the thermal model are compared with the experimental results in Figure 5. A very good agreement is observed for the thermal history in the metal and in the wrap, where the simulation results match very well the experimental results during first 6,000 s of the solidification/cooling process. However, a difference of nearly 100°C is found between the predicted temperature and the experimental data in the ceramic mold. Despite the close location of the thermocouple to the inner surface of the ceramic mold, the temperature registered by this thermocouple hardly achieves 1,100°C, though the melt was poured into the ceramic mold at 1,459°C. The reasons for such discrepancy are discussed in the 'Accuracy of the thermal model' section.

A proper prediction of the liquidus-solidus transition has to be achieved in a reliable thermal model. A deeper analysis of the liquidus-solidus transition is found in Figure 6, which shows a perfect match between simulation and experimental results. The most significant deviation between numerical predictions and experimental results occurs at 300 s after pouring, and the difference is just 4°C.

Since the solidification process of the metal and its thermal history are correctly described, the thermal model can be further utilized to predict the microstructure and defects of the as-cast parts. The next section of this manuscript focuses on the ProCAST model for porosity prediction, which is developed on the basis of the thermal model.

Model for porosity prediction

Description of the model

The ProCAST tool was employed in this work to simulate the development of porosity during investment casting. The physical basis of the model is following. The key variable is the fraction of solid (FS) which extends from FS =0 for liquid to FS =1 for solid, as shown schematically in Figure 7. When the melt solidifies, pockets of liquid are created, surrounded by a mushy zone and then a solid shell. Automatically, the casting is divided into 'regions' having the FS <1. These 'regions' are bounded by isosurfaces. As

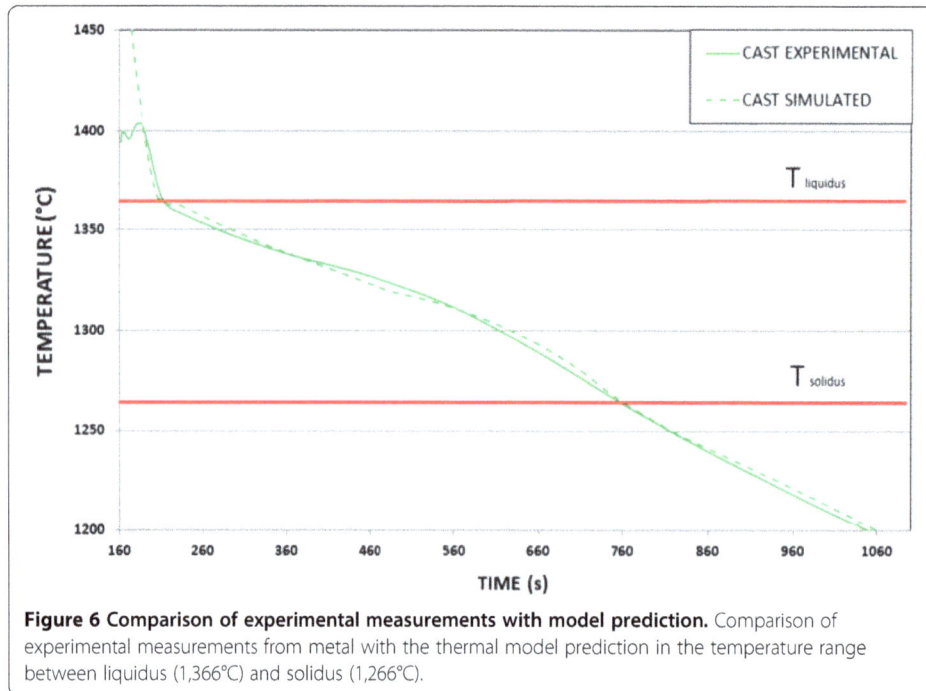

Figure 6 Comparison of experimental measurements with model prediction. Comparison of experimental measurements from metal with the thermal model prediction in the temperature range between liquidus (1,366°C) and solidus (1,266°C).

solidification proceeds and depending upon the complexity of the geometry, the number of 'regions' may increase with time, i.e. one 'region' can be split in more 'regions'. The 'region' disappears once all nodes are completely solidified.

The model is based on comparison of the local FS with a few parameters describing 'key stages' of solidification which determine the porosity in the cast. The first parameter, PI, is a measure of limit of local solid fraction under the surface until piping[a] on the surface can occur. In other words, the model predicts formation of pipe (empty nodes) while FS < PI (Figure 7). In the present calculations, the default value recommended in the ProCAST database PI =0.3 was used, i.e. piping occurs until the solid fraction reaches 30%. No porosity formation takes place in the bulk of the casting while local FS < PI. The second important parameter, PF, is a limit of solid fraction until the liquid can still feed a hot area. PF =0.7 (the default value recommended in the Pro-CAST database) was used in the present simulations. The model predicts the formation of pipe in the form of a shrinkage pore on the surface while PI < FS < PF (Figure 7). If there are no nodes of the 'region' on the free surface having PI < FS < PF, no pipe can be formed and the model predicts macroshrinkage in the bulk of casting (Figure 7). In this case, the macropore nucleates and grows at the highest point of the liquid pocket.

According to the model, microporosity can appear only in the zone having PF < FS <1. The third parameter FL, critical feeding length, is introduced into the model to predict microporosity. The FL value depends upon the size of the mushy zone and thus, the size of the casting. In the present calculations, FL =0.005, following the value recommended in the ProCAST database. Two scenarios are possible in the bulk of casting:

1) There is still some mushy zone (liquid) below PF. Microporosity forms only at the distance higher than FL from the PF isosurface (see zone A for the corresponding situation in Figure 7). The amount of microporosity is equal to the density change

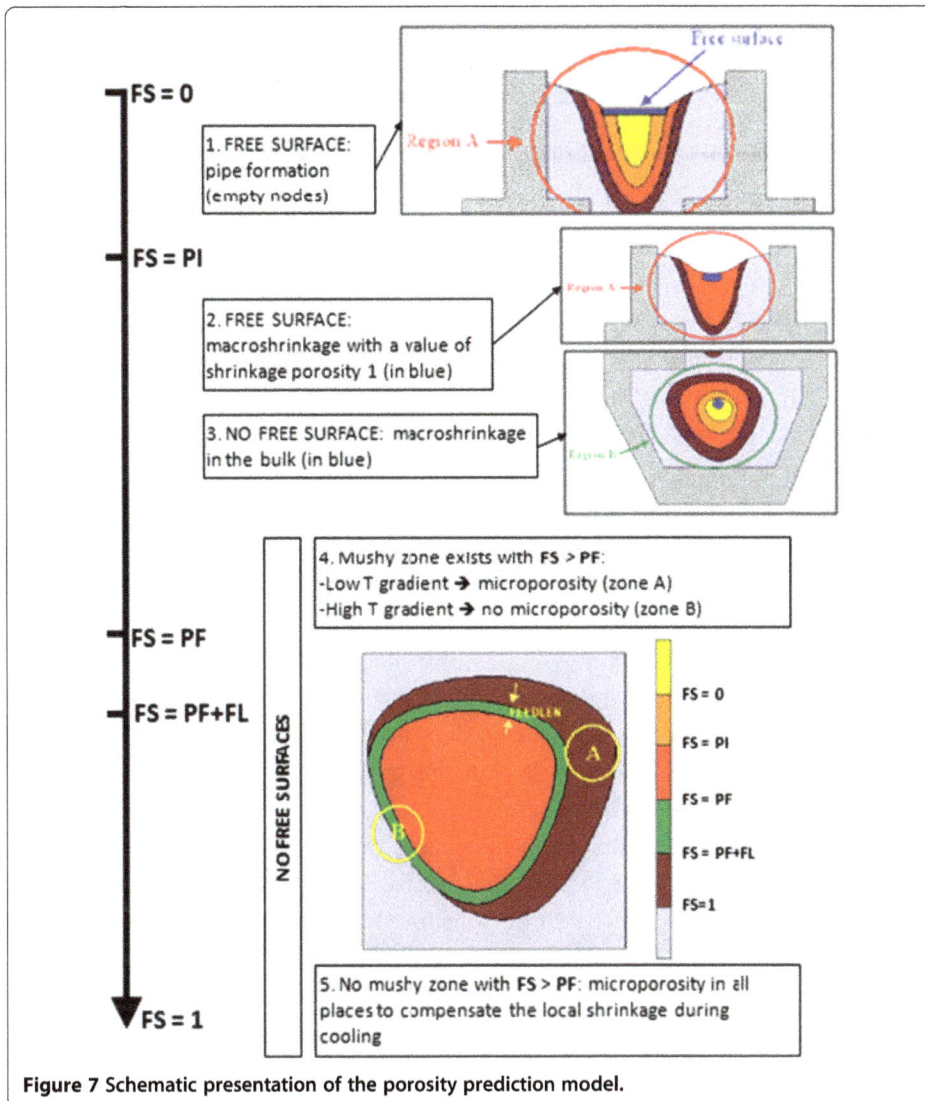

Figure 7 Schematic presentation of the porosity prediction model.

between the local solid fraction and 1. No micropores can form in the case of high-temperature gradients, since the distance between PF and solidus isosurface is smaller than FL (see zone B for the corresponding situation in Figure 7). On the contrary, low-temperature gradient promotes formation of microporosity (see zone A for the corresponding situation in Figure 7).

2) There is no more mushy zone in zones with PF < FS <1. In this case, the parameter FL is not active. In this case, there can be microporosity in the whole region with PF < FS <1 to compensate the shrinkage during cooling. The level of porosity is calculated based on the change of the density for each node as solidification takes place. This variation in the density of the material allows the software to compute the volume corresponding to shrinkage porosity as the limit value of PF is achieved in the nodes.

The modeling results are displayed in ViewCast software. The unit is volume fraction [%]. The results are classified as follows:

- the porosity values below 1% correspond to microporosity;
- the porosity values in the range between 1% and 2.3% correspond to shrinkage porosity;
- the porosity values above 2.3% correspond to the macroporosity.

The present tool was applied for porosity prediction in the as-cast NGVs, and the simulation outcomes are presented in the next section.

Results and discussion

Porosity characterization in the as-cast new generation NGV and experimental validation of the model

Figure 8 shows the porosity predictions (left) and the experimental data (right) of transversal section of a solid vane in the as-cast NGV. The optical micrographs corresponding to the trailing edge (zone a), middle part (zone b) and leading edge (zone c) are also plotted in Figure 8. The analysis of these results shows a good agreement between simulation predictions and experimental results in all zones. The highest level of porosity (≤2.91%) is predicted for the middle part of the solid vane and it is in quantitative agreement with the experimental shrinkage porosity of 3.07% in this area (Figure 8b and Table 3). The average pore size is 22 μm. Analysis of the histogram of pore size distribution shows that the frequency of pores decreases with increasing size and a few macropores with a size up to 196 μm are present in the middle part (Figure 9a). For the leading edge, lower levels of shrinkage porosity (≤2.17%) are predicted by the model, but this prediction overestimates the experimental result, 0.63% (Figure 8a and Table 3). The average pore size slightly decreases to 20.4 μm and the maximum pore size does not exceed 79 μm (Figure 9b). Finally, shrinkage porosity was not predicted in the trailing edge and this is confirmed by experimental study (Figure 8a and Table 3).

Figure 10 illustrates the outcomes of porosity modeling for the longitudinal section of the solid vane along with the optical microscopy images for selected areas. A good correlation between simulation and experimental results is observed. The simulation results show the highest level of macroporosity (≤2.90%) in the red zone (b). The experimental evaluation of porosity in the red zone (b) yields porosity of 4.87% with the average pore size of 113 μm (Table 4). The frequency of pores decreases with increasing size and a few macropores with a size up to 280 μm are present in the hot spot (Figure 11b). Macroporosity of ≤2.83% is expected in the zone (c) according to the model, whereas the experimental results show macroporosity of 4.82% (Table 4). The amount of the large pores decreases in this area (Figure 11c) and the average pore size is reduced to 27 μm, correspondingly (Table 4). A decrease of porosity should take place in zones a and d and shrinkage porosity of ≤2.43% is predicted, though the experimental characterization of these areas shows much lower values of porosity (Table 4).

The simulation results were also validated for the hollow vane of the new generation NGV (Figure 12). Generally lower porosity is predicted at the top (≤2.17%) and bottom (≤2.47%) sections of the hollow vane since the small thickness of the walls hastens the solidification process, thus reducing porosity. The highest porosity (≤2.73%) is expected in the midsection of the hollow vane as it solidifies last. The experimental

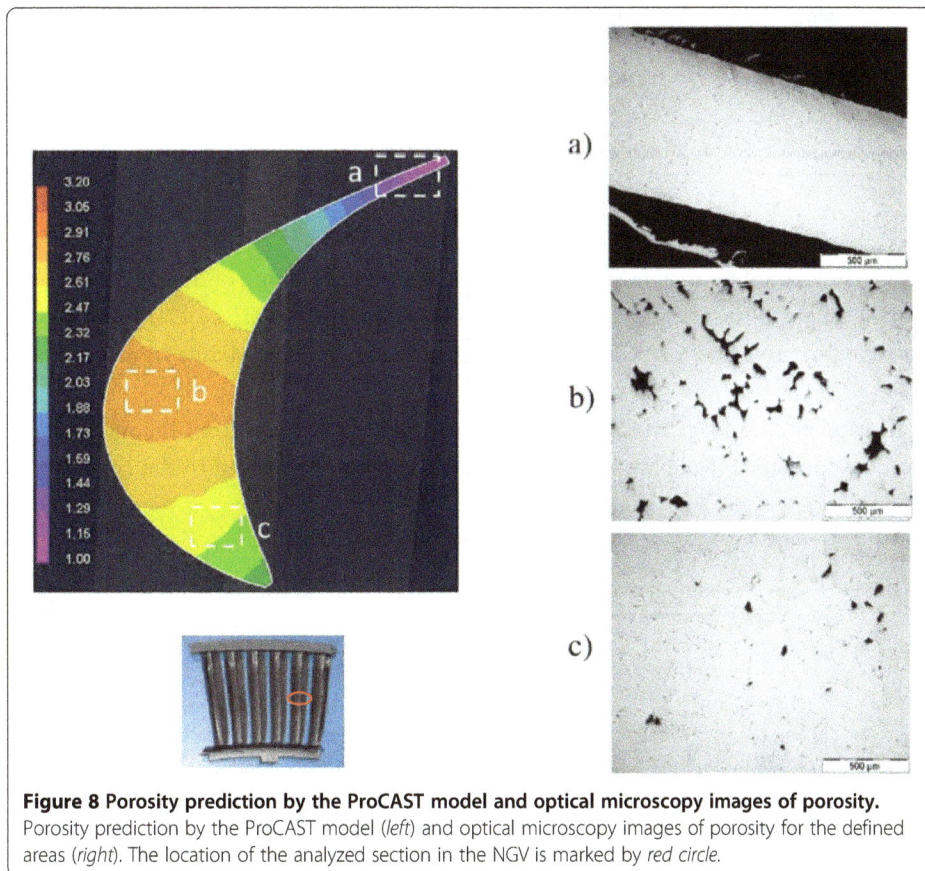

Figure 8 Porosity prediction by the ProCAST model and optical microscopy images of porosity.
Porosity prediction by the ProCAST model (*left*) and optical microscopy images of porosity for the defined areas (*right*). The location of the analyzed section in the NGV is marked by *red circle*.

data of porosity in all these areas follow the trends predicted by the simulation (Figure 12), although the modeling results tend to overestimate the porosity in the hollow vane (Table 5).

Accuracy of the thermal model

The analysis of the simulation results and their comparison with the experimental data clearly show that the thermal history of the metal and wrap is very well described by the thermal model during the solidification and cooling processes (Figure 5). The differences between predictions and experimental measurements in the mold do not seem to be due to the model. The thermal plot measured from the ceramic mold seems to underestimate its real thermal history since the thermocouples record slightly higher temperatures for the insulation wrap compared to the ceramic mold in the time range of 1,300–3,000 s (Figure 5), which cannot be true. This discrepancy can be rationalized on the basis of the shortcomings of the experimental procedure utilized to measure the temperature in the ceramic mold. In particular,

1) Cement was used to fix the thermocouple to the ceramic mold;
2) Some 'air gaps' can appear between the thermocouple and cement due to significant thermal expansion/contraction;

Table 3 The porosity characteristics of the transversal section of the solid vane on Figure 8

	Leading edge	Middle part	Trailing edge
Local porosity fraction [%]	0.63	3.07	No porosity
Average pore size [μm]	20.4	22.0	-

3) There could also be some deviations from the correct positioning of the thermocouple in the ceramic mold during its fixing to the ceramic mold, since this operation has to be performed manually at extreme conditions in limited time.

The 'air gaps' and cement can significantly reduce the heat transfer from the mold to the thermocouple since they have lower thermal conductivity compared to the ceramic

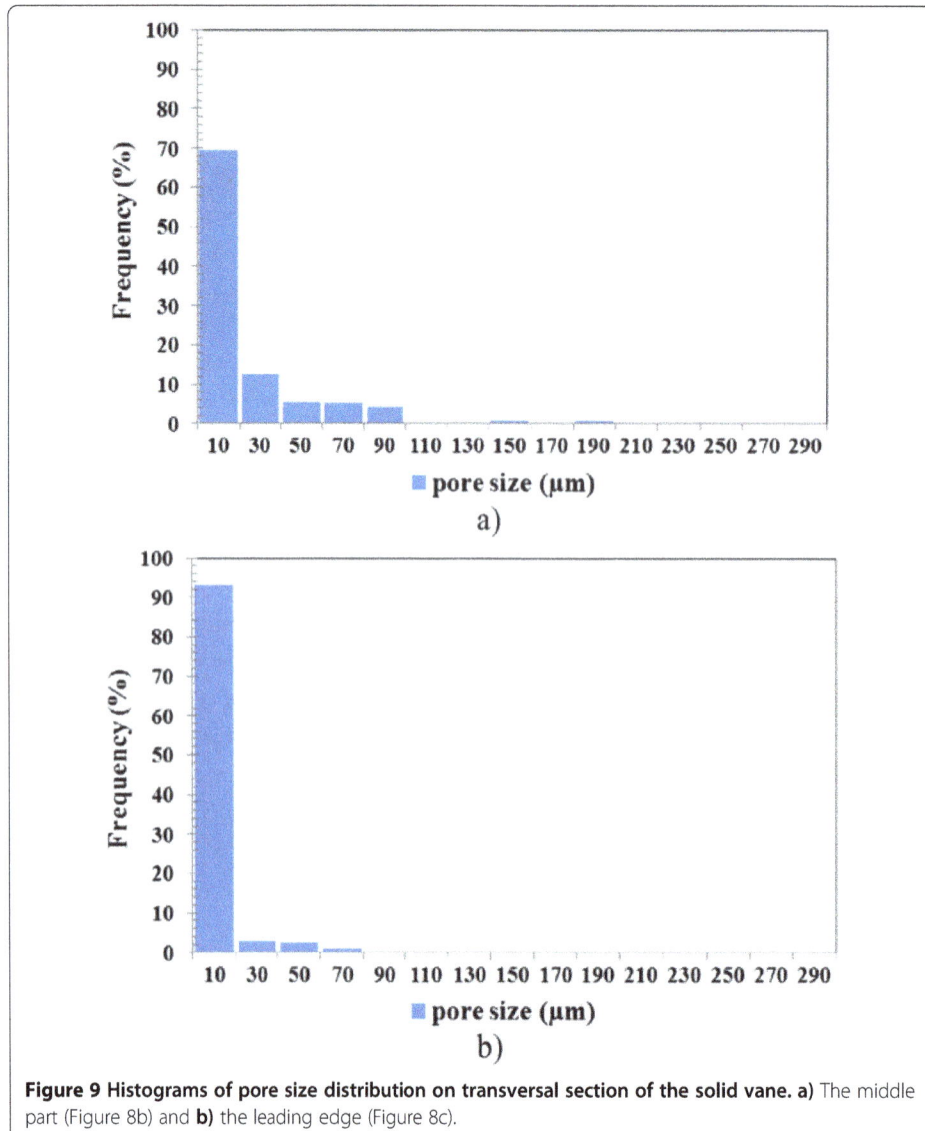

Figure 9 Histograms of pore size distribution on transversal section of the solid vane. a) The middle part (Figure 8b) and **b)** the leading edge (Figure 8c).

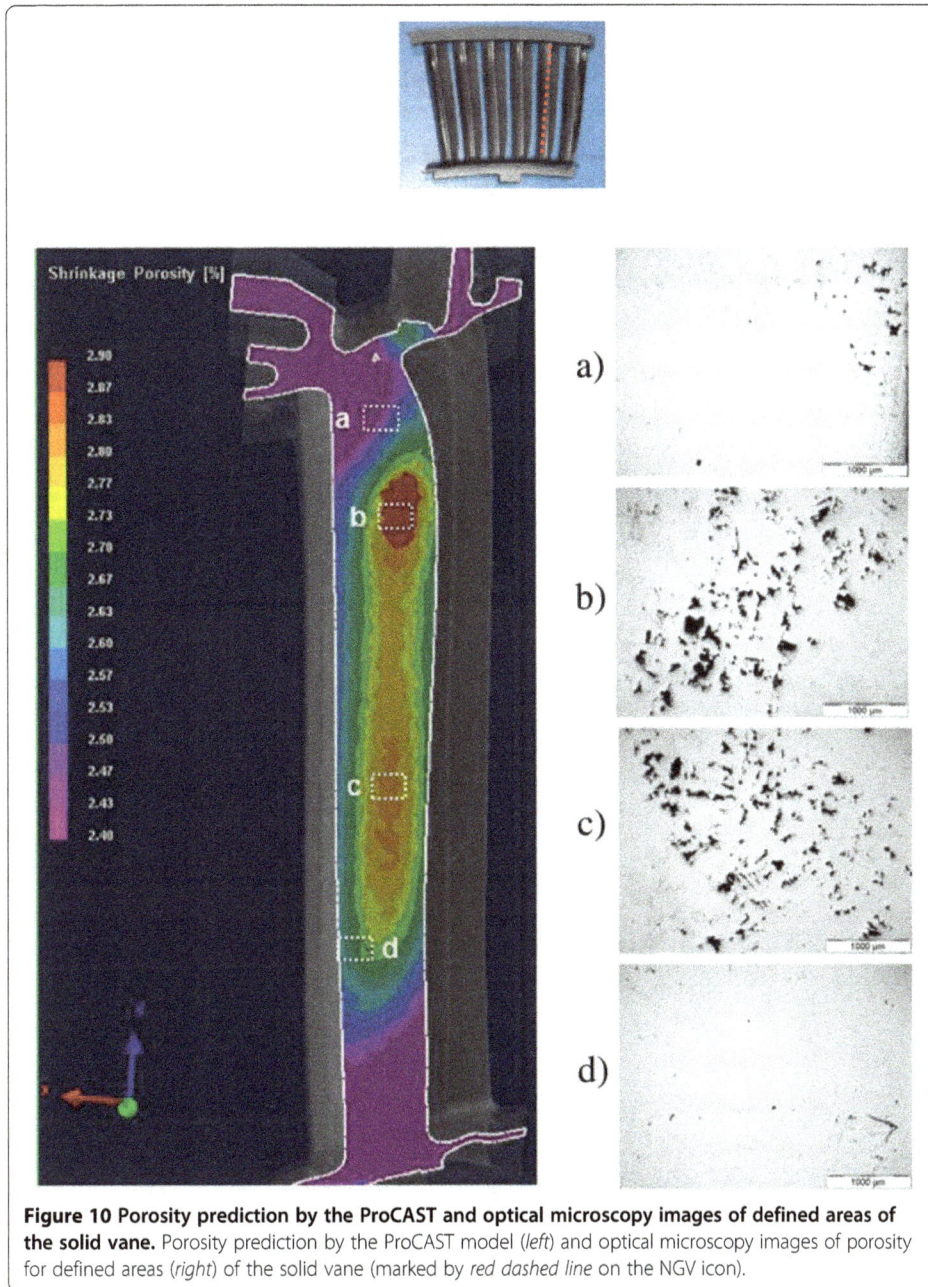

Figure 10 Porosity prediction by the ProCAST and optical microscopy images of defined areas of the solid vane. Porosity prediction by the ProCAST model (*left*) and optical microscopy images of porosity for defined areas (*right*) of the solid vane (marked by *red dashed line* on the NGV icon).

mold. Therefore, the experimental measurements on the mold can yield lower temperatures than the real temperatures, as seen from Figure 5. These shortcomings could also lead to the drop of temperature readings from the thermocouple placed in the ceramic mold at the moment of melt pouring which could result in thermocouple shifting due to thermal expansion of ceramic mold (Figure 5).

Table 4 The porosity characteristics of the longitudinal section of the solid vane on Figure 10

Area	a	b	c	d
Local porosity fraction [%]	0.19	4.87	4.82	0.14
Average pore size [μm]	32	113	27	29

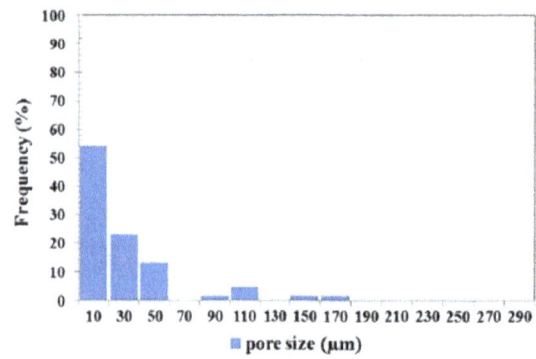

a)

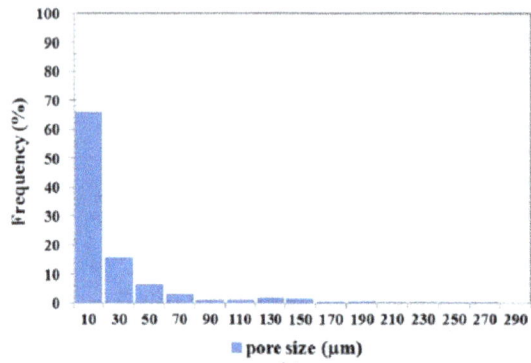

b)

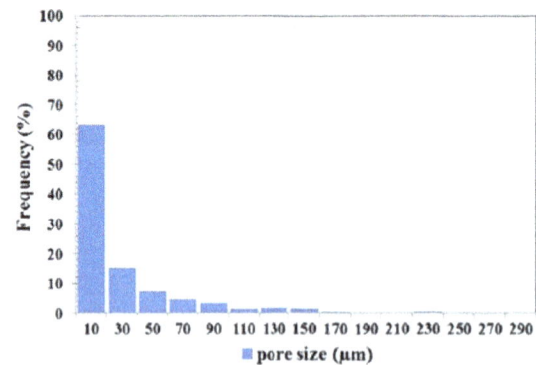

c)

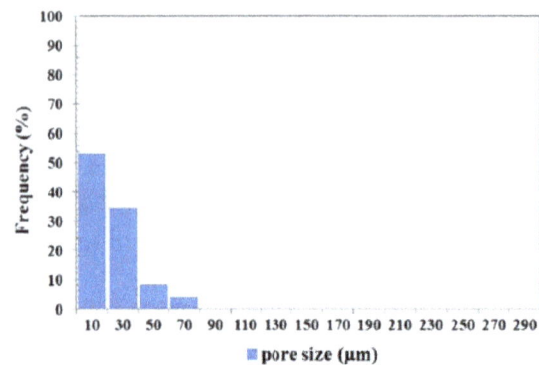

d)

Figure 11 Histograms of pore size distribution on the longitudinal section of the solid vane.
a) Area on Figure 10a, **b)** area on Figure 10b, **c)** area on Figure 10c, and **d)** area on Figure 10d.

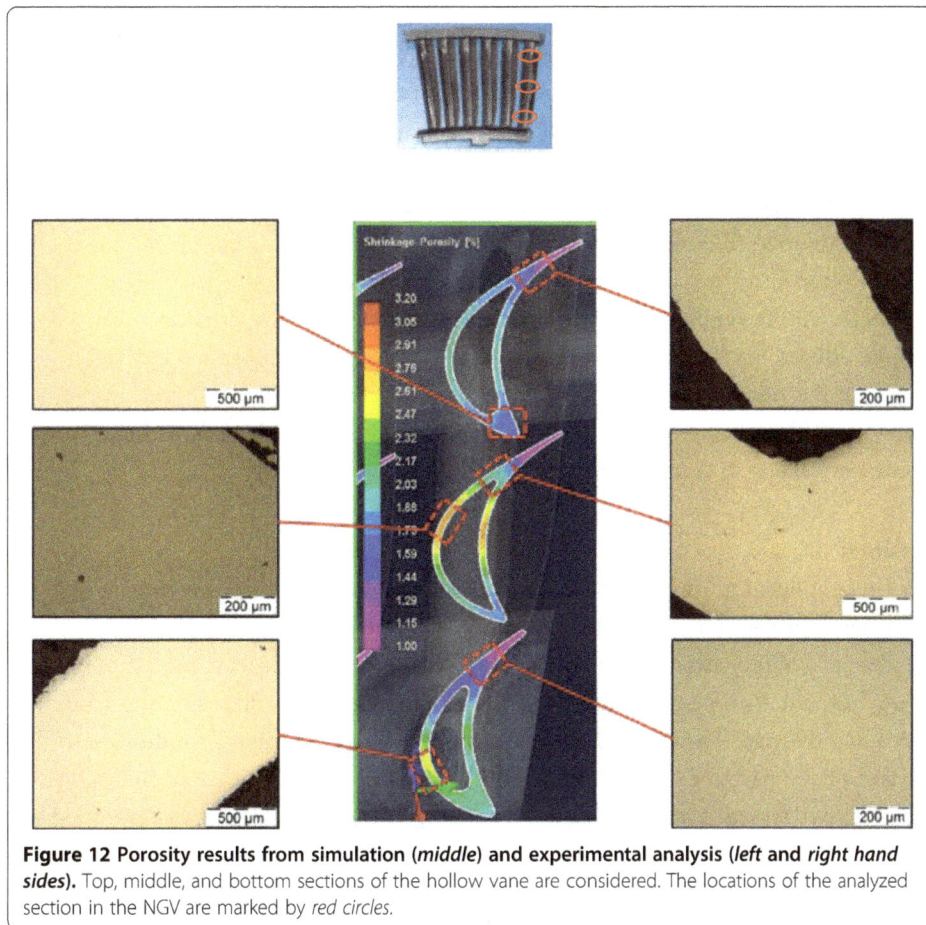

Figure 12 Porosity results from simulation (*middle*) and experimental analysis (*left* and *right hand sides*). Top, middle, and bottom sections of the hollow vane are considered. The locations of the analyzed section in the NGV are marked by *red circles*.

It should be noted that the possible 'air gaps' and the cement were not considered in the thermal model, since it would increase enormously the time required for calculations. Another experimental procedure should be developed for more accurate recording of thermal history in the ceramic mold. Nevertheless, the thermal histories of metal and wrap were accurately predicted as a result of the right selection of thermophysical parameters of the ceramic mold in the thermal model.

Accuracy of the model for porosity prediction

Analysis of the overall porosity prediction shows that the model predicts well the location of hot spots and areas prone to porosity formation throughout the NGV. These areas are located mainly in the solid vanes and this was confirmed by the experimental characterization of the as-cast NGV. A very good match between the simulation predictions and the experimental results was found in many NGV areas. However, the model tends to slightly underestimate porosity in the areas located in the thickest parts of the

Table 5 The porosity characteristics of the transversal section of the hollow vane on Figure 12

Area	Top	Middle	Bottom
Average porosity fraction [%]	0.06	0.08	0.08
Average pore size [μm]	19	17	15

NGV (Figure 10b). This discrepancy could be related to liquid pools which can be formed in those areas during solidification, as reported recently by Kang et al. [22]. Another shortcoming of the model is the overestimation of shrinkage porosity in the thinnest parts of the NGV, which are the hollow vanes with a wall thickness nearly 1 mm. This effect could be explained by formation of skin which can significantly affect the local thermal history of the metal in the thin parts. It should be noted that the rapid skin formation due to freezing of melt with a colder ceramic mold is not taken into account by the model.

All in all, it can be outlined that the simulation tool for porosity prediction can be successfully utilized for further improvement of NGV design. Its application can significantly reduce the number of expensive experimental casting trials which are typically required to find the suitable casting parameters and to develop a manufacturing route for investment casting of complex shape parts at industrial scale.

Conclusions

Investment casting of NGV from Ni-based superalloys was simulated by means of a finite element model. The simulation strategy is targeted to predict the heat exchange during solidification and cooling and the porosity. The casting assembly, consisting of the hollow ceramic mold with NGV-shape interior and insulation wrap, is created and meshed. The thermophysical parameters and boundary conditions are defined for all the parts of the casting assembly, and simulation is carried out using ProCAST. Experimental casting trials are performed for validation of the developed models.

The thermal history of the metal and the insulation wrap during investment casting was accurately predicted. The critical thermal-physical parameters of the casting system were obtained either from the literature or by an inverse simulation procedure by comparing the simulation results for simple casting geometries with experimental data. The hot spots and areas with enhanced porosity which are located in the thickest parts of the NGV were accurately predicted. In addition, the porosity predictions were in good agreement with the experimental results in many NGV areas. The shortcomings of the porosity predictions include a slight underestimation of porosity in some very thick areas and an overestimation of shrinkage porosity in the thinnest parts of the NGV. It is concluded that the developed modeling tool can be successfully utilized for further improvement of NGV design, allowing to minimize the number of casting trials.

Endnote

[a]Piping is the formation of pipes during solidification. Pipes are open-air shrinkage defects which form at the surface of the casting and burrow into the casting.

Competing interests

The authors declare that they have no competing interests.

Authors' contributions

AJT applied thermal model and porosity prediction model and prepared the initial draft of the manuscript. OK and LC provided materials science guidance and expertise in modeling. LM performed experimental casting trials. ECM contributed with the overall development of the main concepts presented in this paper. MR performed quantitative analysis of porosity in the as-cast NGV. SM and IS helped with the validation of the porosity tool and manuscript writing. JL formulated the concept of this work and conceived the workflow, as well as provided materials science guidance. All authors contributed to the manuscript. The final version was prepared by IS and JL and approved by all authors. All authors read and approved the final manuscript.

Acknowledgements
This investigation was carried out in frame of the VANCAST project (EU, FP7, ERA-NET MATERA+). SM and IS acknowledge gratefully the Spanish Ministry of Economy and Competitiveness for financial support through the Ramon y Cajal fellowships. Prof. A. Zryd (Maxwell Technologies SA) and Dr. A. Faes (CSEM SA) are greatly acknowledged for the inverse simulations of experimental casting trials of easy geometry parts as those results constituted the seed for experimental work which had led to this manuscript.

Author details
[1]University of Applied Sciences and Arts Western Switzerland, Sion, Switzerland. [2]CALCOM-ESI, Lausanne, Switzerland. [3]Precicast Bilbao, Bilbao, Spain. [4]IMDEA Materials Institute, C/Eric Kandel 2, Getafe 28906, Madrid, Spain. [5]Department of Materials Science, Polytechnic University of Madrid, Madrid, Spain.

References
1. Razak AMY (2007) Industrial gas turbines: performance and operability. Woodhead Publishing Limited, Cambridge, UK
2. Reed RC (2006) The Superalloys: Fundamentals and Applications. Cambridge University Press, Cambridge, UK
3. Pattnaik S, Karunakar DB, Jha PK (2012) Developments in investment casting process—a review. J Mater Proc Tech 212:2332–2348, doi:10.1016/j.jmatprotec.2012.06.003
4. Anglada E, Meléndez A, Maestro L, Domiguez I (2013) Adjustment of numerical simulation model to the investment casting process. Proc Eng 63:75–83, doi:10.1016/j.proeng.2013.08.272
5. Rafique MMA, Iqbal J (2009) Modeling and simulation of heat transfer phenomena during investment casting. Int J Heat Mass Transf 52:2132–2139, doi:10.1016/j.ijheatmasstransfer.2008.11.007
6. Stefanescu DM (2009) Science and Engineering of Casting Solidification, 2nd edn. Springer Science + Business Media, New York, NY, USA
7. Piwonka TS, Flemings MC (1966) Pore formation in solidification. Trans AIME 236(8):1157–1165
8. Pellini WS (1953) Factors which determine riser adequacy and feeding range. AFS Transactions 61:61–80
9. Niyama E, Uchida T, Morikawa M, Saito S (1981) Predicting shrinkage in large steel castings from temperature gradient calculations. AFS Int Cast Met J 6(2):16–22
10. Carlson KD, Beckermann C (2009) Prediction of shrinkage pore volume fraction using a dimensionless Niyama criterion. Metall Mater Trans A 40:163–175, doi:10.1007/s11661-008-9715-y
11. Kubo K, Pehlke RD (1985) Mathematical modeling of porosity formation in solidification. Metall Mater Trans B 16:359–366, doi: 10.1007/BF02679728
12. Lee PD, Hunt JD (2001) Hydrogen porosity in directionally solidified aluminium copper alloys: a mathematical model. Acta Mater 49:1383–1398, doi:10.1016/S1359-6454(01)00043-X
13. Lee PD, Chirazi A, Atwood RC, Wang W (2004) Multiscale modeling of solidification microstructures, including microsegregation and microporosity, in an Al-Si-Cu alloy. Mater Sci Eng A 365:57–65, doi:10.1016/j.msea.2003.09.007
14. Carlson KD, Lin Z, Beckermann C (2007) Modeling the effect of finite-rate hydrogen diffusion on porosity formation in aluminum alloys. Metall Mater Trans B 38:541–555, doi:10.1007/s11663-006-9013-2
15. Pequet C, Rappaz M, Gremaud M (2002) Modeling of microporosity, macroporosity, and pipe-shrinkage formation during the solidification of alloys using a mushy-zone refinement method: applications to aluminum alloys. Metall Mater Trans A 33:2095–2106, doi:10.1007/s11661-002-0041-5
16. Couturier G, Rappaz M (2006) Effect of volatile elements on porosity formation in solidifying alloys. Model Simul Mater Sci Eng 14(2):253–271, doi:10.1088/0965-0393/14/2/009
17. Couturier G, Rappaz M (2006) Modeling of porosity formation in multicomponent alloys in the presence of several dissolved gases and volatile solute elements. TMS Annual Meeting, San Antonio, TX, USA, pp 143–152
18. Stefanescu DM (2005) Computer simulation of shrinkage related defects in metal castings—a review. Inter J Cast Metal Res 18(3):129–143
19. Lee PD, Chirazi A, See D (2001) Modeling microporosity in aluminum–silicon alloys: a review. J Light Metals 1:15–30, doi:10.1016/S1471-5317(00)00003-1
20. Overfelt RA, Sahai V, Ko YK, Berry JT (1994) Porosity in cast equiaxed alloy 718. In: Loria EA (ed) Proceedings of the TMS Meeting., p 189
21. Monastyrskiy VP (2010) Modeling of porosity formation in Ni-based superalloys. In: Choi JK (ed) Proceedings of the 8th Pacific Rim International Conference on Modeling of Casting and Solidification Process., p 89
22. Kang M, Gao H, Wang J, Ling L, Sun B (2013) Prediction of microporosity in complex thin-wall castings with the dimensionless Niyama criterion. Materials 6:1789–1802
23. Calba L, Lefebvre D (2008) Modeling the investment casting process. ESI-GROUP Resource Center, Paris
24. Harris K, Erickson GL, Schwer RE (1984) MAR-M247 derivations—CM247 LC DS alloy, CMSX single crystal alloys, properties and performance. In: Gell M, Kortovich CS, Bricknell RH, Kent WB, Radvich JF (eds) Proceedings of the 5th International Symposium on Superalloys, TMS., p 221
25. Handbook ASM (2010) Metals Process Simulation, vol 22B. ASM International, Ohio, USA
26. ProCast user Manual & Technical Reference (2007) Version 6.1. ESI software, France
27. Rappaz M, Bellet M, Deville M, Snyder R (2002) Numerical modeling in materials science and engineering. Springer-Verlag, Berlin, Germany
28. Dantzig JA, Rappaz M (2009) Solidification. EPFL-Press, Lausanne, Switzerland
29. Handbook ASM (2008) Casting, vol 15. ASM International, Ohio, USA

30. O'Mahoney D, Browne DJ (2000) Use of experiment and an inverse method to study interface heat transfer during solidification in the investment casting process. Exper Thermal Fluid Sci 22:111–122, doi:10.1016/S0894-1777(00)00014-5

31. Konrad CH, Brunner M, Kyrgyzbaev K, Völkl R, Glatzel U (2011) Determination of heat transfer coefficient and ceramic mold material parameters for alloy IN738LC investment castings. J Mater Proc Tech 211:181–186, doi:10.1016/j.jmatprotec.2010.08.031

32. Santos CA, Quaresma JMV, Garcia A (2001) Determination of transient interfacial heat transfer coefficients in chill mold castings. J Alloys Compd 319:174–186, doi: 10.1016/S0925-8388(01)00904-5

33. Dong Y, Bu K, Dou Y, Zhang D (2011) Determination of interfacial heat-transfer coefficient during investment-casting process of single-crystal blades. J Mater Proc Tech 211:2123–2131, doi:10.1016/j.jmatprotec.2011.07.012

34. Sahai V, Overfelt RA (1995) Contact conductance simulation for alloy 718 investment casting of various geometries. Tran Amer F 103:627–632

35. Yuang XL, Lee PD, Brooks RF, Wunderlich R (2004) The sensitivity of investment casting simulations to the accuracy of thermophysical properties values. In: Proceedings of the International Symposium on Superalloys, TMS., p 951

8

Investment casting of nozzle guide vanes from nickel-based superalloys: part II – grain structure prediction

Agustín Jose Torroba[1], Ole Koeser[2], Loic Calba[2], Laura Maestro[3], Efrain Carreño-Morelli[1], Mehdi Rahimian[4], Srdjan Milenkovic[4], Ilchat Sabirov[4] and Javier LLorca[4,5*]

* Correspondence:
javier.llorca@imdea.org
[4]IMDEA Materials Institute, C/Eric Kandel 2, 28906 Getafe, Madrid, Spain
[5]Department of Materials Science, Polytechnic University of Madrid, 28040 Madrid, Spain
Full list of author information is available at the end of the article

Abstract

The control of grain structure, which develops during solidification processes in investment casting of nozzle guide vanes (NGVs), is a key issue for optimization of their mechanical properties. The main objective of this part of the work was to develop a simulation tool for predicting grain structure in the new generation NGVs made from MAR-M247 Ni-based superalloy. A cellular automata - finite element (CAFE) module is employed to predict the three-dimensional (3D) grain structure in the as-cast NGV. The grain structure in the critical sections of the experimentally cast NGV is carefully analyzed, the experimental results are compared with the modeling outcomes, and the model is calibrated via tuning parameters which govern grain nucleation and growth. The grain structures predicted by the calibrated model show a very good accordance with the real ones observed in the critical sections of the as-cast NGV. It is demonstrated that the calibrated CAFE model is a reliable tool for the foundry industry to predict grain structure of the as-cast NGVs with very high accuracy.

Keywords: Ni-based superalloys; Investment casting; Nozzle guide vanes; Modeling; Cellular automata finite element (CAFE) module; Grain structure

Background

Solidification microstructure is of great importance for controlling the properties and the quality of the nozzle guide vanes (NGVs) produced via investment casting. In the last decades, emergence of accurate simulation capabilities and development of rigorous analytical models have contributed to a better understanding of solidification process and enabled prediction of solidification grain structure.

Phase-field models have attracted considerable interest since the early 90s to describe phase transitions for a wide range of systems [1]. Phase-field models based on the rigorous framework of reversible thermodynamics [2,3] have been developed to describe both the solidification of pure materials [4] and binary alloys [5,6]. They also have been used extensively to simulate numerically dendritic growth into an undercooled liquid [7-10]. These computations provide realistic simulations of dendritic growth, including side arm production and coarsening. Systems with three phases as well as grain structures with an ensemble of grains of different crystallographic orientations have also been modeled by the phase-field method using a vector-valued phase field [11-16]. The shortcoming of phase-field method is the very fine grid size (<1 µm) required to capture the solid liquid boundary layer, limiting the size of domains which can be simulated (e.g., <1 mm^2 in 2D) [17].

Cellular automata (CA) is another technique widely applied in modeling of solidification. The CA technique was originally developed by Hesselbarth and Gbel [18]. It is based on the division of the simulation domain into cells, which contain all the necessary information to represent a given solidification process. Each cell is assigned information regarding the state (solid, liquid, interface, grain orientation, etc.) and the value of the calculated fields (temperature, composition, solid fraction, etc.). In addition, a neighborhood configuration is selected, which includes the cells that can have a direct influence on a given cell. The fields of the cells are calculated by analytical or numerical solutions of the transport and transformation equations. The change of the cell states is calculated through transition rules, which can be analytical or probabilistic. When these rules are probabilistic, the technique is called stochastic. The important feature of the method is that all cells are considered at the same time to define the state of the system in the following time step. Thus, the computational time step can be directly related to the physical time step.

The CA technique is often coupled with finite difference (FD) or finite element (FE) methods that makes it possible to obtain more accurate results. The modeling technique based on combining the CA method for tracking the solid liquid front location with a FD solution of solute diffusion is referred to as CAFD model. This model was successfully used to simulate dendritic growth in a range of alloys [19,20]. The CAFD model solves the solute conservation equation subjected to equilibrium conditions at the solid liquid interface. The model simulates the solutal interaction within the developing dendritic network, predicting when overgrowth or branching will occur. The model was applied to investigate the effect of changing the pulling velocity on directionally solidified dendritic structures in Ni-based superalloys [20].

Rappaz and Gandin [21] started to explore the possibility of coupling FE heat flow computations with two-dimensional (2D) CA calculations describing the mechanisms of nucleation and growth of dendritic grains. The first model was applied only to small specimens of uniform temperature, i.e., it was an isothermal model. One year later, Gandin and Rappaz [22] extended the model to non-uniform temperature situations, and the modeling tool was referred to as cellular automata - finite element (CAFE) module. In 1997, Gandin and Rappaz [23] proposed a three-dimensional (3D) CA algorithm to model the growth of octahedral dendritic grains from the liquid phase. The 3D CAFE module was able to account for different cooling conditions, crystallographic orientations, and growth kinetics parameters. The CAFE model is based on the concept of marginal stability to uniquely define the dendrite tip radius, allowing the analytical solution of Kurz, Giovanola, and Trivedi (KGT) to be applied [24]. CAFE models do not incorporate details of dendritic growth, but they are very useful for simulating grain structures on orders of magnitude larger simulation domains than is possible using the phase-field method. Another advantage of CAFE models is a clear prediction of grain size and shape.

The developed 3D CAFE module found a wide application for simulation of solidification grain structure in casting of complex shape parts. Gandin et al. [25] were first to apply it for grain structure prediction in directionally solidified blades. In particular, the growth competition occurring among columnar grains was directly reproduced, taking into account the crystallographic orientation of the grains and the temperature evolution in representative 3D investment cast parts. Seo et al. [26] successfully applied the model to predict grain structure in turbine blades produced via investment casting of CM247LC Ni-based superalloy. The overall appearance of grain structure in the cast blades was very well reproduced by the CAFE module at various investment casting conditions. Wang et al. [27] applied the CAFE module for simulation of grain selection during single crystal casting of a DD403 Ni-based superalloy with spiral grain selector, and the model was validated experimentally via investment casting using different spiral geometries. It was demonstrated that the CAFE module was a reliable tool for optimization of crystal orientation via manipulation with mold geometry.

The second part of our work aims to develop the CAFE module for grain structure prediction in the new generation NGVs to be produced via investment casting in a real plant process. Along with the thermal model and ProCast model for porosity prediction, it will form a modeling tool for further optimization of the NGV design at industrial scale.

Description of the modeling tool

The CAFE module is a software tool which allows the prediction of structures of castings in which columnar and equiaxed crystals are formed. CAFE is based on a stochastic model which combines grain nucleation and grain growth algorithms with the calculation of the heat transfer by FE as described by Gandin et al. [28]. These algorithms and their parameters are considered below.

Grain nucleation algorithm

The CAFE model assumes that grains may nucleate on the surface of the mold (surface nucleation) or in the bulk (volume nucleation). Both types of nucleation are described by Gaussian distributions as

$$\frac{dn}{d\,\Delta T} \qquad \frac{n_{\max}}{\sigma_{\Delta T} \times \sqrt{2\pi}} \exp\left[-\frac{1}{2}\left(\frac{\Delta T - \Delta T_m}{\sigma \Delta T}\right)^2\right] \qquad\qquad 1$$

where ΔT is the local undercooling, ΔT_m the mean undercooling, $\sigma_{\Delta T}$ the standard deviation, and $n_{\max}$ the maximum nucleation density which can be reached when all the nucleation sites are activated while cooling. These parameters mainly depend on the selected alloy, the shape of the cast part, its volume, and the casting procedure. Their determination is difficult in our case, since no previous references can be found in the literature. Calibration with experimental results is the only possible procedure for their precise estimation. Precicast Bilbao has provided a set of preliminary values from previous experimental work on investment casting of Ni-based superalloys (Table 1), but

Table 1 Initial values of the parameters for the surface and volume nucleation algorithms

Parameter	Surface	Volume
$n_{\max}$	$1 \cdot 10^6$	$1 \cdot 10^9$
ΔT_m	$6\,\mathrm{C}$	$6\,\mathrm{C}$
$\sigma_{\Delta T}$	$1\,\mathrm{C}$	$1\,\mathrm{C}$

their final calibration and validation are still necessary for reaching more accurate results.

Grain growth algorithm

Grain growth kinetics during metal solidification can be determined by the undercooling at dendrite tips. Different theories have been developed for growth kinetics, with the Lipton-Glicksman-Kurz (LGK) model selected as the one to make the linkage between growth velocity and the given undercooling as described by Kurz et al. [24] and Lipton et al. [29]. Thus, the following simplified equation is used by CAFE module for calculations:

$$v \ \Delta T \quad a_2 \Delta T^2 \quad a_3 \Delta T^3 \qquad\qquad 2$$

where v is the dendrite tip velocity, ΔT the undercooling, and a_2 and a_3 are the specific material parameters. The value of the local undercooling (ΔT) is provided by the thermal resolution of the model during ProCAST calculation [28] as the difference between the liquidus temperature and the local temperature in each finite element. The value of the parameters a_2 and a_3 may be calculated by the CAFE module based on the alloy chemical composition and the phase diagram properties, such as liquidus temperature (1,366C for Mar-M247 Ni-based superalloy), Gibbs-Thomson coefficient of the solvent (2 10^{-7} for Ni), chemical concentration of each solute element (Table 2), liquidus slope, and partition coefficient of each element. For multi-component systems, the latter two can only be obtained with the help of thermodynamic databases [30]. Once input data are known, values of parameters a_2 and a_3 are calculated (Table 3).

At the same time, when grains grow, their orientation is randomly determined by the model. Since growth velocity of each grain tip is calculated by undercooling in front of each dendrite tip, the grains better aligned with the maximum undercooling direction will grow faster than those which are not (Figure 1). This effect makes possible the growth of secondary and tertiary side arms on dendrites having favorable orientation as shown in Figure 1.

To obtain good results from calculation, a proper cell size must be defined. If the cell size is too big, poor results will be achieved, but if too small, time calculation will shoot up. A proper cell size is given by the value of the typical secondary dendrite arm spacing measured in the casting, being values between 50 and 100 μm most commonly used [20]. This small size of the cells means that the amount of cells to be considered for the calculation can be extremely high. To tackle this problem, the software first divides the part into blocks, which will be composed by a defined number of cells (Figure 2). The number of cells contained in a single block is chosen by the user in order to improve the calculation. Once the software has divided the part into blocks, each block is given one of three statuses: *stand by*, *active*, or *inactive* (Figure 2). Blocks in *stand by* position are located in the liquid but out of the mushy zone where solidification is taking place. The *active* ones are located in the mushy zone and, therefore,

Table 2 Chemical composition of Mar-M247 Ni-based superalloy

Ni	C	Cr	Co	Mo	W	Ta	Al	Ti	Hf
Base	0.15	8.4	10	0.7	10	3.1	5.5	1.05	1.4

Table 3 Values of parameters for the growth kinetics function of Mar-M247 Ni-based superalloy (CAFE database)

Parameters	Values
a_2	$6.3 \cdot 10^{-7}$
a_3	$3.33 \cdot 10^{-6}$

these blocks enter into calculation by the solver. Finally, the *inactive* blocks correspond to those in the solid zone. The *inactive* blocks have undergone calculation and their results have been stored. This approach is applied in order to reduce the number of blocks that are in calculation simultaneously, so the solution can be achieved faster. This technique is especially useful for parts produced via directional solidification as demonstrated by Qingyan et al. [31], since the number of blocks that are being calculated simultaneously, can be much smaller than in the cases when the solidification process is more distributed.

All parameters defined above are considered by the software to calculate the transition from equiaxed to columnar grains. No specific parameter is used to determine this transition, being described by the grain growth competition, which depends on the mean undercooling in each location of the mushy zone [23].

It is seen that the most important data to perform correct calculations with the CAFE module are provided by the thermal model, so a reliable thermal model well describing the real solidification process is required. A detailed description of the thermal model used in our work can be found in the first part of this manuscript.

Model definition

For the prediction of the NGV grain structure, a reliable thermal model was developed (see part I). This model is imported to the CAFE pre-processor, where parameters necessary to define the nucleation and growth algorithms and the calculation conditions

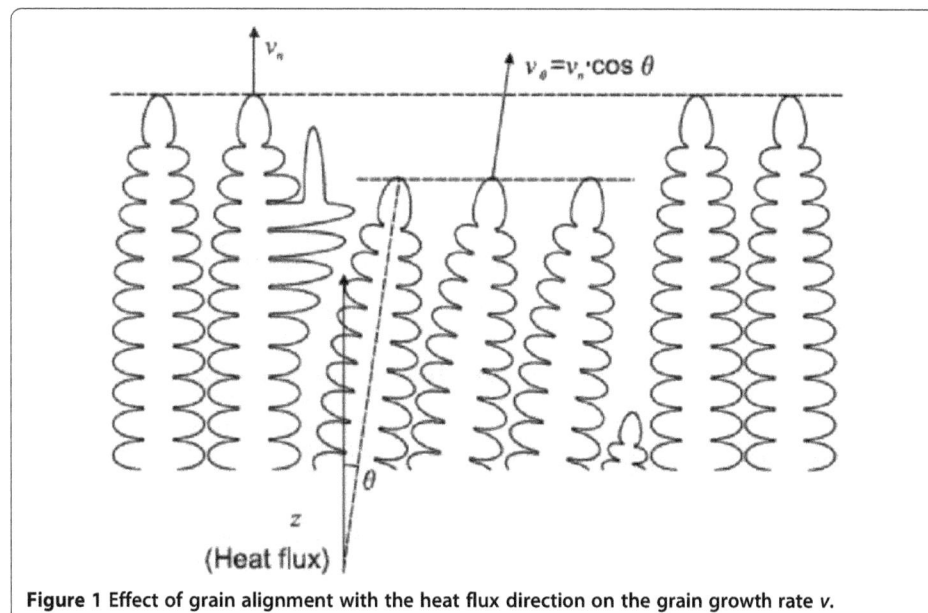

Figure 1 Effect of grain alignment with the heat flux direction on the grain growth rate *v*.

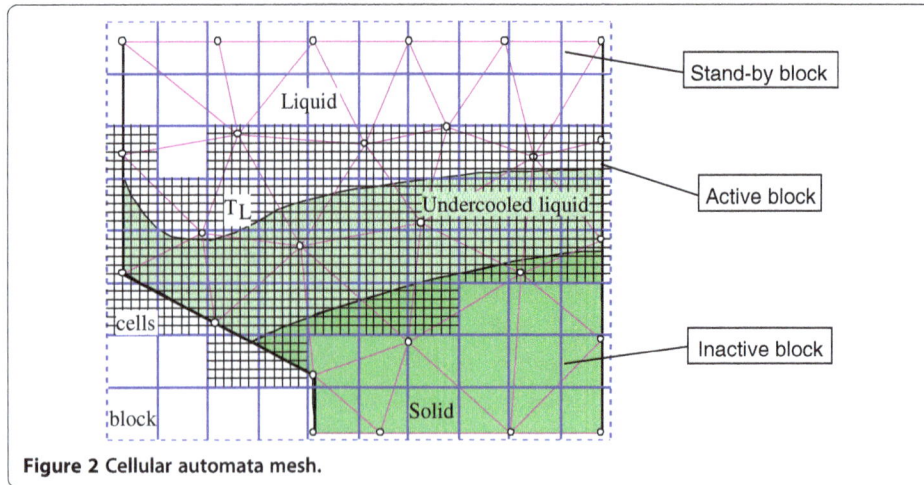

Figure 2 Cellular automata mesh.

are established. For the given example of NGV, the pre-processor is setup as described below.

Domain selection: the model includes not only the cast part but also other elements such as the mold, insulation layer, etc. Therefore, the domain of interest is to be selected. The areas of no interest, such as the feeding system, the pouring cap, etc. can be ignored.

General parameters: the cell size and the number of cells per block are defined at this point. In our case, the selected values are 60 μm for the cell size, and each block is composed of a cube of 10 10 10 cells.

Window definition: the zones for calculations are to be defined. In our case, only critical sections of the NGV are investigated. Modeling results for these sections will be compared with results from experimental analyses of real grain structure. Otherwise, analysis of a whole part would lead to very time-consuming calculations.

Surface nucleation (nucleation algorithm): values that describe Gaussian distribution for surface nucleation (Equation 1) must be defined (Table 1). Since these values depend on the alloy, cast shape and casting procedure, they are calibrated by comparison of modeling and experimental results.

Volume nucleation (nucleation algorithm): no volume nucleation is expected for this part so all parameters for volume nucleation set to zero (Table 4).
Physical data (growth algorithm): defines the growth kinetics of the given alloy (Mar-M247 Ni-based superalloy), parameters a_2 and a_3 for Equation 2 are introduced (Table 3).
Results: after definition of zones for modeling (*window definition*), the CAFE module may request to store additional information such as cuts/planes of interest.

Table 4 Calibrated values of parameters for the surface and volume nucleation algorithms

Parameters	Surface	Volume
n_{max}	$5.5 \, ?10^6$	0
ΔT_m	5.2℃	0
$\sigma_{\Delta T}$	1.2℃	0

Once the CAFE pre-processor is settled, the CAFE solver is run. The obtained grain structure is demonstrated by Visual-Viewer (CALCOM Software).

Methods

Material and experimental procedures

The MAR-M247 Ni-based superalloy was chosen as the material for this investigation. The chemical composition of the material is presented in Table 2. Preparation of the ceramic molds and investment casting process was described in detail in the first part of the manuscript. The as-cast NGV was cut into smaller specimens for analysis of grain structure. The selected areas for grain structure evaluation are shown below in the Conclusions section. The specimens were ground and polished to a mirror-like surface using standard metallographic technique. The polished specimens were etched using a chemical solution consisting of 25 g $FeCl_3$, 60 ml HCl, and 25 ml H_2O to reveal grain structure.

The optical microscope OLYMPUS BX51 (Olympus Corporation, Shinjuku-ku, Japan) was used for characterization of grain structure. At least three images were taken from each area of interest. Quantitative analysis of grain structure (grain size, standard deviation of grain size, and aspect ratio) was performed using ANALYSIS software. The grain size was measured as an equivalent circle diameter due to complex shape of some grains. Aspect ratio was calculated as a ratio of grain length to its width, as specified in the ANALYSIS software.

Results and discussion

Modeling vs. experimental

After first comparison between experimental and calculated results, the parameters governing the nucleation algorithm in the CAFE module were calibrated. The final grain structure predicted by the CAFE module for four critical NGV sections (transversal section of a hollow vane, transversal and longitudinal sections of a solid vane, and longitudinal section of the bottom platform) is compared with the real grain structure of the as-cast NGV. The comparison is based on (1) visual comparison of grain structure for critical sections of the NGV and (2) comparison of average grain size, standard deviation of the grain size, and aspect ratio of grains for these NGV sections.

The final calibrated values for the surface and volume nucleation algorithms are presented in Table 4. As calibrated values will remain the same for the full cast part, the prediction of the grain structure at any location of the piece will be possible. It should be noted that the final values of parameters for surface nucleation are close to the original ones proposed by Precicast Bilbao (Table 1). Nevertheless, even such a small difference can significantly affect the final modeling result.

Grain structure on transversal section of a hollow vane

The location of the selected section is marked by a red circle in the NGV icons in Figure 3, where the general view of the grain structure predicted by CAFE (Figure 3b) and grain structure of the real NGV (Figure 3a) are illustrated. An accurate comparison of the modeling and experimental results can be made analyzing optical microscopy images taken at higher magnification from leading and trailing edges of the hollow vane and its middle part (Figure 4). Qualitative and quantitative analysis shows that the

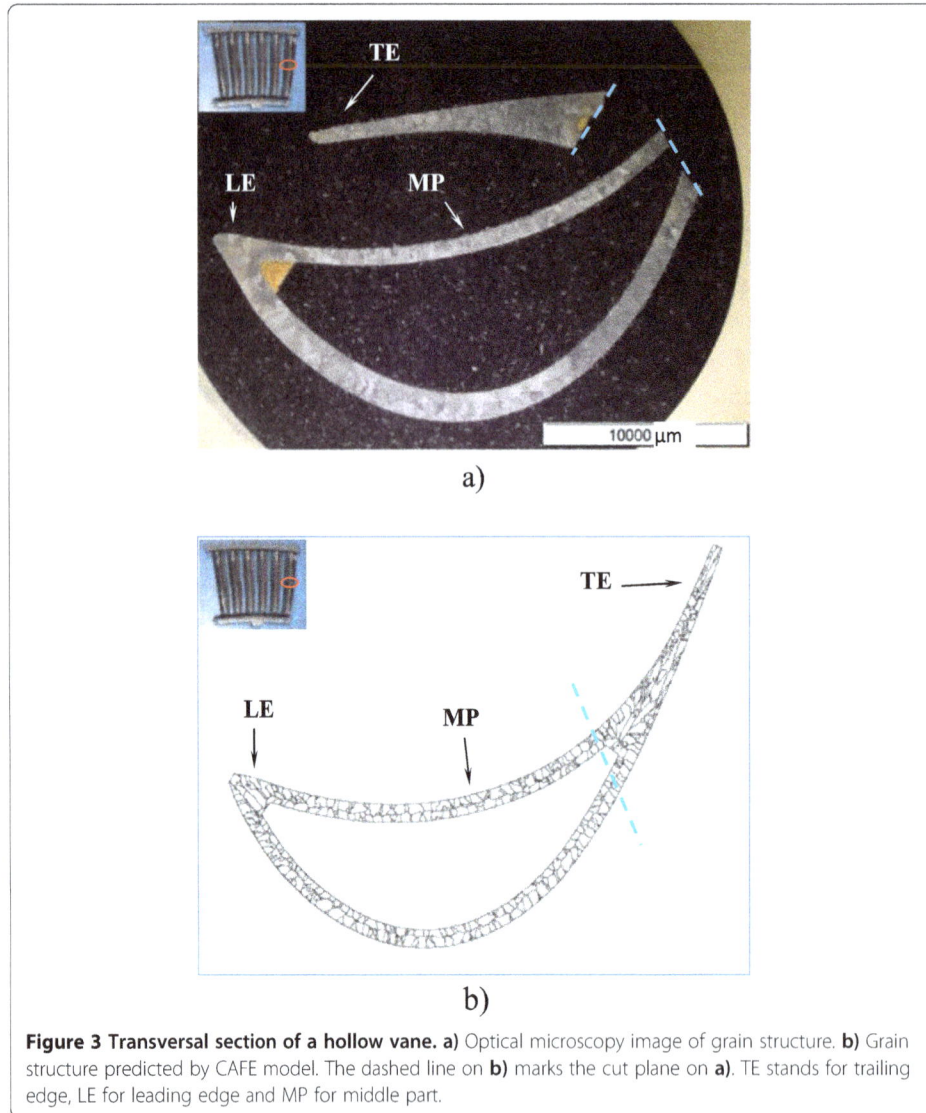

Figure 3 Transversal section of a hollow vane. a) Optical microscopy image of grain structure. **b)** Grain structure predicted by CAFE model. The dashed line on **b)** marks the cut plane on **a)**. TE stands for trailing edge, LE for leading edge and MP for middle part.

modeling results perfectly match the grain structure of the hollow vane in all areas of the hollow vane. Fast solidification occurs in the thin walls (Figure 4b) resulting in the formation of homogeneous grain structure consisting of small equiaxed grains. Both equiaxed and columnar grains are observed in the leading edge (Figure 4a) and trailing edge (Figure 4c) of the hollow vane due to a more complex character of the local heat flux during solidification. The average grain size, its standard deviation, and aspect ratio from modeling and experimental results have nearly the same values in all areas of the hollow vane (Table 5).

Grain structure on transversal section of a solid vane

The analyzed transversal section of the solid vane is marked by a red circle on the NGV icons (Figure 5). A very good agreement between modeling and experimental results is observed (Figure 5). The fast solidification in the thin trailing edge leads to formation of small equiaxed grains (Figure 6a). As the mushy zone moves towards the middle part of the solid vane, the condition of directional solidification is achieved resulting in

a)

b)

c)

Figure 4 Microstructure of transversal section of a hollow vane (Figure 3) at higher magnifications.
Optical microscopy images (left) are compared with CAFE prediction (right): **a)** leading edge, **b)** middle part, and **c)** trailing edge.

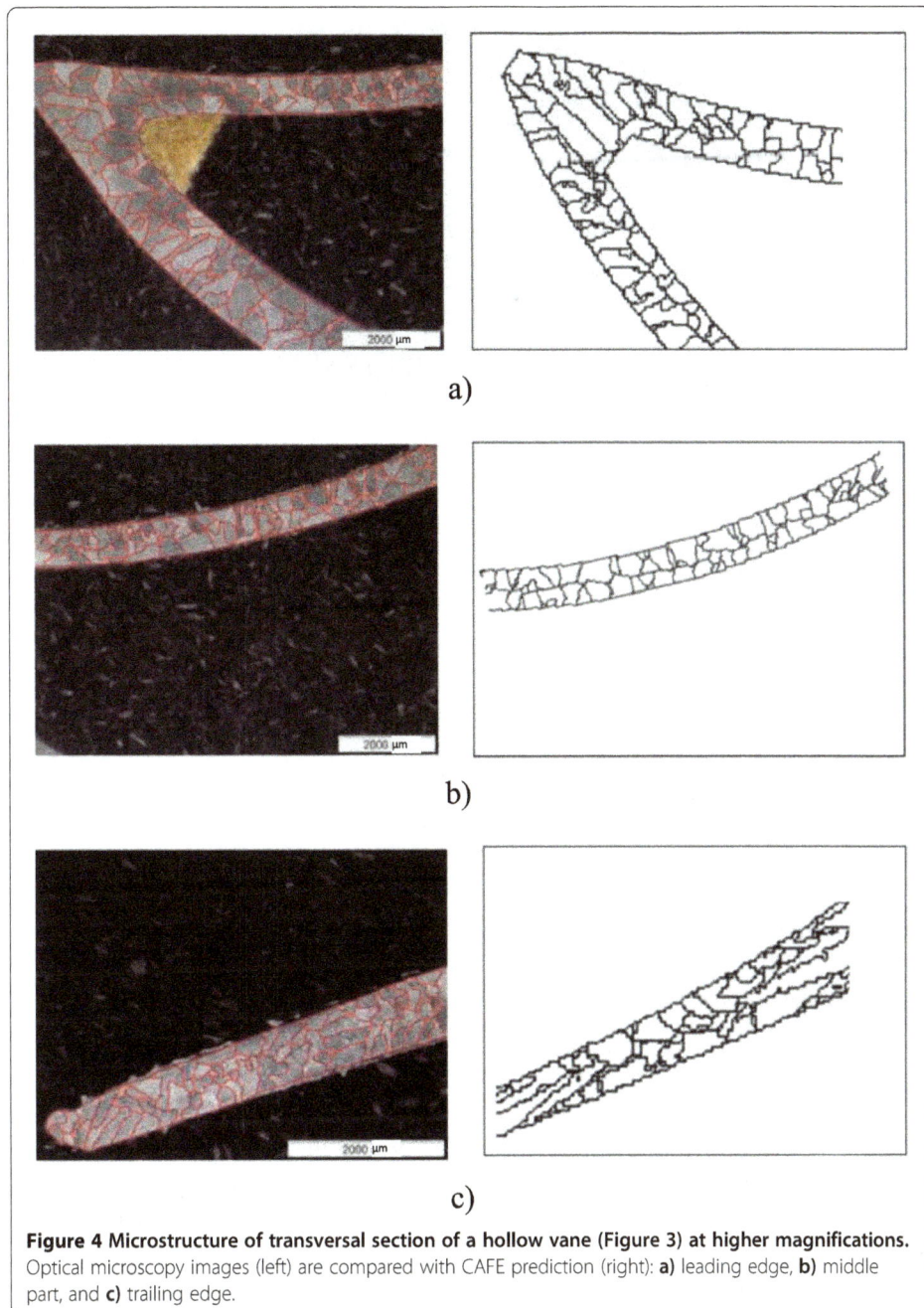

Table 5 Comparison of grain structure predicted by CAFE model with experimental results for transversal section of the hollow vane (Figures 3 to 4)

	Leading edge		Middle part		Trailing edge	
	Experiment	Model	Experiment	Model	Experiment	Model
Average grain size (µm)	756	732	550	541	269	271
Standard deviation (µm)	347	362	226	228	170	165
Aspect ratio	1.7	1.6	1.5	1.4	1.9	2.0

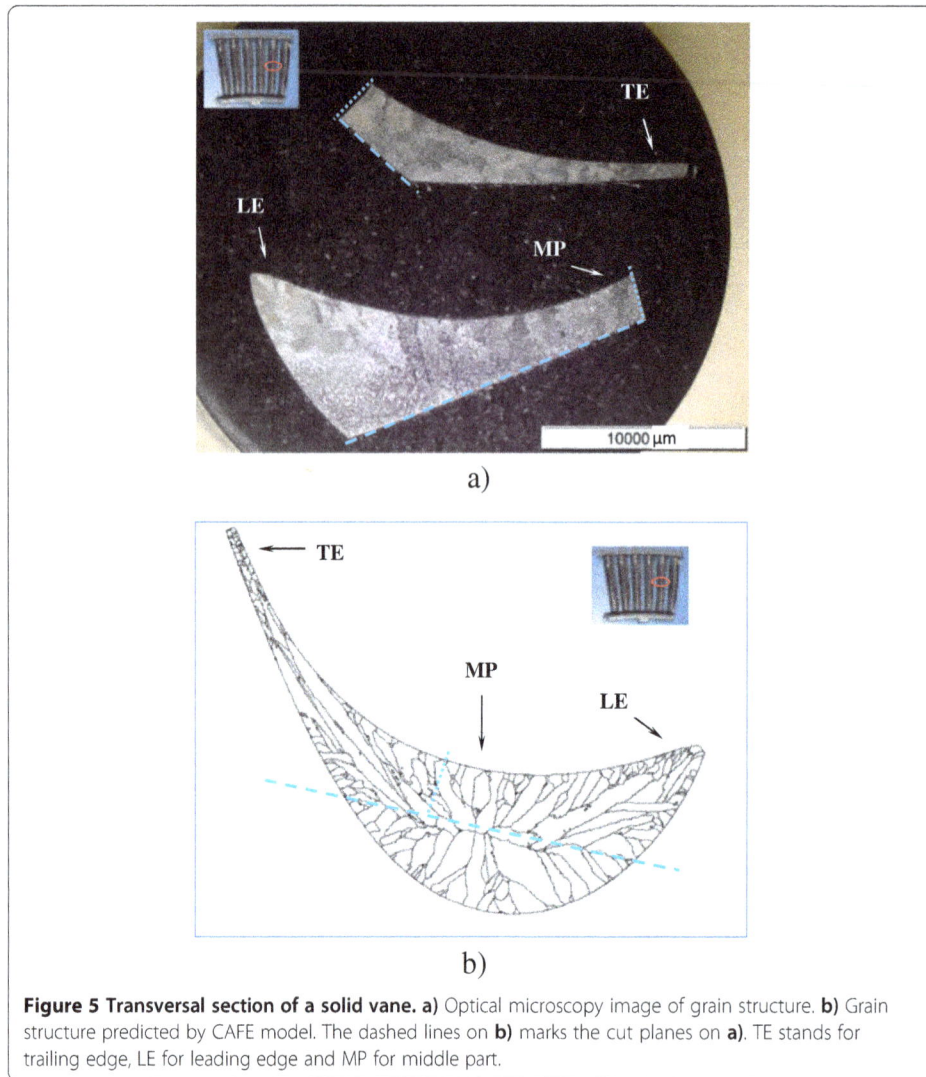

Figure 5 Transversal section of a solid vane. a) Optical microscopy image of grain structure. **b)** Grain structure predicted by CAFE model. The dashed lines on **b)** marks the cut planes on **a)**. TE stands for trailing edge, LE for leading edge and MP for middle part.

transition from equiaxed to columnar grains (Figure 6b). A similar effect is observed on the leading edge of the solid vane, where equiaxed grains initially appear on the edge and transform into columnar grains when solidification proceeds into middle part (Figure 6c).

Average grain size and its standard deviation from the CAFE prediction are in a good accordance with those experimentally measured on the transversal section of the solid vane (Table 6). However, the model predicts somewhat higher aspect ratio in all parts of the transversal section of the solid vanes. Nevertheless, the visual comparison of the predicted and real grain structures shows a good match also for grain shape (Figure 6).

Grain structure on longitudinal section of a solid vane

For analysis of grain structure on the longitudinal section of a solid vane, it was cut along its axis. The cut plane is marked by a blue line on Figure 5b. The location of the analyzed section is marked by a red circle in the NGV icon in Figure 7a, where grain structure of the real NGV is also presented. The grain boundaries have been marked by a red line for easier identification. The grain structure predicted by CAFE model

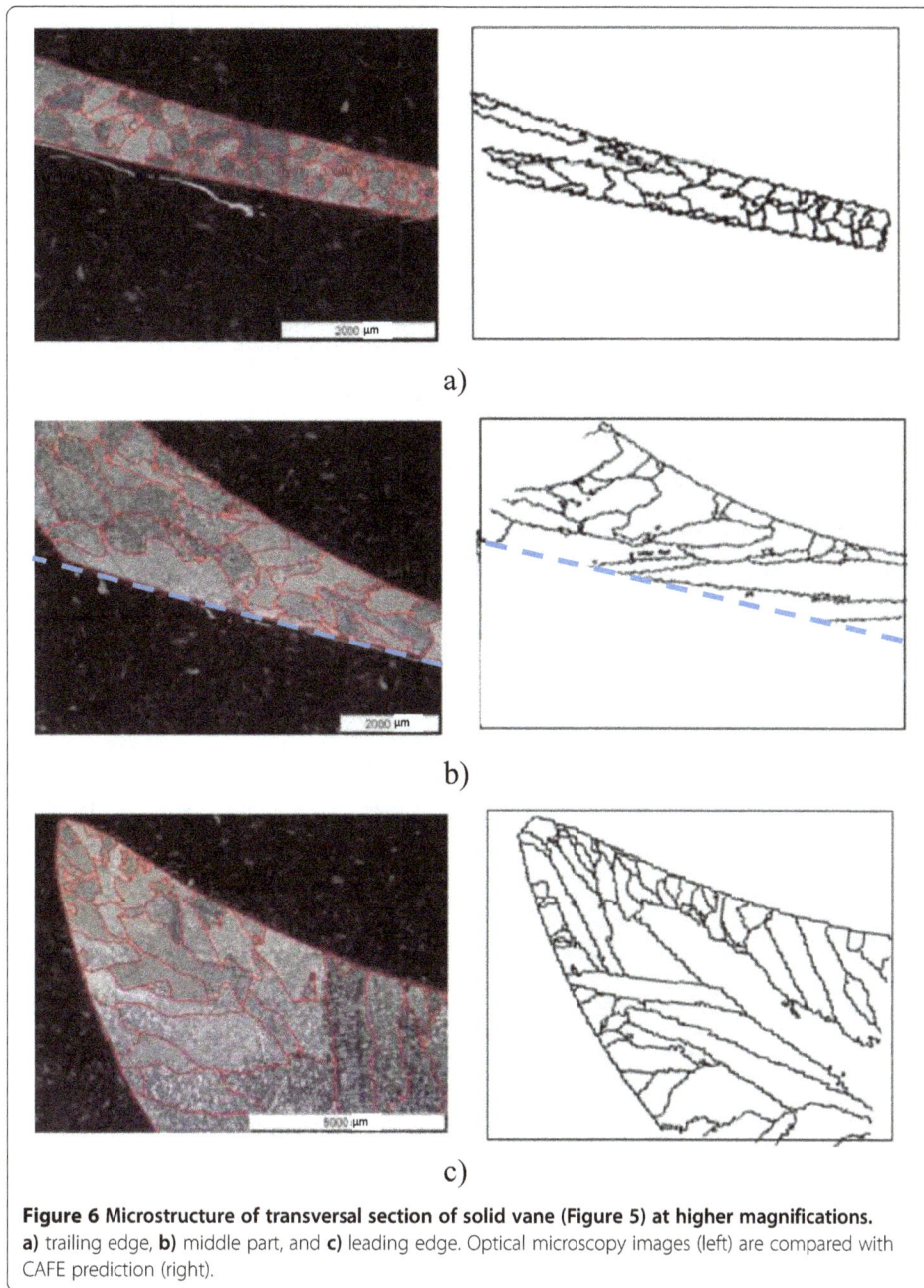

a)

b)

c)

Figure 6 Microstructure of transversal section of solid vane (Figure 5) at higher magnifications.
a) trailing edge, **b)** middle part, and **c)** leading edge. Optical microscopy images (left) are compared with CAFE prediction (right).

Table 6 Comparison of grain structure predicted by CAFE model with experimental results for transversal section of solid vane (Figures 5 to 6)

	Leading edge		Middle part		Trailing edge	
	Experiment	Model	Experiment	Model	Experiment	Model
Average grain size (µm)	1,560	1,345	785	869	281	397
Standard deviation (µm)	813	508	451	361	213	138
Aspect ratio	2.2	2.5	2.5	3.0	1.6	2.0

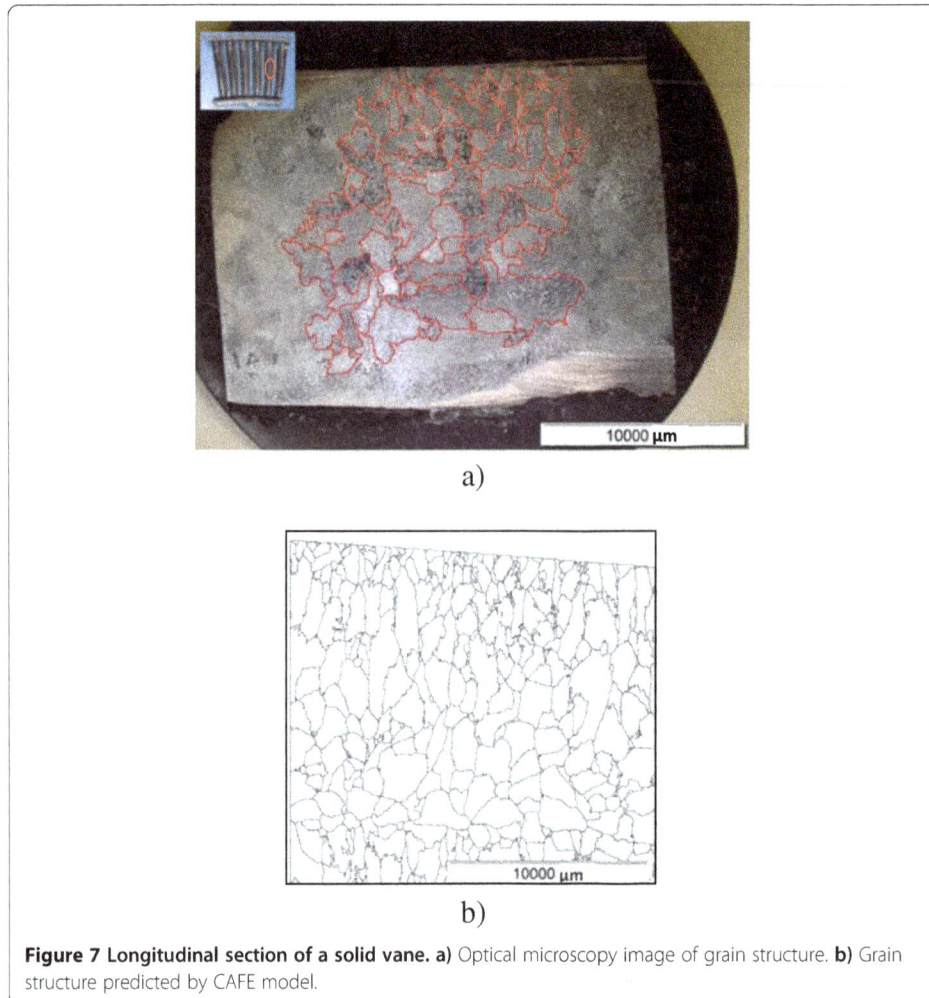

Figure 7 Longitudinal section of a solid vane. a) Optical microscopy image of grain structure. **b)** Grain structure predicted by CAFE model.

(Figure 7b) is similar to the experimental results. Microstructure with small grains close to the surface of solid vane (on the top of images), where solidification begins, is gradually transformed into microstructure with coarse grains when the mushy zone moves towards the central part of the solid vanes which is solidified in the last stage. It should be noted that the grain structure does not vary along the vane axis (horizontal axis in Figure 7).

Quantitative analysis shows that the CAFE model underestimates the average grain size, though the values of grain size from the model and experimental measurements overlap if standard deviation is taken into account (Table 7). The CAFE model provides a somewhat higher average value of grain aspect ratio which fits better to the grains

Table 7 Comparison of grain structure predicted by CAFE model with experimental results for longitudinal section of solid vane (Figure 7)

	Experiment	Model
Average grain size (μm)	2,153	1,458
Standard deviation (μm)	1,179	738
Aspect ratio	1.5	2.2

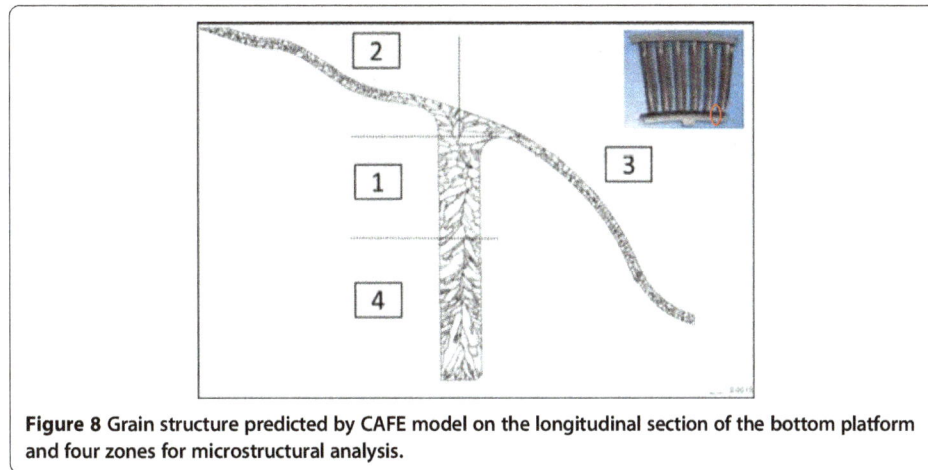

Figure 8 Grain structure predicted by CAFE model on the longitudinal section of the bottom platform and four zones for microstructural analysis.

close to the solid vane surface (top part of the images in Figure 7). Nevertheless, it can be concluded that there is a good match between the modeling and experimental results.

Grain structure on longitudinal section of a bottom platform

The location of the selected section in the bottom platform is marked by a red circle in the NGV icon in Figure 8. For convenience, the longitudinal section has been divided into four zones as shown in Figure 8, which also illustrates the grain structure predicted by the CAFE module. The modeling results show a good match with the real grain structure on the longitudinal section of the bottom platform in all four zones (Figure 9). Again, fast solidification in very thin elements (zones 2 and 3) leads to formation of homogeneous microstructure consisting of small equiaxed grains as seen from the optical microscopy image and grain structure generated by the model (Figure 9). On the contrary, columnar grains prevail in the microstructure of the thicker element (zones 1 and 4).

The average grain size measured from optical microscopy images in zones 1 and 2 are in a very good accordance with the model prediction (Table 8). The model tends to underestimate the average grain size in zones 3 and 4, though the ranges of grain size from model and experiment overlap (Table 8). The shape of grains in the modeled

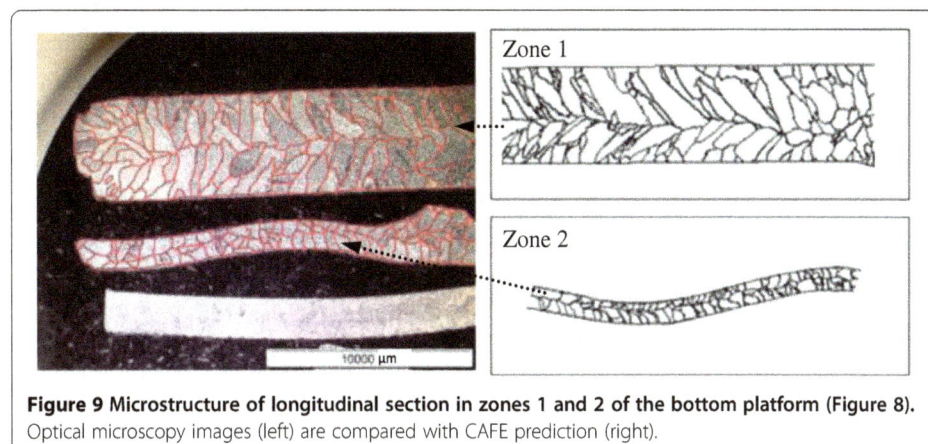

Figure 9 Microstructure of longitudinal section in zones 1 and 2 of the bottom platform (Figure 8). Optical microscopy images (left) are compared with CAFE prediction (right).

Table 8 Comparison of grain structured predicted by CAFE model with experimental results for longitudinal section of the bottom platform (Figures 8 to 9)

| | Zone 1 | | Zone 2 | | Zone 3 | | Zone 4 | |
	Exp.	Model	Exp.	Model	Exp.	Model	Exp.	Model
Average grain size (μm)	1,742	1,500	672	575	1,008	675	1,610	1,070
Standard deviation (μm)	1,458	866	300	287	713	337	189	418
Aspect ratio	1.8	2.0	2.4	2.2	2.3	2.0	2.3	2.1

grain structure is very similar to that observed in the optical microscopy images in all four zones (Figure 9). From Table 8, it is seen that the CAFE model predicts well the aspect ratio of grains in all analyzed zones of the bottom platform.

The comparison of the modeled grain structure with the real one formed in the critical sections of the as-cast NGV clearly shows that the CAFE module is a very useful tool for prediction of grain structure in the complex shape parts manufactured from Ni-based superalloys via investment casting. It is able to predict with high accuracy the size, shape, and orientation of grains throughout the complex shape part. The thermal model and models for porosity and grain structure prediction constitute a tool for further improvement of NGV design. This tool can provide the optimum parameters for investment casting at low cost in a quick manner. It should be also noted that thus obtained modeling results on porosity and grain structure could potentially be used for modeling of mechanical and functional properties of various sections of NGVs. So the integrated modeling tools could be developed in the future, which will dramatically minimize or even eliminate the number of experimental casting trials.

Conclusions

A CAFE module was employed to predict the 3D grain structure in NGVs manufactured from Ni-based superalloys via investment casting. The grain structure of the critical sections in the experimentally cast NGV was carefully analyzed, the experimental results were compared with preliminary grain structure prediction, and the model was calibrated via tuning parameters in the algorithms describing grain nucleation and growth.

It is demonstrated that the calibrated CAFE model is a reliable tool for the foundry industry to predict grain structure in the new design NGVs with high accuracy. Microstructure consisting of small equiaxed grains is predicted in the trailing edge and thin walls of the vanes, where fast solidification occurs. Grain growth follows the heat flux directions described by the thermal model, so larger grains appear in the thicker sections. The predicted grain size is always in the range of grain sizes measured in the real as-cast NGV, though the CAFE module provides smaller standard deviation. Transition from equiaxed to columnar grains is correctly predicted in the bottom platform of the NGV.

It is outlined that the calibrated CAFE module is a useful and reliable tool for the foundry industry to predict grain structure of the as-cast NGVs.

Competing interests
The authors declare that they have no competing interests.

Authors contributions
AJT applied the CAFE model and created the initial draft of the manuscript. OK and LC provided the materials science guidance and expertise in modeling. LM performed the experimental casting trials. ECM contributed with the overall development of the main concepts presented in this paper. MR performed the quantitative analysis of grain structure in the as-cast NGV. SM and IS helped with the validation of the developed grain structure prediction tool and manuscript writing. JL formulated the concept of this work and conceived the workflow, as well as provided materials science guidance. All authors read and approved the final manuscript.

Acknowledgements
This investigation was carried out in frame of the VANCAST project (EU, FP7, ERA-NET MATERA+). SM and IS acknowledge gratefully the Spanish Ministry of Economy and Competitiveness for financial support through the Ramon y Cajal fellowships.

Author details
[1]University of Applied Sciences and Arts Western Switzerland, 1950 Sion, Switzerland. [2]CALCOM-ESI, 1015 Lausanne, Switzerland. [3]Precicast Bilbao, 48901, Barakaldo, Bilbao, Spain. [4]IMDEA Materials Institute, C/Eric Kandel 2, 28906 Getafe, Madrid, Spain. [5]Department of Materials Science, Polytechnic University of Madrid, 28040 Madrid, Spain.

References

1. Janssens KGF, Raabe D, Miodownik Y, Kozeschnik MA, Nestler B (2007) Computational Materials Engineering. Elsevier Academic Press, Burlington, MA, USA
2. Penrose O, Fife PC (1990) Thermodynamically consistent models of phase-field type for the kinetic of phase transitions. Phys D 43:44 62, doi:10.1016/0167-27890167-2789(90)90015-H
3. Wang SL, Sekerka RF, Wheeler AA, Coriell SR, Murray BT, Braun RJ, McFadden GB (1993) Thermodynamically-consistent phase-field models for solidification. Phys D 69:189 200, doi:10.1016/0167-27890167-2789(93)90189-8
4. Caginalp G, Fife PC (1986) Phase field methods of interfacial boundaries. Phys Rev B 33:7792 7794, doi:10.1103/PhysRevB.33.7792
5. Lowen H, Bechoefer J, Tuckerman LS (1992) Crystal growth at long times: critical behavior at the crossover from diffusion to kinetics-limited regimes. Phys Rev A 45:2399 2415, doi:10.1103/PhysRevA.45.2399
6. Warren JA, Boettinger WJ (1994) Prediction of dendritic growth and microsegregation patterns in a binary alloy using the phase-field method. Acta Metall Mater 43:689 703, doi:10.1016/0956-7151(94)00285-P
7. Kobayashi R (1993) Modeling and numerical simulations of dendritic crystal-growth. Phys D 63:410 423, doi:10.1016/0167-2789(93)90120-P
8. Wheeler AA, Murray BT, Schaefer RJ (1993) Computation of dendrites using a phase field model. Phys D 66:243 262, doi:10.1016/0167-27890167-2789(93)90242-S
9. Wang SL, Sekerka RF (1996) Algorithms for phase eld computations of the dendritic operating state at large su-. percoolings. J Comp Phys 127:110 117. doi:10.1006/jcph.1996.0161
10. Provatas N, Goldenfeld N, Dantzig J (1998) Efficient computation of dendritic microstructures using adaptive mesh refinement. Phys Rev Lett 80:3308 3311, doi:10.1103/PhysRevLett.80.3308
11. Chen LQ, Young W (1994) Computer simulation of the domain dynamics of a quenched system with a large number of nonconserved order parameters: the grain-growth kinetics. Phys Rev B 50:15752 15756, doi:10.1103/PhysRevB.50.15752
12. Chen LQ (1995) A novel computer simulation technique for modeling grain growth. Scr Metall Mater 32:115 120, doi:10.1016/S0956-716X(99)80022-3
13. Steinbach I, Pezzolla F, Nestler B, Seesselberg M, Schmitz GJ, Rezende J (1996) A phase field concept for multiphase systems. Phys D 94:135 147, doi:10.1016/0167-27890167-2789(95)00298-7
14. Nestler B, Wheeler AA (1998) Anisotropic multi-phase-field model: interfaces and junctions. Phys Rev E 57:2602 2609, doi:10.1103/PhysRevE.57.2602
15. Kobayashi R, Warren JA, Carter WC (1998) Vector-valued phase field model for crystallization and grain boundary formation. Phys D 119:415 423, doi:10.1016/S0167-2789(98)00026-8
16. Garcke H, Nestler B, Stoth B (1998) On anisotropic order parameter models for multi-phase systems and their sharp interface limits. Phys D 115:87 108, doi:10.1016/S0167-2789(97)00227-3
17. Boettinger WJ, Warren JA (1996) The phase-field method: simulation of alloy dendritic solidification during recalescence. Metall Mater Trans A 27:657 669, doi:10.1007/BF02648953
18. Hesselbarth HW, Göbel IR (1991) Simulation of recrystallization by cellular automata. Acta Metall 39:2135 2143. doi:10.1016/0956-7151(91)90183-2
19. Wang W, Kermanpur A, Lee PD, McLean M (2003) Simulation of dendritic growth in the platform region of single crystal superalloy turbine blades. J Mater Sci 38:4385 4391, doi:10.1023/A:1026303720544
20. Wang W, Lee PD, McLean M (2003) A model of solidification microstructures in nickel based superalloys: predicting primary dendrite spacing selection. Acta Mater 51:2971 2987, doi:10.1016/S1359-6454(03)00110-1
21. Rappaz M, Gandin CA (1993) Probabilistic modelling of microstructure formation in solidification processes. Acta Metall Mater 41:345 360, doi:10.1016/0956-7151(93)90065-Z
22. Gandin CA, Rappaz M (1994) A coupled finite element-cellular automaton model for the prediction of dendritic grain structures in solidification processes. Acta Metall Mater 42:2233 2246, doi:10.1016/0956-7151(94)90302-6
23. Gandin CA, Rappaz M (1997) A 3D cellular automaton algorithm for the prediction of dendritic grain growth. Acta Mater 45:2187 2195, doi:10.1016/S1359-6454(96)00303-5
24. Kurz W, Giovanola B, Trivedi R (1986) Theory of microstructural development during rapid solidification. Acta Metall 34:823 830, doi:10.1016/0001-6160(86)90056-8

25. Gandin CA, Rappaz M, Desbiolles JL, Lopez R, Swierkosz M, Thevoz PH (1997) 3D modeling of dendritic grain structures in turbine blade investment cast parts. In: Loria EA (ed) Proceedings of the TMS Meeting. TMS, p 121
26. Seo SM, Kim IS, Jo CY, Ogi K (2007) Grain structure prediction of Ni-base superalloy castings using the cellular automaton-finite element method. Mater Sci Eng A 449 451:713 716, doi:10.1016/j.msea.2006.02.400
27. Wang N, Liu L, Gao S, Zhao X, Huang T, Zhang J, Fu H (2014) Simulation of grain selection during single crystal casting of a Ni-base superalloy. J Alloys Compd 586:220 229, doi:10.1016/j.jallcom.2013.10.036
28. Gandin CA, Desbiolles JL, Rappaz M, Thevoz P (1999) A three-dimensional cellular automation-finite element model for the prediction of solidification grain structures. Metall Mater Trans A 30:3153 3165, doi:10.1007/s11661-999-0226-2
29. Lipton J, Glicksman ME, Kurz W (1987) Equiaxed dendrite growth in alloys at small supercooling. Metall Trans A 18:341 345, doi:10.1007/BF02825716
30. ProCast user Manual & Technical Reference (2007) Technical Reference (2007). Version 6.1. ESI software, France
31. Qingyan X, Baicheng L, Dong P, Jing Y (2012) Progress on modeling and simulation of directional solidification of superalloy turbine blade casting. Res Develop 2:69 77

9

Experimental measurement of surface strains and local lattice rotations combined with 3D microstructure reconstruction from deformed polycrystalline ensembles at the micro-scale

Paul A Shade[*], Michael A Groeber, Jay C Schuren and Michael D Uchic

* Correspondence:
paul.shade.1@us.af.mil
Air Force Research Laboratory,
Materials and Manufacturing
Directorate, 2230 10th Street,
Wright-Patterson AFB, OH 45433, USA

Abstract

This article describes a new approach to characterize the deformation response of polycrystalline metals using a combination of novel micro-scale experimental methodologies. An in-situ scanning electron microscope (SEM)-based tension testing system was used to deform micro-scale polycrystalline samples to modest and moderate plastic strains. These tests included measurement of the local displacement field with nm-scale resolution at the sample surface. After testing, focused ion beam serial sectioning experiments that incorporated electron backscatter diffraction mapping were performed to characterize both the internal 3D grain structure and local lattice rotations that developed within the deformed micro-scale test samples. This combination of experiments enables the local surface displacements and internal lattice rotations to be directly correlated with the underlying 3D polycrystalline microstructure, and such information can be used to validate and guide further development of modeling and simulation methods that predict the local plastic deformation response of polycrystalline ensembles.

Keywords: Micro-tensile test; Plastic deformation; Microstructure

Background

Many structural components are fabricated from polycrystalline materials, and the desire to both optimize the performance and extend the lifetime of metallic alloys has fostered the development of advanced micromechanical modeling and simulation tools that can accurately predict the deformation response of polycrystalline ensembles. Experimental and computational techniques working toward this goal have been the subject of numerous studies, and have evolved with increasing fidelity at decreasing length scales. One example of many approaches to address this need is crystal plasticity finite element modeling (CP-FEM) focused on explicitly representing the morphology and local crystallographic orientations of polycrystalline microstructures [1]. These models can predict the development of intra- and inter-granular gradients in the deformation field, as well as the evolution of grain morphology and local lattice rotations, and yet at the same time have known limitations such as the inability to accurately account for length scale effects [2-4].

Experimental validation of such methods is critical to guide their further development and implementation. However, due to experimental and computational challenges, validation studies which compare experimental data to simulations which explicitly incorporate the experimental microstructure have been historically limited. These have largely involved studies where only the surface microstructure of a mechanical test specimen has been experimentally determined and subsequently used as input for either 2D or quasi-3D simulations [5-7], or to approximate the 3D microstructure of a simplified material (i.e., very large grain materials where the sub-surface microstructure is assumed to be columnar) [8-12]. St-Pierre collected 2D electron backscatter diffraction (EBSD) scans of the surface microstructure of a tensile sample and used microstructure statistics to generate a 3D mesh of the tensile sample with the experimental surface and a realistic sub-surface virtual microstructure [13]. Musienko utilized successive electropolishing on a post-deformation tensile specimen combined with EBSD scans to determine the 3D microstructure from a small volume in a region-of-interest near the specimen surface, which was subsequently meshed and simulated to compare to the tensile experiment [14].

In the present study, we demonstrate a new methodology for generating mechanical test datasets combined with explicit microstructure representation of the entire test specimen. We have employed in-situ SEM-based micro-scale tensile testing [15-18] combined with surface strain mapping to track the evolution of surface strains throughout the mechanical test [19-23]. Micro-scale test volumes are amenable to 3D serial sectioning in focused ion beam-scanning electron microscopes (FIB-SEM), and performing such experiments while incorporating EBSD mapping allows for capturing the post-deformation microstructure, including local lattice rotations [24-27]. The combination of all of these techniques allows the collection of rich datasets for model development and validation studies; these efforts are described in other publications [28].

Methods, results and discussion

Sample preparation

The material selected for this work was a 99.0% purity annealed Ni foil with a nominal thickness of 50 μm. The foil contained no appreciable texture and was comprised of equiaxed grains with an average diameter of approximately 10 μm. Micro-tensile samples were fabricated from the foil by implementing a stencil mask technique [29]. This technique involves using standard microelectronics processing methods to produce high aspect-ratio stencil masks from a Si wafer. Once fabricated, the stencil masks are placed on top of the foil and the mask-and-foil are co-sputtered using a broad ion beam milling system. This ultimately transfers the pattern of the stencil mask into the foil, creating an array of test structures. For the present experiments, the Si wafer was 200 μm in thickness and the pattern consisted of an array of tensile samples integrally attached to the bulk substrate. Milling was conducted with a Gatan Precision Etching Coating System, operated with a 6 kV Ar^+ broad ion beam for approximately 40 hours.

After completing the stencil mask procedure described above, the Ni tensile samples required further micro-machining to remove both tapered sidewalls and a thin coating of re-deposited material. An FEI Nova 600 Dual Beam FIB-SEM was used to perform these tasks. First, the top and bottom surfaces of the samples were milled to remove any re-deposited material and also to thin the specimens to the desired thickness. Surface striations, a.k.a. "curtaining", invariably developed during this process because of the relatively large dimensions of the specimens for ion milling which required the use of relatively large beam currents for the final microfinishing step (> 5 nA), as can be seen in Figure 1. Following this, an automated procedure was developed and implemented to remove the taper from the specimen sidewalls. This procedure combined motorized microscope stage movements with image recognition-optimized placement of milling patterns to iteratively cross-section mill the perimeter of the sample while maintaining a biased back-tilt of 1° relative to the sidewall. This allowed nearly perfectly orthogonal sidewalls to be produced.

Three samples were fabricated with a final specimen geometry consisting of a rectangular cross section, a gage width of 21 μm, a thickness of 38 μm, and a gage length of 80 μm. Images of the samples prior to testing can be seen in Figure 1. The flat sample surfaces were conducive to collecting EBSD patterns before and after mechanical testing, and also for making surface strain measurements throughout the mechanical test. The choice of material, grain size and specimen dimensions allowed for roughly 200 grains to be included in the gage volume. This allowed the experiment to include a sufficient number of grains such that the results would be relevant to interrogating a polycrystalline response, yet have the total number of grains be small enough such that

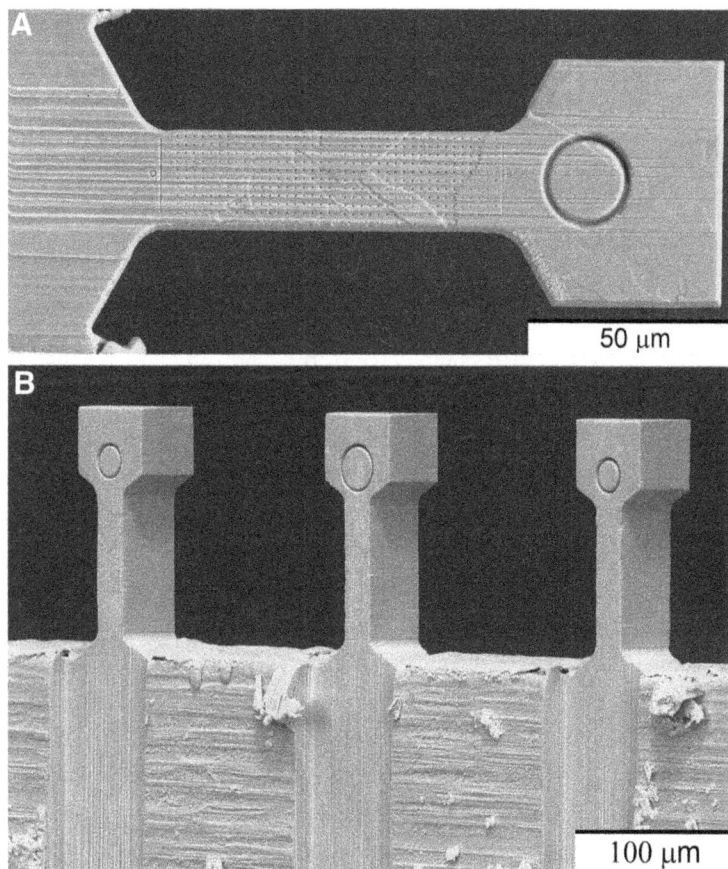

Figure 1 Representative SEM images of the Ni micro-tensile samples prior to testing. (A) is a higher magnification image of one of the samples, highlighting the presence of a grid of points which were used as markers for surface strain analysis. **(B)** is an oblique view of all three samples.

the test could still be directly simulated using a CP-FEM model instantiated with the explicit 3D microstructure [28]. Furthermore, the specimen dimensions are appropriate for micro-tensile testing and post-test 3D-EBSD serial sectioning in the FIB-SEM.

Mechanical testing

In-situ SEM-based micro-tensile testing was conducted using a custom-built mechanical testing device [18,30,31]. Selected details of the device construction have been reported elsewhere [31]. The device is displacement-controlled using a piezoelectric actuator, and load is measured with a strain-gage-based load cell. The local sample displacements are calculated from SEM images by tracking the positional change of distinct features on the specimen surface. An alignment flexure ensures linear motion of the loading train [32,33], and the samples are precisely positioned for testing by attaching the bulk substrate to a piezoelectric controlled x-y-z micro-positioning stage. The FIB-SEM was used to manufacture a tensile grip into the end of a SiC fiber which was 8 mm in length and 0.1 mm in diameter, and attached at the other end to the load cell. The 80:1 aspect ratio of the SiC fiber enables the tensile grip to have an extremely low lateral stiffness, thus minimizing the lateral constraints imposed on the specimen during mechanical testing [18,31]. As a result, the imposed boundary conditions are different from that in a traditional tensile test.

The mechanical tests were conducted in-situ in an FEI Sirion SEM equipped with a 4 pi image acquisition system. The recorded images had dimensions that were 6000 × 2000 pixels, corresponding to a pixel size of ~ 2 nm, and contained 16 bit depth. The tests were conducted in a quasi-static manner, where the samples underwent sequential periods of loading at a constant actuator voltage ramp rate (open-loop displacement rate control) separated by periods in which the actuator was held at a constant voltage to facilitate the collection of high resolution SEM images. While the voltage ramp rate of the actuator was held constant during the loading portions of each test, i.e., the actuator displaced a constant amount for each loading segment, the displacement achieved by the sample varied due to the high compliance of the load cell. This is illustrated in Figure 2, which shows a plot of stress and strain rate vs strain for the three tested samples, where it can be seen that the strain rates were initially very low during elastic loading and increased rapidly until reaching a more stable rate once each sample started to plastically deform. Note that the strain rate values reported were calculated by measuring the change in engineering strain between two images and dividing by the period of time over which the voltage values supplied to the actuator were increased. Therefore, the strain rate values reported are actually an upper limit estimate, as they do not account for the possibility of creep in the sample while the load is held static during image collection. For strain rate sensitive materials this mode of testing may significantly affect the results, however, the impact is expected to be minimal for Ni polycrystals. Due to the relatively high elastic modulus (~ 200 GPa) and low yield stress (~ 100 MPa) of the Ni foil, total elastic strain values for the samples corresponded to displacement values of only about 2 pixels, and therefore it was not possible to accurately measure elastic modulus values from these experiments.

The three samples were tested to different strain levels (~ 1.1, 2.5, and 11.9% axial engineering strain), as shown in Figure 2. Despite the limited number of grains within

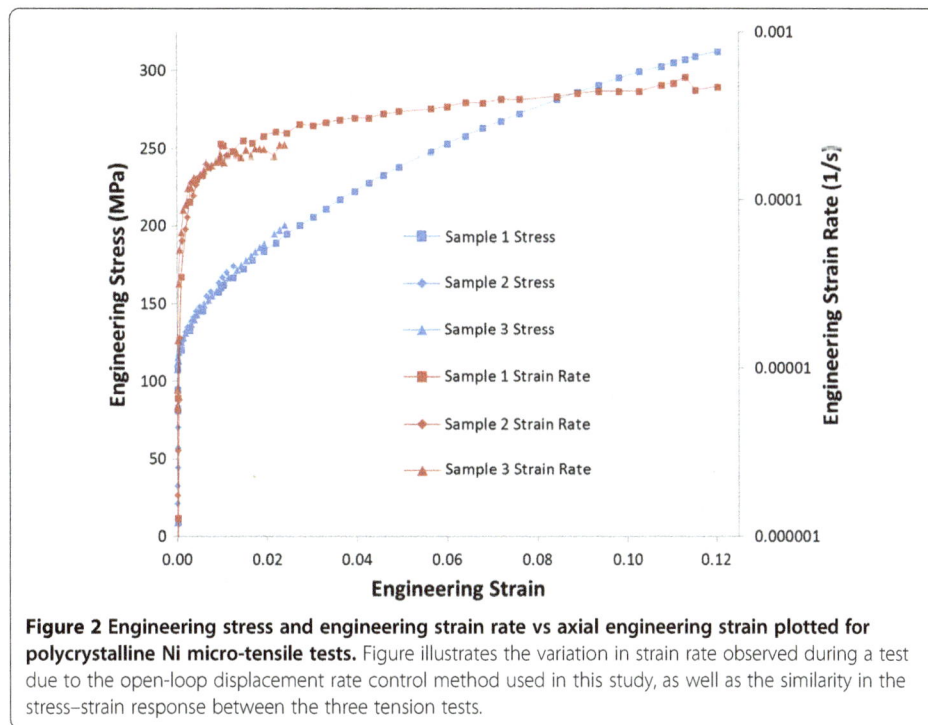

Figure 2 Engineering stress and engineering strain rate vs axial engineering strain plotted for polycrystalline Ni micro-tensile tests. Figure illustrates the variation in strain rate observed during a test due to the open-loop displacement rate control method used in this study, as well as the similarity in the stress–strain response between the three tension tests.

the gage volume and expected variation in local grain configurations among the three samples, the three engineering stress–strain curves are very similar. This agreement highlights the potential limitation of using the global stress–strain response as a sole validation metric, and thus other measures are required for interrogating the plastic deformation behavior of polycrystalline ensembles.

Surface strain mapping

The evolution and distribution of surface strains is typically a direct output of modeling tools such as CP-FEM, and these quantities are also measurable using modern digital image correlation (DIC) methods [8,10,12-14]. Both random and regular patterns can be used for DIC analysis, and for the present study a regular grid of points was milled onto the top surface of the micro-tensile samples prior to testing using the FIB-SEM. The markers (points) were circular with an approximate diameter of 30 nm and a point-to-point spacing of 2.3 μm. An example of this marker pattern can be seen in Figure 1A.

The distortion of the grid throughout an experiment was measured from the individual images and used to determine local surface strains, following a methodology similar to that described by Biery et al. [23]. First, marker positions in each image were determined using a script that quickly found rough marker coordinates by performing a binary segmentation with a threshold intensity value that highlighted the markers, and then calculated the centroid of the resulting cluster of pixels at each marker. Refined coordinates were subsequently determined with sub-pixel accuracy by calculating the peak positions of a 2D Gaussian fit around each marker in the original non-segmented images. Marker positions in the image prior to testing were taken as a reference, and second order polynomial fits were calculated that mapped the positions of a central marker and the nearest surrounding markers in the reference image to those in the distorted image. Strain values were then determined from the coefficients of the polynomial fits following equations 1–3 from Biery et al. [23].

Figure 3 shows images of the deformed grid in the strain mapping region combined with von Mises effective strain plots, where the von Mises effective strain was calculated using equation 1 from Wu et al. [34]. Note the strongly heterogeneous nature of the strain distribution in all three samples, where some regions remain nearly undeformed while others contain strains which are on the order of double the average value. For example, the sample deformed to 1.1% axial strain had local axial strain values that ranged from –0.2 to 2.3% with a standard deviation of 0.4%. The difference is magnified in the higher strain samples, where the sample deformed to 2.5% axial strain had local axial strain values that ranged from –0.3 to 5.0% with a standard deviation of 1.2%, and the sample deformed to 11.9% axial strain had local axial strain values that ranged from –0.8 to 30.8% with a standard deviation of 6.1%. Videos which show the evolution of the axial (XX), transverse (YY), and in-plane shear (XY) surface strain distribution, along with the surface strain mapping region and stress–strain plot,

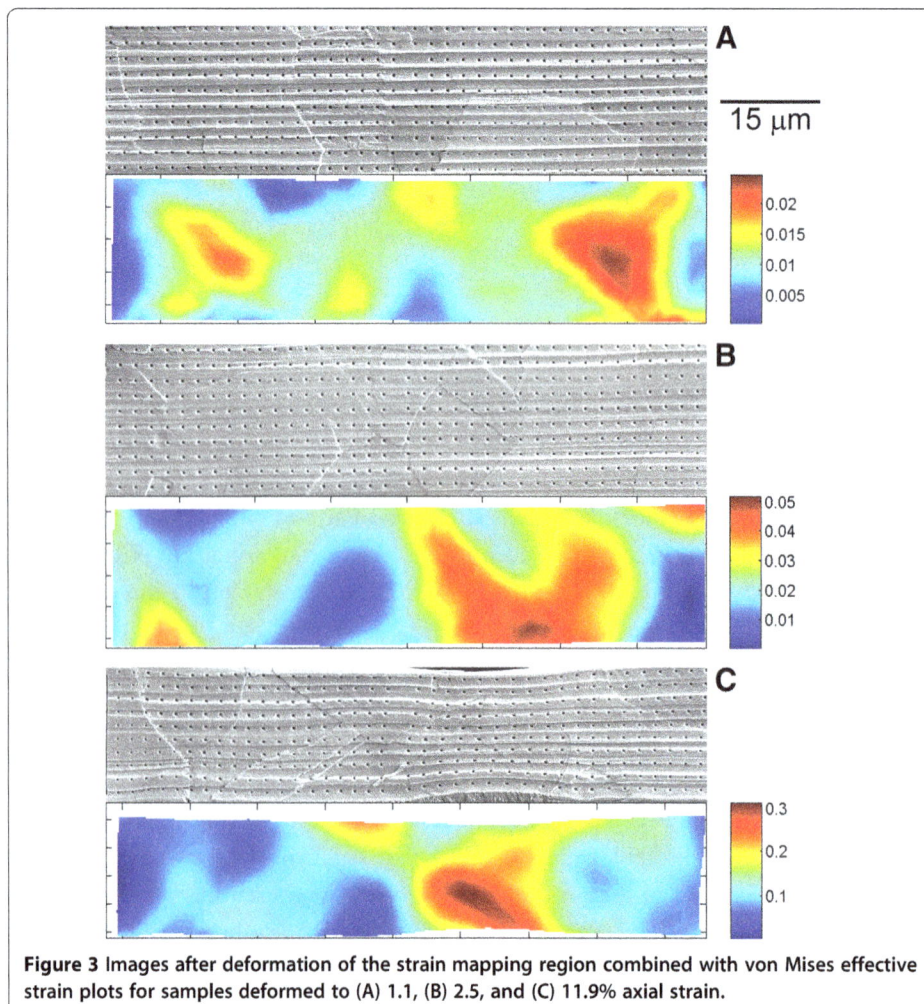

Figure 3 Images after deformation of the strain mapping region combined with von Mises effective strain plots for samples deformed to (A) 1.1, (B) 2.5, and (C) 11.9% axial strain.

for all three samples are available in the online version of this manuscript (Additional files 1, 2 and 3).

3D-EBSD serial sectioning

The internal microstructures of the deformed samples were characterized following mechanical testing by 3D-EBSD serial sectioning using the aforementioned FIB-SEM equipped with a TSL Hikari EBSD detector. The 3D-EBSD serial sectioning process has been described in detail elsewhere [24-27]. Briefly, the process consisted of repeated cross-section milling of the sample using the ion beam, followed by repositioning the sample via tilt, rotation and translation of the 5-axis microscope stage to collect a variety of images or crystallographic (EBSD) maps for each section. This process has been fully automated with the development of custom codes that utilize FEI RunScript software to control the FIB-SEM, and AutoIt automation software to initiate the collection of EBSD maps and facilitate communication between the FIB-SEM control computer and the EBSD acquisition system.

Prior to collection of the 3D-EBSD data, a ~ 3 μm thick layer of Pt was deposited onto the specimen surface and a series of fiducial markers for fine scale positioning were milled into the bulk substrate near the tensile sample. These fiducial markers are used in conjunction with image matching algorithms to optimize the kinematic position of the sample prior to ion milling or data acquisition, and also allow for precise placement of ion milling patterns to minimize the effects of sample drift or minor variability in sample positioning. Cross-section milling was conducted with a 30 kV Ga^+ ion beam at a current of 6.5 nA. The cross-sections were milled with a 1° back-tilt to compensate for taper of the cross-section surface, and the incremental section thickness was approximately 250 nm. Crystallographic orientation information was captured for each section by collecting EBSD maps, using a 20 kV accelerating voltage and 250 nm pixel size. The grain structure was also imaged using ion-induced secondary electron (ISE) images using a 30 kV Ga^+ ion beam and a current of 0.1 nA, resulting in an in-plane resolution of ~ 50 nm. These ISE images display significant channeling contrast for polycrystalline grain structures [24,35] and were collected at two different tilt values to increase the probability of having at least one ISE image where all neighboring grain pairs displayed visibly different gray-scale intensities. However, due to the long time duration of these experiments, the intensity of the ISE images varied dramatically over the course of the serial sectioning experiment (which we attributed to slowly-varying changes in the current delivered to the sample by the ion column), and as a result a number of the ISE images had poor contrast that prevented the use of image processing methods to segment the internal grain structure. Examples of the various data collected for each section are shown in Figure 4. Each 3D-EBSD dataset consisted of approximately 400 sections and required approximately 6 days of collection time.

Additionally, for one of the samples the raw EBSD patterns were saved for every pixel within a scan. Enabling this option significantly slows the acquisition process, in part because of the requirement to not use pattern binning and additionally to allow time for the computer to store large quantities of image data. As such, the raw EBSD pattern data was collected with a reduced resolution of 1 μm voxels by using an in-plane pixel size of 1 μm and only collecting this data for every fourth cross-section. The resolution

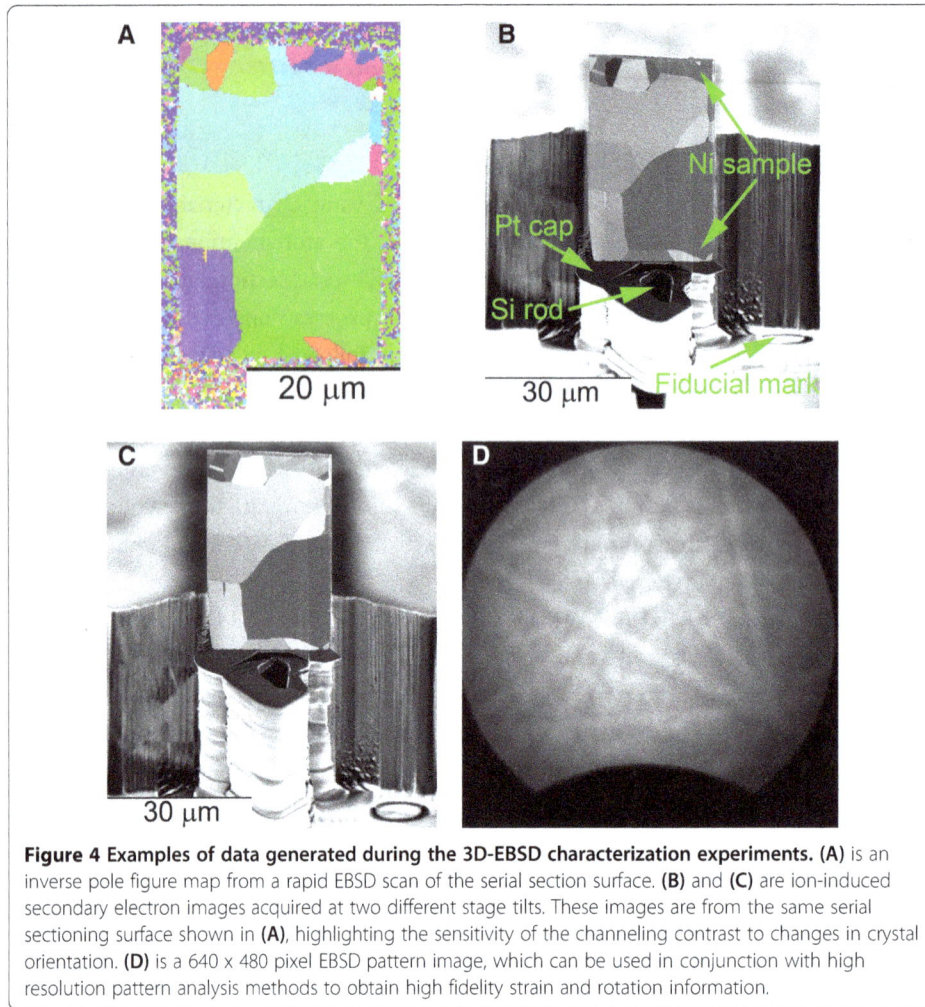

Figure 4 Examples of data generated during the 3D-EBSD characterization experiments. (A) is an inverse pole figure map from a rapid EBSD scan of the serial section surface. **(B)** and **(C)** are ion-induced secondary electron images acquired at two different stage tilts. These images are from the same serial sectioning surface shown in **(A)**, highlighting the sensitivity of the channeling contrast to changes in crystal orientation. **(D)** is a 640 x 480 pixel EBSD pattern image, which can be used in conjunction with high resolution pattern analysis methods to obtain high fidelity strain and rotation information.

of the pattern images was 640 × 480, an example of which can be seen in Figure 4D. A single crystal Si rod was extracted from a wafer and placed on top of the sample using an Omniprobe micro-manipulator to be used as a pattern center reference (as can be seen in Figure 4B and C), however, differences between Si and Ni diffraction pattern intensities made it difficult to find a set of camera parameters optimized for both and as such the pattern quality in this experiment was insufficient for this application. Currently, the raw EBSD patterns are not being used, however, in the future we hope to use this data to extract residual strains along with more precise crystallographic orientations [36].

The series of individual 2D EBSD maps were aligned and reconstructed [37,38] using DREAM.3D software [39] to produce a 3D volume. The procedure used to register, segment and reconstruct the 3D EBSD data in DREAM.3D is the following. After importing the original series of EBSD scans, the sample volume was identified from the empty space surrounding the sample by using a multiple threshold criteria. For the data sets shown, we have selected threshold values for both image quality [40] and confidence index of the EBSD data that correctly identified most of the voxels associated with the sample volume, with some misclassified voxels internal to the sample as well as in the empty space surrounding the sample. Next, gross section-to-section translations were removed by calculating the center-of-mass of the sample in each section, and aligning these coordinates to a common reference line. For these experiments, the common reference line corresponds to the tensile axis, which was normal to the serial sectioning plane. Following this procedure, the 3D reconstruction of the sample volume contained some visible alignment artifacts due to erroneous data points affecting the center-of-mass calculation, most noticeably at the gage-to-grip transition between the sample and the substrate. In this region of the sample, the nearby surfaces of the substrate that are in the view field of the EBSD scan (but not part of the cross-section surface) generate indexed EBSD data. These erroneous points were removed using a combination of automatic and manual filters to identify these voxels as empty space. Section-to-section translations were re-calculated using the center-of-mass alignment procedure, producing 3D volumes that contained minimal registration artifacts, as shown in Figure 5. Note that prior to performing the serial sectioning experiment on the sample shown in Figure 5A, two longitudinal and three slanted lines were FIB-milled into the top surface of the microsample before deposition of the protective platinum cap. These linear features are clearly visible in the 3D reconstruction, highlighting the accuracy of the data registration procedure.

After completing the registration process, the internal grain structure was segmented in DREAM.3D using a disorientation criteria, where sample voxels were iteratively grouped into fields (grains) when the disorientation between neighboring voxels was less than a user-defined angular threshold, here 5 degrees. This segmentation resulted in the definition of the internal grain structure, however, additional clean-up steps were required to re-assign internal data points that were deemed as erroneous, often the result of indexing errors or from identification as bad data via the original multi-threshold criteria. These features were removed from the data volume and the corresponding voxels were re-assigned using minimum size filters in DREAM.3D, where these filters were set to an ad-hoc threshold size of 16 voxels, corresponding to a volume of 0.25 μm^3. Lastly, a combination of a 1 voxel erosion/dilation morphological filter and a surface smoothing filter were used to eliminate one-voxel wide lines and trenches on the surface of the sample. This latter filter operates by iteratively examining the coordination number of all surface voxels, and altering them by either removing voxels that have a high coordination number with the empty space, or by performing the reverse by filling empty space voxels that have a high coordination number with the sample. The 3D reconstruction of the sample deformed to 2.5% axial strain, along with engineering stress – engineering strain data and SEM images from the micro-tensile test used to calculate surface strain maps, have been made publically available [41].

After data clean-up, additional calculations were performed in DREAM.3D on the data volumes, and these metrics can be assigned to each of the fields (grains) and/or voxels that comprise the tensile volume. For example, the inverse pole figure (IPF) coloring relative to the tensile axis, is shown in Figure 5 (also included in the online version of this manuscript as Additional files 4 and 5). Figure 6 shows multiple 3D reconstructions of one of the tensile samples (which achieved a total strain of 11.9%), demonstrating additional metrics that can be calculated and displayed, in concert with other quantities that are measured as part of the EBSD acquisition process such as Image Quality [40]. Specifically, the following metrics are shown in Figure 6: Schmid factor for each grain (assuming a state of uniaxial tension), the average disorientation

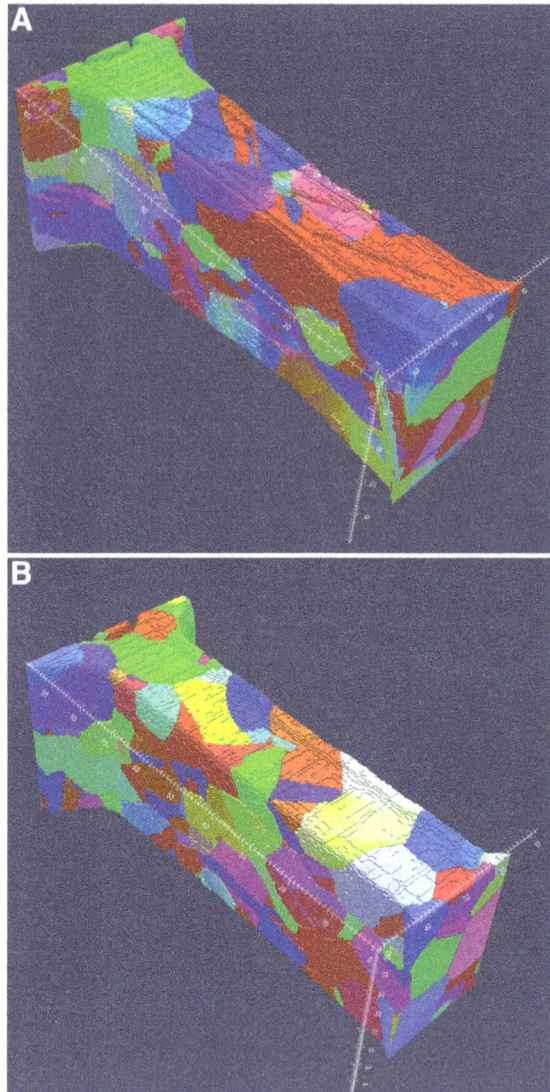

Figure 5 3D reconstructions of two of the deformed Ni polycrystalline micro-tensile samples. Voxel colors represent the local inverse pole figure orientation relative to the tensile axis, and the units listed on the 3D scale bar are in micrometers. The sample in **(A)** was deformed to 11.9% axial plastic strain. Note the presence of two longitudinal and three slanted lines which were FIB-milled into the top surface of the microsample prior to performing the serial sectioning experiment, highlighting the accuracy of the registration procedure. The sample in **(B)** was deformed to 2.5% axial plastic strain.

of each voxel relative to its local neighborhood (Kernel Average Misorientation, 1st nearest neighbor shell), the voxel disorientation relative to a reference grain orientation, here the average orientation of each grain (Grain Reference Misorientation), the Manhattan Distance for each voxel relative to the grain boundary network, IPF coloring, and Image Quality. This data is also displayed online as a movie in Additional file 6. These examples clearly demonstrate the fidelity and complexity with which the internal crystallographic structure of the test volume can be characterized after testing, and, coupled to the surface strain maps and test volumes with controlled boundary conditions, provide a rich palette of data to link the evolution of surface deformations to both the far-field stress state and the underlying microstructure.

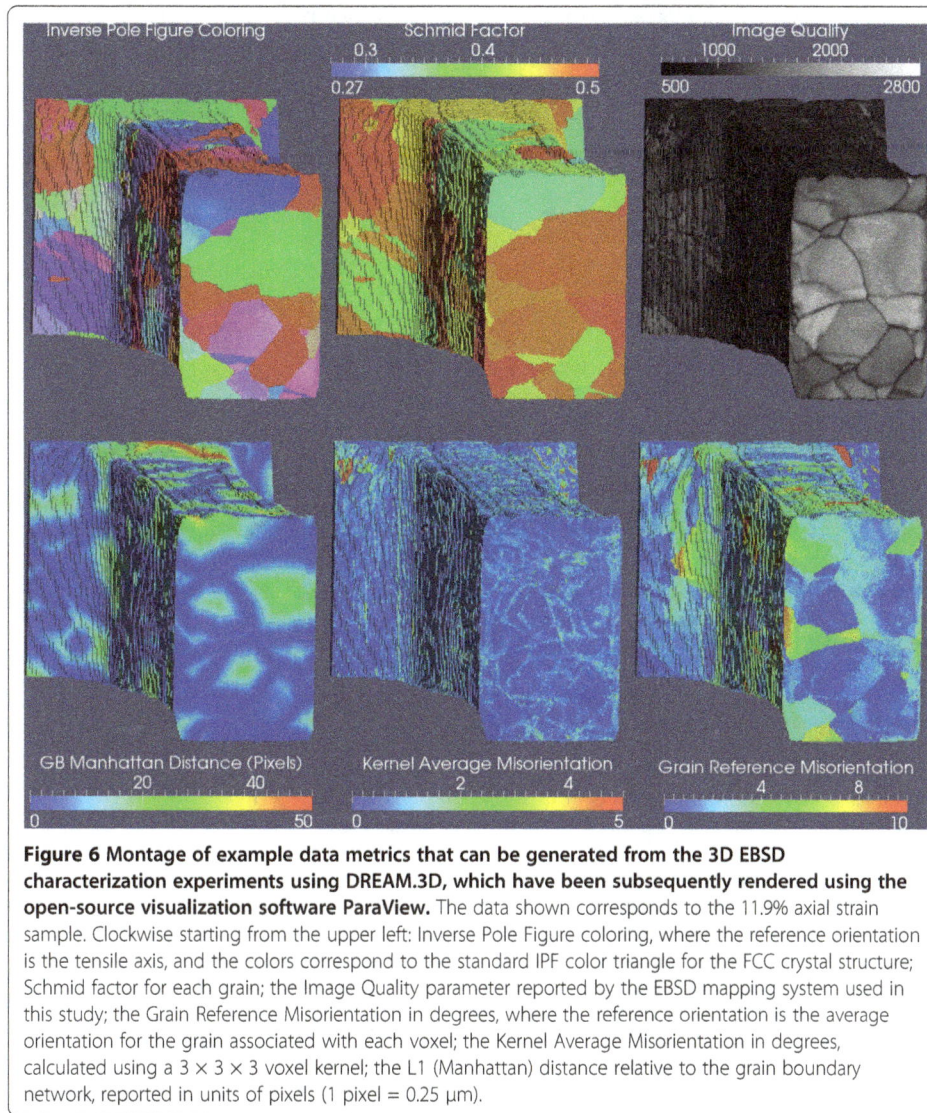

Figure 6 Montage of example data metrics that can be generated from the 3D EBSD characterization experiments using DREAM.3D, which have been subsequently rendered using the open-source visualization software ParaView. The data shown corresponds to the 11.9% axial strain sample. Clockwise starting from the upper left: Inverse Pole Figure coloring, where the reference orientation is the tensile axis, and the colors correspond to the standard IPF color triangle for the FCC crystal structure; Schmid factor for each grain; the Image Quality parameter reported by the EBSD mapping system used in this study; the Grain Reference Misorientation in degrees, where the reference orientation is the average orientation for the grain associated with each voxel; the Kernel Average Misorientation in degrees, calculated using a 3 × 3 × 3 voxel kernel; the L1 (Manhattan) distance relative to the grain boundary network, reported in units of pixels (1 pixel = 0.25 µm).

Conclusion

In the present study, we demonstrated a new methodology for generating high-fidelity mechanical test data sets combined with explicit 3D microstructure representation of the entire test specimen, with the intent to couple this data to simulations for model validation and development. This was accomplished utilizing a micro-scale mechanical test specimen, so that the test volumes were amenable to examination via an established 3D microstructure characterization technique, 3D-EBSD serial sectioning with a FIB-SEM. Future studies may collect similar data on larger (mm-scale) samples by utilizing emerging destructive [42] and nondestructive [43] microstructure characterization techniques.

Surface strain distributions and internal lattice rotations were measured, and will serve as metrics from which to compare to simulations in future validation studies. One caveat to using this data for validation studies is that only the microstructure from the deformed specimen can be measured, as 3D-EBSD serial sectioning is a destructive process. Hence, some assumptions will have to be made in terms of assigning initial orientations to the individual grains (removing internal lattice rotations due to deformation), and also the initial grain morphology (since the measured microstructure will be distorted due to the deformation). The usefulness of this technique for validation studies is therefore likely best at lower total strain values.

Availability of supporting data

The 3D reconstruction of the sample deformed to 2.5% axial strain, along with engineering stress – engineering strain data and SEM images from the micro-tensile test used to calculate surface strain maps, have been made publically available [41].

Additional files

Additional file 1: Video showing evolution of surface strains during micro-tension test of the sample deformed to 1.1% axial plastic strain. Top left is the axial (XX) strain component; middle left is the transverse (YY) strain component; bottom left is the in-plane shear (XY) strain component; top right shows the engineering stress versus axial engineering strain curve; bottom right shows the deforming sample.

Additional file 2: Video showing evolution of surface strains during micro-tension test of the sample deformed to 2.5% axial plastic strain. Top left is the axial (XX) strain component; middle left is the transverse (YY) strain component; bottom left is the in-plane shear (XY) strain component; top right shows the engineering stress versus axial engineering strain curve; bottom right shows the deforming sample.

Additional file 3: Video showing evolution of surface strains during micro-tension test of the sample deformed to 11.9% axial plastic strain. Top left is the axial (XX) strain component; middle left is the transverse (YY) strain component; bottom left is the in-plane shear (XY) strain component; top right shows the engineering stress versus axial engineering strain curve; bottom right shows the deforming sample.

Additional file 4: 3D reconstruction of the sample deformed to 11.9% axial plastic strain. Voxel colors represent the local inverse pole figure orientation relative to the tensile axis, and the units listed on the 3D scale bar are in micrometers.

Additional file 5: 3D reconstruction of the sample deformed to 2.5% axial plastic strain. Voxel colors represent the local inverse pole figure orientation relative to the tensile axis, and the units listed on the 3D scale bar are in micrometers.

Additional file 6: Montage of example data metrics that can be generated from the 3D EBSD characterization experiments using DREAM.3D, which have been subsequently rendered using the open-source visualization software ParaView. The data shown corresponds to the 11.9% axial strain sample. Clockwise starting from the upper left: Inverse Pole Figure coloring, where the reference orientation is the tensile axis, and the colors correspond to the standard IPF color triangle for the FCC crystal structure; Schmid factor for each grain; the Image Quality parameter reported by the EBSD mapping system used in this study; the Grain Reference Misorientation in degrees, where the reference orientation is the average orientation for the grain associated with each voxel; the Kernel Average Misorientation in degrees, calculated using a 3 × 3 × 3 voxel kernel; the L1 (Manhattan) distance relative to the grain boundary network, reported in units of pixels (1 pixel = 0.25 μm).

Competing interests
The authors declare that they have no competing interests.

Authors' contributions
The micro-tensile samples were fabricated by PS and MU. The mechanical test was conducted by PS, MG, and MU. The surface strain analysis was conducted by PS and JS. The 3D EBSD serial section data was collected by PS, MG, and MU. The 3D EBSD reconstructions were conducted by MG and MU. All authors have read and approved the final manuscript.

Acknowledgements
The authors would like to thank Dr. R. Wheeler (UES Inc., MicroTesting Solutions LLC) who developed the micro-testing device used in these experiments, and Adam Shiveley (UES, Inc.) for help setting up communication between various instrument computers. The authors would also like to acknowledge useful discussions with Drs. D.M. Dimiduk (Air Force Research Laboratory), T.J. Turner (Air Force Research Laboratory), and Y.S. Choi (UES Inc.). The authors acknowledge support from the Air Force Office of Scientific Research (AFOSR, program managers Dr. Joan Fuller and Dr. Ali Sayir) and the Materials & Manufacturing Directorate of the Air Force Research Laboratory.

References

1. Roters F, Eisenlohr P, Hantcherli L, Tjahjanto DD, Bieler TR, Raabe D (2010) Overview of constitutive laws, kinematics, homogenization and multiscale methods in crystal plasticity finite-element modeling: Theory, experiments, applications. Acta Mater 58:1152–1211

2. Lim H, Lee MG, Kim JH, Adams BL, Wagoner RH (2011) Simulation of polycrystal deformation with grain and grain boundary effects. Int J Plast 27:1328–1354

3. Uchic MD, Dimiduk DM, Florando JN, Nix WD (2004) Sample dimensions Influence strength and crystal plasticity. Science 305:986–989

4. McDowell DL (2008) Viscoplasticity of heterogeneous metallic materials. Mater Sci Eng R 62:67–123

5. Becker R, Panchanadeeswaran (1995) Effects of grain interactions on deformation and local texture in polycrystals. Acta Metall Mater 43:2701–2719

6. Bhattacharyya A, El-Danaf E, Kalidindi SR, Doherty RD (2001) Evolution of grain-scale microstructure during large strain simple compression of polycrystalline aluminum with quasi-columnar grains: OIM measurements and numerical simulations. Int J Plast 17:861–883

7. Cheong KS, Busso EP (2006) Effects of lattice misorientations on strain heterogeneities in FCC polycrystals. J Mech Phys Solids 54:671–689

8. Heripre E, Dexet M, Crepin J, Gelebart L, Roos A, Bornert M, Caldemaison D (2007) Coupling between experimental measurements and polycrystal finite element calculations for micromechanical study of metallic materials. Int J Plast 23:1512–1539

9. Kalidindi SR, Bhattacharyya A, Doherty RD (2004) Detailed analyses of grain-scale plastic deformation in columnar polycrystalline aluminium using orientation image mapping and crystal plasticity models. Proc Roy Soc Lond A 460:1935–1956

10. Rehrl C, Volkert B, Kleber S, Antretter T, Pippan R (2012) Crystal orientation changes: A comparison between a crystal plasticity finite element study and experimental results. Acta Mater 60:2379–2386

11. Turner TJ, Semiatin SL (2011) Modeling large-strain deformation behavior and neighborhood effects during hot working of a coarse-grain nickel-base superalloy. Model Simul Mater Sci Eng 19:065010

12. Zhao Z, Ramesh M, Raabe D, Cuitino AM, Radovitzky R (2008) Investigation of three-dimensional aspects of grain-scale plastic surface deformation of an aluminum oligocrystal. Int J Plast 24:2278–2297

13. St-Pierre L, Heripre E, Dexet M, Crepin J, Bertolino G, Bilger N (2008) 3D simulations of microstructure and comparison with experimental microstructure coming from O.I.M. analysis. Int J Plast 24:1516–1532

14. Musienko A, Tatschl A, Schmidegg K, Kolednik O, Pippan R, Cailletaud G (2007) Three-dimensional finite element simulation of a polycrystalline copper specimen. Acta Mater 55:4121–4136

15. Gianola DS, Eberl C (2009) Micro- and nanoscale tensile testing of materials. JOM 61:24–35

16. Kiener D, Grosinger W, Dehm G, Pippan R (2008) A further step towards an understanding of size-dependent crystal plasticity: In situ tension experiments of miniaturized single-crystal copper samples. Acta Mater 56:580–592

17. Kim JY, Greer JR (2009) Tensile and compressive behavior of gold and molybdenum single crystals at the nano-scale. Acta Mater 57:5245–5253

18. Wheeler R, Shade PA, Uchic MD (2012) Insights gained through image analysis during in-situ micromechanical experiments. JOM 64:58–65

19. Peters WH, Ranson WF (1982) Digital imaging techniques in experimental stress analysis. Opt Eng 21:427–432

20. Sutton MA, Wolters WJ, Peters WH, Ranson WF, McNeill SR (1983) Determination of displacements using an improved digital correlation method. Image Vision Comput 1:133–139

21. Sutton MA, Li N, Joy DC, Reynolds AP, Li X (2007) Scanning electron microscopy for quantitative small and large deformation measurements Part I: SEM imaging at magnifications from 200 to 10,000. Exp Mech 47:775–787

22. Wissuchek DJ, Mackin TJ, DeGraef M, Lucas GE, Evans AG (1996) A simple method for measuring surface strains around cracks. Exp Mech 36:173–179

23. Biery N, De Graef M, Pollock TM (2003) A method for measuring microstructural-scale strains using a scanning electron microscope: Applications to γ-titanium aluminides. Metall Mater Trans A 34:2301–2313

24. Uchic MD, Groeber M, Wheeler R, Scheltens F, Dimiduk DM (2004) Augmenting the 3D characterization capability of the dual beam FIB-SEM. Microsc Microanal 10(Suppl 2):1136–1137

25. Groeber MA, Haley BK, Uchic MD, Dimiduk DM, Ghosh S (2006) 3D reconstruction and characterization of polycrystalline microstructures using a FIB-SEM system. Mater Charact 57:259–273

26. Uchic MD, Groeber MA, Dimiduk DM, Simmons JP (2006) 3D microstructural characterization of nickel superalloys via serial-sectioning using a dual beam FIB-SEM. Scripta Mater 55:23–28

27. Zaafarani N, Raabe D, Singh RN, Roters F, Zaefferer S (2006) Three-dimensional investigation of the texture and microstructure below a nanoindent in a Cu single crystal using 3D EBSD and crystal plasticity finite element simulations. Acta Mater 54:1863–1876

28. Turner TJ, Shade PA, Groeber MA, Schuren JC (2013) The influence of microstructure on surface strain distributions in a nickel micro-tension specimen. Model Simul Mater Sci Eng 21:015002

29. Shade PA, Kim SL, Wheeler R, Uchic MD (2012) Stencil mask methodology for the parallelized production of microscale mechanical test samples. Rev Sci Instrum 83:053903

30. Uchic MD, Dimiduk DM, Wheeler R, Shade PA, Fraser HL (2006) Application of micro-sample testing to study fundamental aspects of plastic flow. Scripta Mater 54:759–764

31. Shade PA, Wheeler R, Choi YS, Uchic MD, Dimiduk DM, Fraser HL (2009) A combined experimental and simulation study to examine lateral constraint effects on microcompression of single-slip oriented single crystals. Acta Mater 57:4580–4587

32. Jones RV (1951) Parallel and rectilinear spring movements. J Sci Instrum 28:38–41

33. Jones RV, Young IR (1956) Some parasitic deflexions in parallel spring movements. J Sci Instrum 33:11–15

34. Wu A, De Graef M, Pollock TM (2006) Grain-scale strain mapping for analysis of slip activity in polycrystalline B2 RuAl. Phil Mag 86:3995–4008

35. Orloff J, Utlaut M, Swanson L (2003) High resolution focused ion beams: FIB and its applications. Kluwer Academic/Plenum, New York

36. Wilkinson AJ, Meaden G, Dingley DJ (2006) High-resolution elastic strain measurement from electron backscatter diffraction patterns: New levels of sensitivity. Ultramicroscopy 106:307–313

37. Bhandari Y, Sarkar S, Groeber M, Uchic MD, Dimiduk DM, Ghosh S (2007) 3D polycrystalline microstructure reconstruction from FIB generated serial sections for FE analysis. Comput Mater Sci 41:222–235

38. Ghosh S, Bhandari Y, Groeber M (2008) CAD-based reconstruction of 3D polycrystalline alloy microstructures from FIB generated serial sections. CAD 40:293–310

39. DREAM.3D. http://dream3d.bluequartz.net/

40. Schwartz AJ, Kumar M, Adams BL (2000) Electron backscatter diffraction in materials science. Kluwer Academic/Plenum, New York

41. Shade PA, Groeber MA, Schuren JC, Uchic MD (2013) 3D microstructure reconstruction of polycrystalline nickel micro-tension test.. http://hdl.handle.net/11115/152

42. Uchic M, Groeber M, Shah M, Callahan P, Shiveley A, Scott M, Chapman M, Spowart J (2012) An automated multi-modal serial sectioning system for characterization of grain-scale microstructures in engineering materials. In: De Graef M, Poulsen HF, Lewis A, Simmons J, Spanos G (ed) 1st International Conference on 3D Materials Science, pp 195–202

43. Suter RM, Hennessy D, Xiao C, Lienert U (2006) Forward modeling method for microstructure reconstruction using x-ray diffraction microscopy: single crystal verification. Rev Sci Instrum 77:123905

Permissions

The contributors of this book come from diverse backgrounds, making this book a truly international effort. This book will bring forth new frontiers with its revolutionizing research information and detailed analysis of the nascent developments around the world.

We would like to thank all the contributing authors for lending their expertise to make the book truly unique. They have played a crucial role in the development of this book. Without their invaluable contributions this book wouldn't have been possible. They have made vital efforts to compile up to date information on the varied aspects of this subject to make this book a valuable addition to the collection of many professionals and students.

This book was conceptualized with the vision of imparting up-to-date information and advanced data in this field. To ensure the same, a matchless editorial board was set up. Every individual on the board went through rigorous rounds of assessment to prove their worth. After which they invested a large part of their time researching and compiling the most relevant data for our readers.

The editorial board has been involved in producing this book since its inception. They have spent rigorous hours researching and exploring the diverse topics which have resulted in the successful publishing of this book. They have passed on their knowledge of decades through this book. To expedite this challenging task, the publisher supported the team at every step. A small team of assistant editors was also appointed to further simplify the editing procedure and attain best results for the readers.

Apart from the editorial board, the designing team has also invested a significant amount of their time in understanding the subject and creating the most relevant covers. They scrutinized every image to scout for the most suitable representation of the subject and create an appropriate cover for the book.

The publishing team has been an ardent support to the editorial, designing and production team. Their endless efforts to recruit the best for this project, has resulted in the accomplishment of this book. They are a veteran in the field of academics and their pool of knowledge is as vast as their experience in printing. Their expertise and guidance has proved useful at every step. Their uncompromising quality standards have made this book an exceptional effort. Their encouragement from time to time has been an inspiration for everyone.

The publisher and the editorial board hope that this book will prove to be a valuable piece of knowledge for researchers, students, practitioners and scholars across the globe.

List of Contributors

Stephen R Niezgoda
Materials Science and Technology Division, Los Alamos National Laboratory, Los Alamos NM 87545, USA
Department of Materials Science and Engineering, Department of Mechanical and Aerospace Engineering, The Ohio State University, Columbus, OH 43218, USA

Anand K Kanjarla
Department of Metallurgical and Materials Engineering, Indian Institute of Technology Madras, Chennai, 600036, India

Surya R Kalidindi
George W. Woodruff School of Mechanical Engineering, Georgia Institute of Technology, Atlanta, GA 30332, USA

Lifei Du
Key Laboratory of Space Applied Physics and Chemistry-Ministry of Education, School of Science, Northwestern Polytechnical University, Xi'an 710072, China

Rong Zhang
Key Laboratory of Space Applied Physics and Chemistry-Ministry of Education, School of Science, Northwestern Polytechnical University, Xi'an 710072, China

Ayman A Salem
Materials Resources LLC, Dayton, OH 45402, USA

Joshua B Shaffer
Materials Resources LLC, Dayton, OH 45402, USA

Daniel P Satko
Materials Resources LLC, Dayton, OH 45402, USA

S Lee Semiatin
Air Force Research Laboratory, Materials and Manufacturing Directorate, Wright-Patterson AFB, OH 45433, USA

Surya R Kalidindi
Georgia Institute of Technology, Atlanta, GA 30332, USA

Jonathan D Madison
Computational Materials & Data Science, Sandia National Laboratories, 87185 Albuquerque, NM, USA

Larry K Aagesen
Materials Science & Engineering, University of Michigan, 48109 Ann Arbor, MI, USA

Victor WL Chan
Materials Science & Engineering, University of Michigan, 48109 Ann Arbor, MI, USA

Katsuyo Thornton
Materials Science & Engineering, University of Michigan, 48109 Ann Arbor, MI, USA

Joy H Forsmark
Materials Research Department, Ford Motor Company, Research and Innovation Center, MD3182, P.O Box 2053, Dearborn, MI 48121, USA

Jacob W Zindel
Materials Research Department, Ford Motor Company, Research and Innovation Center, MD3182, P.O Box 2053, Dearborn, MI 48121, USA

Larry Godlewski
Materials Research Department, Ford Motor Company, Research and Innovation Center, MD3182, P.O Box 2053, Dearborn, MI 48121, USA

Jiang Zheng
Department of Materials Science and Engineering, University of Michigan, 2300 Hayward St., Ann Arbor, MI 48109, USA

John E Allison
Department of Materials Science and Engineering, University of Michigan, 2300 Hayward St., Ann Arbor, MI 48109, USA

Mei Li
Materials Research Department, Ford Motor Company, Research and Innovation Center, MD3182, P.O Box 2053, Dearborn, MI 48121, USA

Michael Yeager
Department of Mechanical Engineering and Center for Composite Materials, University of Delaware, Newark, DE 19716, USA

Suresh G Advani
Department of Mechanical Engineering and Center for Composite Materials, University of Delaware, Newark, DE 19716, USA

Agustin Jose Torroba
University of Applied Sciences and Arts Western Switzerland, 1950 Sion, Switzerland

Ole Koeser
CALCOM-ESI, 1015 Lausanne, Switzerland

Loic Calba
CALCOM-ESI, 1015 Lausanne, Switzerland

Laura Maestro
Precicast Bilbao, 48901, Barakaldo, Bilbao, Spain

Efrain Carreo-Morelli
University of Applied Sciences and Arts Western Switzerland, 1950 Sion, Switzerland

Mehdi Rahimian
IMDEA Materials Institute, C/Eric Kandel 2, 28906 Getafe, Madrid, Spain

Srdjan Milenkovic
IMDEA Materials Institute, C/Eric Kandel 2, 28906 Getafe, Madrid, Spain

Ilchat Sabirov
IMDEA Materials Institute, C/Eric Kandel 2, 28906 Getafe, Madrid, Spain

Javier LLorca
IMDEA Materials Institute, C/Eric Kandel 2, 28906 Getafe, Madrid, Spain
Department of Materials Science, Polytechnic University of Madrid, 28040 Madrid, Spain

Paul A Shade
Air Force Research Laboratory, Materials and Manufacturing Directorate, 2230 10th Street, Wright-Patterson AFB, OH 45433, USA

Michael A Groeber
Air Force Research Laboratory, Materials and Manufacturing Directorate, 2230 10th Street, Wright-Patterson AFB, OH 45433, USA

Jay C Schuren
Air Force Research Laboratory, Materials and Manufacturing Directorate, 2230 10th Street, Wright-Patterson AFB, OH 45433, USA

Michael D Uchic
Air Force Research Laboratory, Materials and Manufacturing Directorate, 2230 10th Street, Wright-Patterson AFB, OH 45433, USA

www.ingramcontent.com/pod-product-compliance
Lightning Source LLC
Chambersburg PA
CBHW050454200326
41458CB00014B/5180